Peptide Analysis Protocols

Methods in Molecular Biology

John M. Walker, Series Editor

36. **Peptide Analysis Protocols,** edited by ***Ben M. Dunn and Michael W. Pennington,*** 1994
35. **Peptide Synthesis Protocols,** edited by ***Michael W. Pennington and Ben M. Dunn,*** 1994
34. **Immunocytochemical Methods and Protocols,** edited by ***Lorette C. Javois,*** 1994
33. ***In Situ* Hybridization Protocols,** edited by ***K. H. Andy Choo, 1994***
32. **Basic Protein and Peptide Protocols,** edited by ***John M. Walker, 1994***
31. **Protocols for Gene Analysis,** edited by ***Adrian J. Harwood, 1994***
30. **DNA–Protein Interactions,** edited by ***G. Geoff Kneale, 1994***
29. **Chromosome Analysis Protocols,** edited by ***John R. Gosden, 1994***
28. **Protocols for Nucleic Acid Analysis by Nonradioactive Probes,** edited by ***Peter G. Isaac, 1994***
27. **Biomembrane Protocols:** *II. Architecture and Function,* edited by ***John M. Graham and Joan A. Higgins, 1994***
26. **Protocols for Oligonucleotide Conjugates,** edited by ***Sudhir Agrawal, 1994***
25. **Computer Analysis of Sequence Data:** *Part II,* edited by ***Annette M. Griffin and Hugh G. Griffin, 1994***
24. **Computer Analysis of Sequence Data:** *Part I,* edited by ***Annette M. Griffin and Hugh G. Griffin, 1994***
23. **DNA Sequencing Protocols,** edited by ***Hugh G. Griffin and Annette M. Griffin, 1993***
22. **Optical Spectroscopy, Microscopy, and Macroscopic Techniques,** edited by ***Christopher Jones, Barbara Mulloy, and Adrian H. Thomas, 1994***
21. **Protocols in Molecular Parasitology,** edited by ***John E. Hyde, 1993***
20. **Protocols for Oligonucleotides and Analogs,** edited by ***Sudhir Agrawal, 1993***
19. **Biomembrane Protocols:** *I. Isolation and Analysis,* edited by ***John M. Graham and Joan A. Higgins, 1993***
18. **Transgenesis Techniques,** edited by ***David Murphy and David A. Carter, 1993***
17. **Spectroscopic Methods and Analyses,** edited by ***Christopher Jones, Barbara Mulloy, and Adrian H. Thomas, 1993***
16. **Enzymes of Molecular Biology,** edited by ***Michael M. Burrell, 1993***
15. **PCR Protocols,** edited by ***Bruce A. White, 1993***
14. **Glycoprotein Analysis in Biomedicine,** edited by ***Elizabeth F. Hounsell, 1993***
13. **Protocols in Molecular Neurobiology,** edited by ***Alan Longstaff and Patricia Revest, 1992***
12. **Pulsed-Field Gel Electrophoresis,** edited by ***Margit Burmeister and Levy Ulanovsky, 1992***
11. **Practical Protein Chromatography,** edited by ***Andrew Kenney and Susan Fowell, 1992***
10. **Immunochemical Protocols,** edited by ***Margaret M. Manson, 1992***
9. **Protocols in Human Molecular Genetics,** edited by ***Christopher G. Mathew, 1991***

Earlier volumes are still available. Contact Humana for details.

Methods in Molecular Biology • 36

Peptide Analysis Protocols

Edited by

Ben M. Dunn

University of Florida, Gainesville, FL

and

Michael W. Pennington

Bachem Bioscience, King of Prussia, PA

Humana Press **Totowa, New Jersey**

999 Riverview Drive, Suite 208
Totowa, New Jersey 07512

This publication is printed on acid-free paper. ∞
ANSI Z39.48-1984 (American National Standards Institute) Permanence of Paper for Printed Library Materials

Printed in the United States of America. 10 9 8 7 6 5 4 3 2 1

Library of Congress Cataloging in Publication Data

Peptide analysis protocols/edited by Ben M. Dunn and Michael W. Pennington
p. cm.—(Methods in molecular biology; 36)
Companion v. to: Peptide synthesis protocols.
Includes index.
ISBN 0-89603-274-4 (acid-free paper)
1. Peptides—Analysis—Laboratory manuals. I. Dunn, Ben M. II. Pennington, Michael W. III. Peptide synthesis protiocols. IV. Series: Methods in molecular biology (Totowa, N.J.); 36
QP552.P4P43 1994
574.19'2456—dc20 94-22663
CIP

Preface

As the technology base for the preparation of increasingly complex peptides has improved, the methods for their purification and analysis have also been improved and supplemented. Peptide science routinely utilizes tools and techniques that are common to organic chemistry, protein chemistry, biophysical chemistry, enzymology, pharmacology, and molecular biology. A fundamental understanding of each of these areas is essential for interpreting all of the data that a peptide scientist may see.

The purpose of *Peptide Analysis Protocols* is to provide the novice with sufficient practical information necessary to begin developing useful analysis and separation skills. Understanding and developing these skills will ultimately yield a scientist with broadened knowledge and good problem-solving abilities. Although numerous books that address different specialties, such as HPLC, FAB-MS, CE, and NMR, have been written, until now no single volume has reviewed all of these techniques with a focus on "getting started" in separation and analysis of peptides.

This volume will also provide those who already possess practical knowledge of the more advanced aspects of peptide science with detailed applications for each of these protocols. Because the chapters have been written by researchers active in each of the fields that they discuss, a great deal of information on and insight into solution of real problems that they have encountered is presented. Examplary results are clearly demonstrated and discussed. For more advanced investigations, supplementary experiments are often suggested.

Additionally, *Peptide Analysis Protocols* contains several applications and extensions of peptide science. Thus, we include chapters on such topics as the construction of peptide libraries, recombinant peptide synthesis, immunochemical methods used for peptides, the kinetic analysis of proteases using peptide substrates, and *de novo* protein design and synthesis. These represent state-of-the-art technologies and applications that make this volume complete when used in combination with its companion work in this series, *Peptide Synthesis Protocols*.

It is hoped that after reading *Peptide Analysis Protocols* and applying the techniques described, the reader can develop successful separations, analyses, and applications of each of his or her peptides.

We wish to thank the authors for their excellent contributions and cooperation.

Ben M. Dunn
Michael W. Pennington

Contents

Contents of the companion volume:

Peptide Synthesis Protocols

(Note: Cross references to these chapters will include the notation PSP)

Contributors

KARL-HEINZ ALTMANN • *Section de Chimie, Universite de Lausanne, Lausanne, Switzerland. Present Address: Ciba-Geigy AG, Basel, Switzerland*

PHILIP C. ANDREWS • *Department of Biological Chemistry, University of Michigan, Ann Arbor, MI*

DANIEL M. BOLLAG • *Merck, Sharpe, and Dohme, West Point, PA*

MICHAEL E. BYRNES • *Bachem Bioscience, King of Prussia, PA*

J. MARK CARTER • *Cytogen, Princeton, NJ*

WAYNE L. CODY • *Department of Chemistry, Parke-Davis Pharmaceutical Research, Division of Warner-Lambert Co., Ann Arbor, MI*

DAN L. CRIMMINS • *Core Protein/Peptide Facility, Howard Hughes Medical Institute, Washington University School of Medicine, St. Louis, MO*

P. R. DAS • *Schering-Plough Research Institute, Kenilworth, NJ*

RUTH DAVENPORT • *Department of Biochemistry and Molecular Biology, College of Medicine, University of Florida, Gainesville, FL*

BEN M. DUNN • *Department of Biochemistry and Molecular Biology, College of Medicine, University of Florida, Gainesville, FL*

MARK G. HINDS • *NMR Laboratory, Biomolecular Research Institute, Parkville, Australia*

PAUL D. HOEPRICH, JR. • *Applied Biosystems Inc., Foster City, CA*

DANIELE M. LEONARD • *Department of Chemistry, Parke-Davis Pharmaceutical Research, Warner-Lambert Co., Ann Arbor, MI*

MANFRED MUTTER • *Section de Chimie, Universite de Lausanne, Lausanne, Switzerland*

UDO NIRENBERG • *Bachem Feinchemikalien AG, Bubendorf, Switzerland*

Raymond S. Norton • *NMR Laboratory, Biomolecular Research Institute, Parkville, Australia*

Jan Pohl • *Microchemical Facility, Winship Cancer Center, Emory University School of Medicine, Atlanta, GA*

B. N. Pramanik • *Schering-Plough Research Institute, Kenilworth, NJ*

Tomi K. Sawyer • *Department of Chemistry, Parke-Davis Pharmaceutical Reserach, Warner-Lambert Co., Ann Arbor, MI*

Paula E. Scarborough • *Department of Pathology, Duke University, Durham, NC*

Verena Steiner • *Section de Chimie, Universite de Lausanne, Lausanne, Switzerland. Present Address: Ciba-Geigy AG, Basel, Switzerland*

Jay S. Stout • *BioNebraska Inc., Lincoln, NE*

Wieslaw Swietnicki • *Department of Biochemistry and Molecular Biology, College of Medicine, University of Florida, Gainesville, FL*

Gabriele Tuchscherer • *Section de Chimie, Universite de Lausanne, Lausanne, Switzerland*

Gino Van Heeke • *Protein Expression Laboratory, Interdisciplinary Center for Biotechnology Research, University of Florida, Gainesville, FL; BioNebraska, Inc. Lincoln, NE*

Fred W. Wagner • *BioNebraska Inc., Lincoln, NE; Department of Biochemistry, University of Nebraska, Lincoln, NE*

Thomas E. Wheat • *Analytical Technology Inc., Boston, MA*

Chapter 1

Gel-Filtration Chromatography

Daniel M. Bollag

1. Introduction

Gel filtration chromatography is a method for separating proteins and peptides based on their size *(1)*. The chromatographic matrix consists of porous beads, and the size of the bead pores defines the size of macromolecules that may be fractionated. Those proteins or peptides that are too large to enter the bead pores are "excluded," and thus elute from the column first (Fig. 1). Since large molecules do not enter the beads, they have less volume to pass through, which is why they are the first to elute from the column. Smaller macromolecules that enter some, but not all of the pores are retained slightly longer in the matrix and emerge from the column next. Finally, small molecules filter through most of the pores, and they elute from the column with an even larger elution volume. This method is also called gel permeation, molecular sieve, gel-exclusion, and size-exclusion chromatography. Since no binding is required and harsh elution conditions can be avoided, gel-filtration chromatography rarely inactivates enzymes, and often is used as an important step in peptide or protein purification (*see* Note 1).

The chief limitations of gel-filtration chromatography are that the separation may be slow and that the resolution of the emerging peaks is limited (*see* Note 2). The speed of sample elution is limited primarily by the requirement for a long, narrow column in order to permit sufficient component separation, although the procedure may be accelerated by the use of matrices permitting faster flow rates and by the use of pumps or high-pressure chromatography equipment if the matrix can tolerate the added pressure. The resolution is limited since the sample

From: *Methods in Molecular Biology, Vol. 36: Peptide Analysis Protocols*
Edited by: B. M. Dunn and M. W. Pennington

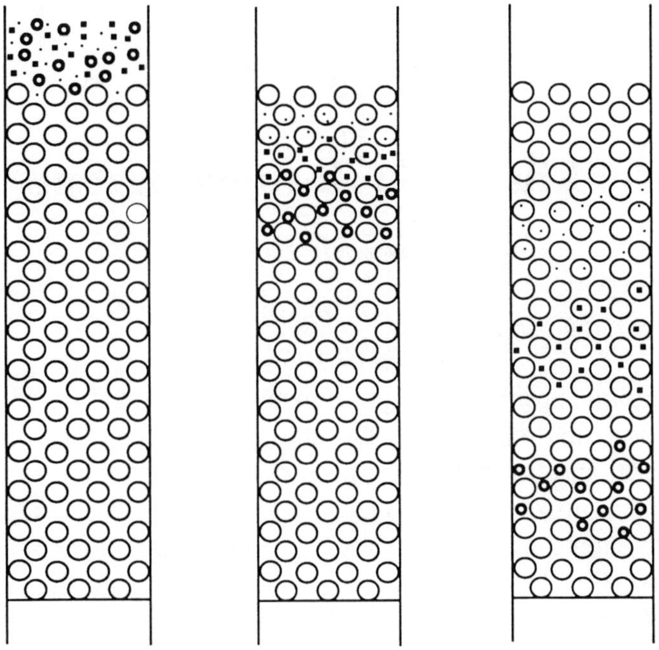

Fig. 1. Schematic representation of gel-filtration chromatography. Molecules of different size in the left frame are separated according to size during migration through the gel-filtration matrix as shown in the middle and right frames.

does not bind to the matrix. Therefore, careful selection of the matrix fractionation range is essential, and gel-filtration chromatography is frequently used as a separation step when only a small number of contaminants remain.

Gel-filtration chromatography separates proteins and peptides based on their diameter during chromatography. Thus, gel filtration allows an estimation of the molecular weight of a protein or multiprotein complex *(2)*. However, a molecular-weight estimation is based on the assumption that the protein is generally globular in shape. Separation on the basis of size may also permit an approximation of a dissociation constant for a protein–protein or protein–ligand interaction *(3)*. In addition, gel-filtration chromatography may be used for sample desalting or for changing the buffer of the sample (*see* Note 1). The versatility of gel-filtration chromatography has made this separation technique an extremely useful and popular tool for protein or peptide purification and analysis.

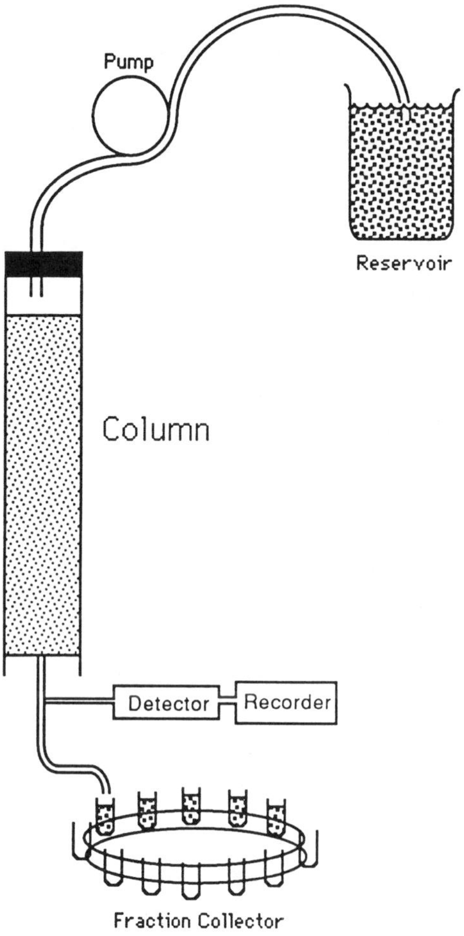

Fig. 2. Schematic representation of chromatography equipment.

2. Materials

The equipment required for gel-filtration chromatography is very simple, but a more sophisticated laboratory system may be preferable to save time and provide more reproducible results. The heart of a gel-filtration chromatography setup (Fig. 2) is the column, which generally consists of a glass cylinder containing a column support. Columns for gel filtration are generally long and narrow, but the diameter should be at least 10 mm, so that anomalous effects from the protein and buffer interactions with the column wall can be avoided. Adaptors for the top and

Table 1
Gel-Filtration Chromatography[a]

Name	Fractionation range, kDa	Linear flow rate, cm/h
BioGel P-6	1–6	10
BioGel P-60	3–60	5
BioGel P-100	5–100	5
Sephacryl S-100 HR	1–100	
Sephacryl S-200 HR	5–250	15
Sephacryl S-300 HR	10–1500	15
Sephadex G-25	1–5	
Sephadex G-50	1.5–30	5
Sephadex G-100	4–150	5
Sephadex G-200	5–600	2
Sepharose CL-6B	10–4000	18

[a]The fractionation range defines the approximate protein and peptide molecular weights that can be separated with the matrix. The linear flow rate can be converted into a volumetric flow rate (mL/min) by multiplying by the cross-sectional area (πr^2) of the column.

bottom of the column allow homogeneous and efficient delivery of sample or buffer to the column matrix. Tubing from the filtration column should be narrow bore to keep remixing of the separated components to a minimum. A reservoir for the buffer to be delivered to the column can be connected via a pump that can control the column flow rate. A UV wavelength detector monitors the absorbance of the eluting sample, and the signal can be sent to a recorder or a personal computer for analysis. The eluting sample may be directed to a fraction collector that sequentially collects aliquots of the eluant either according to time or volume. All of this equipment can be purchased as individual components or as an integrated high-pressure chromatography system, depending on the needs of the user.

The column matrix for gel filtration must be chosen carefully to allow the best resolved separation of the component of interest from the contaminants. The matrix should be chosen so that the sample molecular weight falls in the middle of the matrix fractionation range or so that contaminating components are well resolved from the desired component. Table 1 provides information for selection of the proper matrix type; suppliers such as Bio-Rad and Pharmacia can be consulted for further information. Coarser matrices offer faster flow rates, which may, however, lead to reduced resolution of peaks. A coarse matrix will thus be better for such uses as desalting a protein or exchanging the buffer of the sample, whereas a fine matrix is preferred for separations.

If the matrix is not supplied as a preswollen slurry, the dry powder needs to be swollen in buffer. Swelling is generally carried out by gently swirling the matrix in buffer. Using a magnetic stirrer may cause the matrix particles to be broken into "fine" particles, which can cause irregularities in column packing and may also reduce the column flow rate. Thus, agitation by a rotary shaker or occasionally swirling the matrix by hand is recommended for swelling. Swelling can be carried out at room temperature or by boiling, which speeds the hydration process significantly; matrix manufacturers should be consulted for swelling information. Fines are removed by swirling the slurry containing the gel-filtration matrix and, after most of the matrix particles have settled, pouring off the supernatant. This procedure is repeated several times. A preswollen gel may only require reequilibration in the appropriate buffer. Degassing of the matrix is important to reduce the likelihood that air bubbles will form in the column. To degas the gel-filtration matrix, apply a vacuum to the matrix solution for up to an hour while agitating the matrix slurry.

3. Methods

3.1. Packing the Column

1. The chromatography matrix is first prepared and degassed as a thick slurry (the buffer supernatant should comprise only 25% of the matrix volume). Space below the column support should be filled with buffer so no bubbles will form.
2. Add a small amount of buffer, and close the outlet after a small amount of buffer has been allowed to flow out.
3. Then, in a single step, the slurry is poured down a glass rod into the column or along the side of a column that is temporarily tilted slightly, and the column outlet is opened. If necessary, a column extension or funnel is attached to the column in order to permit packing of the matrix in a single operation; otherwise, uneven beds can form. Care must be taken to be sure that air bubbles are not trapped as the matrix packs or the column will have to be repacked. If bubbles develop early during packing, they can be removed by gently stirring the matrix.
4. Once the matrix has been poured, it is possible to connect the pump and attach the reservoir (but do not exceed the maximum pressure recommended for the matrix). Two- or three-column bed volumes should pass through the packed matrix to stabilize and equilibrate the column.

3.2. Checking the Column

Initially after packing the column, a visual inspection for air bubbles is necessary, since bubbles will cause mixing during chromatography that will reduce the resolution substantially. As a more rigorous test of column packing, 0.2% blue dextran (1% of the column bed volume) can be loaded on the column and should travel through the matrix as a well-defined, horizontal band. If the column is well packed, the blue dextran should elute in no more than twice the volume that was applied.

3.3. Sample Application

1. The sample should ideally be fairly concentrated (10–20 mg/mL), and the sample solution should be less than twice as viscous as the elution buffer or peaks may become too broad. Sample volume should be 1–5% of the column bed volume; a larger volume may lead to poor resolution, whereas a sample volume smaller than 1% of the bed volume will not generally improve the separation.
2. The elution buffer should be chosen to preserve the protein's activity and should contain a low ionic strength buffer (e.g., 20–100 m*M*) to minimize nonspecific ionic or hydrophobic interactions.
3. The buffer in the column should be eluted until the buffer reaches the top of the matrix surface. Then the outlet should be closed. Remember that the chromatography matrix must never be allowed to run dry.
4. The sample is gently layered on top of the matrix, taking care not to disturb the packed matrix.
5. Open the outlet, and allow the buffer to drain until the liquid level again reaches the matrix surface. Then close the outlet.
6. Add a small amount of buffer to the column, and run the buffer just into the column in order to wash the remaining sample into the matrix.
7. Finally, refill the column with buffer, and attach the pump and reservoir. At this point, elution of the sample may begin.

3.4. Column Elution

The buffer is simply run through the column until the peaks of interest have been eluted. Recoveries are typically over 85%. Slower flow rates generally yield better resolution, so some adjustments for optimal separation may be necessary.

3.5. Column Regeneration and Storage

1. Following elution of the sample, the gel-filtration matrix should be regenerated to remove any of the remaining sample components. For most

matrices, regeneration is carried out by washing 0.2*M* NaOH or nonionic detergent through the column, and then reequilibrating with the appropriate buffer for the next experiment.

2. If the column will be stored overnight or longer before the next use, it is advisable to maintain the gel-filtration matrix in a solution containing an inhibitor of microbial growth. For most applications, a buffer containing 0.02% sodium azide is effective for preventing the growth of microorganisms. Other inhibitors include 0.01–0.02% trichlorobutanol or 0.002% hibitane (but do not use hibitane with Sepharose).
3. Finally, some matrices should not be stored in solutions of very high or low pH.

4. Notes

1. Gel-filtration chromatography, aside from its utility in protein and peptide purification, can also be employed for exchanging the buffer in which a macromolecule is found. Since the original sample buffer passes through a matrix, such as Sephadex G-25, much more slowly than a polypeptide, the protein or peptide can be eluted with a new buffer that has been used for column equilibration and elution. In this fashion, an ion-exchange chromatography fraction can be exchanged into a lower salt buffer ("desalting") or a sample can be separated from low-molecular-weight contaminants, such as nucleotides or metals. This separation is a very distinct one, so the sample may be as large as 30% of the column bed volume without affecting the separation. Some matrix suppliers now offer spin columns, which allow desalting or nucleotide removal by passing the sample through the filtration matrix in a rapid centrifugation step.
2. Poor peak resolution may be the result of:
   a. Improper selection of matrix: Use a matrix with a fractionation range that brackets the molecular weight of the desired protein (i.e., the molecular weight is in the middle of the separation range). Be aware that a nonglobular or denatured protein elutes differently from a globular protein.
   b. Wrong matrix grade: A finer matrix grade may be available that offers better resolution, although separation times will be longer.
   c. Column is too short: A longer column will allow better resolution: resolution increases as the square root of column length.
   d. Flow rate is too high: A faster flow rate reduces resolution.
   e. Large dead space before elution fractions are collected: Dead space is the region at the bottom of the chromatography column that allows the temporary accumulation of eluent before fraction collection occurs. If this space is large, protein peaks will remix, reducing

the resolution. A well-designed column will contain minimal dead space.

   f. The sample volume is too large: For a good separation, the sample should be between 1 and 5% of the column bed volume.
   g. The column is poorly packed: Uneven column packing or air bubbles trapped in the column matrix cause irregular flow patterns leading to poorer separation.

3. Skewed protein peaks may be the result of:
   a. Poor sample application: It is possible to practice sample application using blue dextran as described in Section 3.2.
   b. Protein adsorption to the matrix: Matrix adsorption can be suspected when peaks tail off slowly; adding a stronger ionic strength salt may reduce these undesired interactions. In addition, changing the buffer pH or composition may improve the situation.
4. A low flow rate can be traced to:
   a. Plugged filters or tubing: Such a situation can sometimes be remedied by adding some detergent or denaturant to the buffer or by reversing the buffer flow through the column. Otherwise, the column must be dismantled, cleaned, and repacked.
   b. A clogged matrix surface: If a residue has formed on top of the matrix, scrape off and remove the top layer of the matrix, then stir the top centimeter of the remaining matrix, and allow to settle slowly.
   c. A pump is poorly functioning.
   d. A matrix that is incompletely swollen, is compressed, or contains too many "fines": If this is the case, the column must be repacked.
   e. Microbial growth in the matrix: A new chromatography column must be prepared.
5. Poor recovery of the sample might be caused by:
   a. Sample precipitation: Too little or too much salt can result in precipitation of the protein and poor entry into the column.
   b. Adsorption effects: *See* Note 3b.
   c. Elution conditions that are too harsh: This may release a necessary cofactor or damage the component of interest.
   d. Microbial growth: *See* Note 4e.
   e. Proteolysis: Include protease inhibitors in buffer.
   f. Slight adsorption of the sample to the matrix and very slow elution as a peak that cannot be distinguished from the background: A nonionic detergent may disrupt this interaction without damaging the macromolecule.
   g. Dissociation from a complex or necessary cofactor during elution: Mixing fractionated aliquots may reactivate the sample.

References

1. Stellwagen, E. (1990) Gel filtration. *Methods Enzymol.* **182**, 317–328.
2. Preneta, A. Z. (1989) Separation on the basis of size: gel permeation chromatography, in *Protein Purification Methods: A Practical Approach* (Harris, E. L. V. and Angal, S., eds.), IRL, Oxford, pp. 293–305.
3. Pharmacia Fine Chemicals (1991) *Gel Filtration: Principles and Methods*. Uppsala, Sweden.

Chapter 2

Ion-Exchange Chromatography

Daniel M. Bollag

1. Introduction

Ion-exchange chromatography allows the separation of proteins and peptides by taking advantage of their net charge. These macromolecules can also be concentrated by ion exchange either on a column or as a batch procedure (*see* Note 5). Although procedures for separating peptides or proteins vary according to each individual molecule, many basic rules apply to all ion-exchange purifications, and these generalized procedures will be described in this chapter.

The key determinant for adsorption to an ion-exchange matrix is the charge of a peptide or protein *(1)*. Thus, a protein has an affinity for an anion-exchange matrix (such as DEAE-Sepharose) if the protein has an overall negative charge (Fig. 1), and conversely, a cation-exchange matrix binds a positively charged protein. Because of the ionization state of surface amino acids, the net charge of a protein or peptide varies with the pH of the buffer (Fig. 2). The pH is referred to as the protein's isoelectric point (pI) when the total number of positive charges on a protein equals the number of negative charges—in other words, when the protein's net charge is zero. A protein is negatively charged at a pH above its pI and positively charged at a pH below its pI. As seen in Fig. 2, a protein becomes more highly charged as the pH moves further away from the protein's isoelectric point. For most separations, a pH that is 1 U from the pI of the protein is best for achieving the reversible binding required in ion-exchange chromatography.

From: *Methods in Molecular Biology, Vol. 36: Peptide Analysis Protocols*
Edited by: B. M. Dunn and M. W. Pennington Copyright ©1994 Humana Press Inc., Totowa, NJ

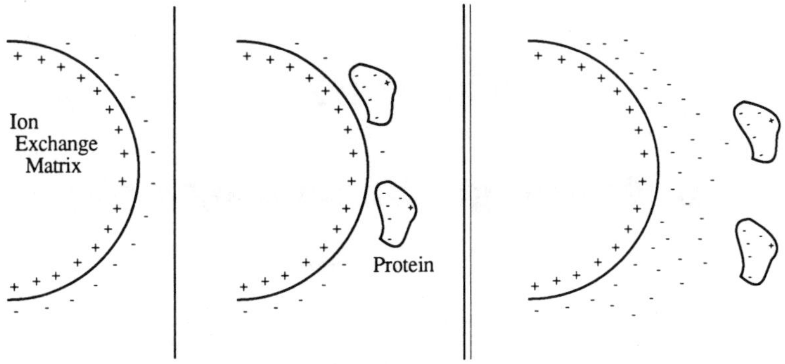

Fig. 1. Schematic example of ion-exchange chromatography. Left frame, an ion-exchange matrix with negative counterions. Center frame, negatively charged protein attached to ion-exchange matrix. Right frame, high counterion concentration has caused protein to detach from matrix.

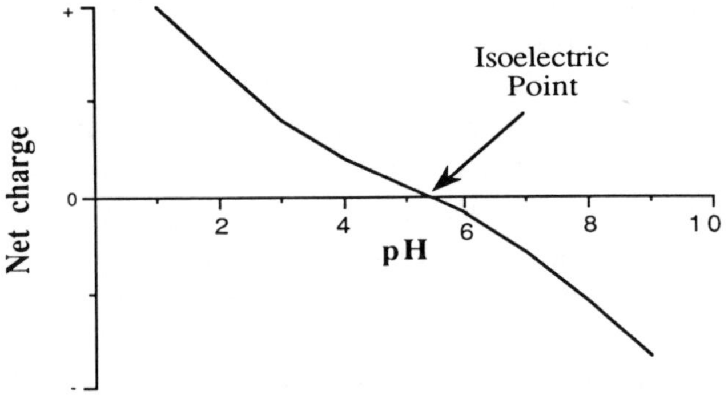

Fig. 2. Example of a protein's overall charge as a function of pH.

Ion-exchange chromatography proceeds in two steps: binding of the protein or peptide to the matrix followed by its elution. In an example of anion exchange (Fig. 1), an anion-exchange matrix is initially positively charged and in equilibrium with a negatively charged counterion (e.g., Cl^-). When the negatively charged protein or peptide of interest is applied to the column, the macromolecule displaces the chloride counterion and remains bound to the matrix. To elute the macromolecule, a higher concentration of counterion (e.g., 1*M* Cl^-) is added to the column (*see* Note 4). The protein is displaced by the strong competition of the concentrated counterion and is eluted from the column. The differing affinities of various proteins for an ion-exchange matrix provide a sensitive method for their separation on the basis of charge.

Selecting the best ion-exchange matrix for the separation is important. An ion-exchange matrix is derivatized with a functional group that defines the matrix as an anion- or cation-exchange matrix. Anion exchangers are derivatized with positively charged groups, whereas cation exchangers contain negatively charged groups. Most anion-exchange matrices are substituted with a diethylamino ethyl (DEAE) group (for example, DEAE-Sephadex or DEAE-Sepharose) or a quaternary amine (Mono Q). Cation-exchange matrices generally contain a carboxymethyl (CM) group (thus, CM-Sephadex or CM-Sepharose) or a sulfomethyl group (Mono S). Those groups that are weakly basic (DEAE) or weakly acidic (CM) bind proteins or peptides with relatively low affinity, such that the interactions can be disrupted without overly harsh conditions.

Key factors in deciding which matrix to use include the pH stability, swelling properties, capacity, and flow properties of the ion-exchange matrix. If extremes of pH are to be used during chromatography, a matrix that resists breaking down under such conditions should be chosen. Special attention must be paid to a matrix that is soft and easily compressed if the column is to be run under pressure or if changes in ionic strength can cause significant matrix swelling (for example, with Sephadex) (*see* Note 3). The capacity of the matrix is an estimate of how much protein can be bound per unit volume of matrix, and familiarity with the capacity will help in determining what volume of matrix should be used for separating or concentrating the sample of interest. Only 10–20% of the available capacity should be used for applications where high resolution of components is required. For example, 1 mL of DEAE-Sepharose CL-6B matrix can bind up to 100 mg of hemoglobin, although for high-resolution separation, a total of only 10–20 mg of hemoglobin should be applied/mL of this matrix. In addition, when speed is important during the fractionation procedure, the flow rate afforded by a matrix becomes an important factor, although in general a faster flow rate results in lower resolution of elution peaks. The most widely used matrices are crosslinked dextrans, such as Sephadex and Sephacryl, crosslinked agaroses, such as Sepharose and Bio-Gel A, beaded agaroses (Sepharose Fast Flow and Sepharose CL-6B), beaded celluloses (Sephacel), and crosslinked polyacrylamides, such as Bio-Gel P. Manufacturers (especially Pharmacia-LKB and Bio-Rad) should be consulted for information concerning individual matrices.

Purification of proteins or peptides is best achieved by utilizing separation steps that depend on different properties of the macromolecules. For example, ion-exchange chromatography may be followed by gel-filtration chromatography (*see* previous chapter) to take advantage of the differences in particle size, and may then be followed by affinity chromatography in which separation is based on specific interactions with a ligand. By carefully planning a purification to exploit different properties of a protein or peptide, a high level of purification should be possible with minimal loss of the sample.

2. Materials

2.1. Chromatography Equipment

Equipment for ion-exchange chromatography can range from a simple homemade apparatus to sophisticated automated instruments that improve the speed and reproducibility of separations. The minimal basic materials required for chromatography are a column attached to a fraction collector. A pump, gradient maker, buffer reservoir, detector, and recorder may be attached to the basic equipment (Fig. 3). The column dimensions should be determined by the application: a short, wide column is most commonly used for ion-exchange chromatography when speed is desired, whereas a longer, narrower column will allow better separation of components. The column is composed of a cylinder, usually made of glass, with a flat porous supporting material at the bottom on which the ion-exchange matrix rests. Some manufacturers also supply adaptors for the top and bottom of the column, which facilitate sample loading and connect the column to the detector and reservoir. An important feature of a column outlet is a minimal amount of dead space to prevent sample mixing after separation on the column. The fraction collector permits samples to be fractionated according to time or volume. These basic components are the heart of the chromatographic separation hardware.

Increased flexibility and automation are provided by additional equipment. A buffer reservoir eliminates the need for manual addition of buffer during the fractionation process, and the introduction of a pump permits the regulation of buffer flow. A gradient maker is required for gradient elutions. UV wavelength detector and a chart recorder can be attached to the column outlet allowing an initial reading of the column elution profile. A concise discussion of chromatographic equipment, including suggestions for suppliers, is provided in ref. 2, pp. 186–189.

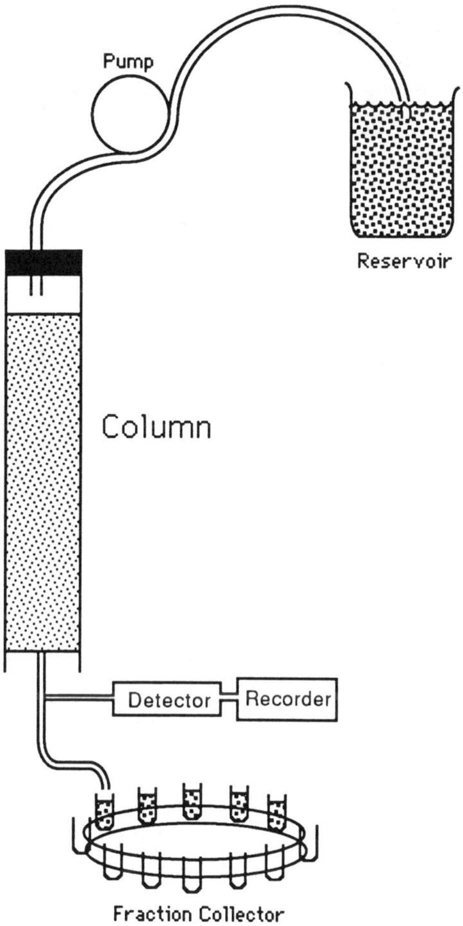

Fig. 3. Schematic representation of chromatography equipment.

2.2. Column and Matrix

Preparing an ion-exchange matrix for chromatography involves swelling the matrix, removing fine particles, packing the column, and equilibrating the matrix prior to sample application. For most applications involving enzymes, it is advisable to handle the sample at 4°C in order to reduce the loss of enzyme activity. If a chromatographic procedure is to be run in the cold, it is necessary to pour, store, and run the column at 4°C, since changes in temperature may cause bubble formation.

1. Many matrices are supplied as swollen gels, and those that are ordered as dry powders should be hydrated by incubating with the experimental buffer at 100°C for one to several hours or at room temperature for several hours to several days according to the manufacturer's instructions. Five to 50 mL of swollen gel are generally obtained for each gram of dry matrix material. During swelling, the buffer should be changed several times. The matrix is best agitated by gentle swirling, since mechanical agitation with a magnetic stir bar may break the matrix into smaller particles ("fines").
2. Fines may cause uneven column packing and can substantially reduce the flow rate. Prior to packing the column, fines should be removed by swirling the matrix slurry, allowing the slurry to settle, and decanting the supernatant to remove the fines. This removal procedure should be repeated several times. Finally, an estimation of the appropriate matrix volume to use will depend on the protein-binding capacity of the matrix as well as the required separation resolution and speed, as described in Section 1.
3. Column packing and equilibration must be done carefully to minimize problems with flow rates and column reproducibility. Before adding the ion-exchange matrix, the column should be prepared by removing air from the dead space at the bottom of the column. Add a small amount of degassed buffer to the column, allow buffer to flow through the column outlet (thus pushing out any air bubbles), and close the column outlet. To reduce the possibility of trapping air bubbles in the matrix, the matrix should be degassed prior to packing. Adequate degassing for most applications is achieved by applying a vacuum to the matrix solution for up to an hour. Agitating the matrix slurry during degassing reduces the possibility of air bubbles remaining lodged between matrix particles.
4. A thick slurry (the matrix should comprise 75% of the slurry volume) is poured down a glass rod into the column or down the side of a slightly tilted column so that no air bubbles are trapped in the matrix as it settles. Once the column is straightened, the column outlet is opened and more buffer added as the matrix packs. At this point, a column adaptor can be attached, and the column may be connected to the buffer reservoir.
5. To equilibrate the matrix, pass several column volumes of buffer through the column. The pH and conductivity of the buffer should be the same before application and after elution from the column. Alternatively, the matrix can be equilibrated by washing on a Buchner funnel prior to packing the column. A well-poured column that is carefully equilibrated and maintained will allow excellent, reproducible separations for many experiments.

2.3. Sample

For ion-exchange chromatography, the sample initially should be in a buffer of low ionic strength (below 50 m*M*). If the binding characteristics of the protein of interest are known, it is advisable to apply the sample in a buffer with an ionic strength slightly below that required for sample elution. This procedure is useful to eliminate more rapidly those contaminants with a lower binding affinity. If the sample to be applied is turbid, filtration is an important step to prevent clogging of the column (a filter pore size of 0.45 mm is recommended). As mentioned above, the buffer pH is critical in defining the affinity of a protein for the ion-exchange matrix. Most proteins are negatively charged at pH 8, so an anion-exchange matrix with pH 8 buffer is appropriate for many applications. If the isoelectric point of the protein is not known, a small-scale experiment may be helpful in determining the protein's binding profile for various pH ranges *(3)*.

3. Methods

3.1. Sample Application

1. The sample solution is applied after the ion-exchange matrix has been packed in the column and equilibrated with the starting buffer. Be particularly careful not to allow the column matrix to run dry during sample application and chromatography, since this may change the binding properties of the matrix or cause protein denaturation.
2. Allow the buffer to drain until it reaches the bed surface, and close the column outlet. Gently apply sample solution to the bed surface using a pipet, taking care not to disturb the bed surface or agitate the sample. Then, open the column outlet, allow the sample solution to enter into the column until the liquid reaches the bed surface, and reclose the outlet.
3. Gently add some starting buffer to the bed surface, allow the buffer to enter the column, and close the column outlet again. This step serves to wash the sample residue on the column walls into the ion-exchange matrix.
4. Finally, add starting buffer gently to the column, and attach the column to the reservoir. At this point, column washing and elution may begin.

3.2. Column Elution

1. Once the sample has been loaded on the column, the ion-exchange matrix should be washed with the starting buffer in order to elute any unbound material. Typically, three to ten column bed volumes of buffer are used for washing the column, but a more reliable indication of how long to wash can be obtained by monitoring the eluent optical density or protein con-

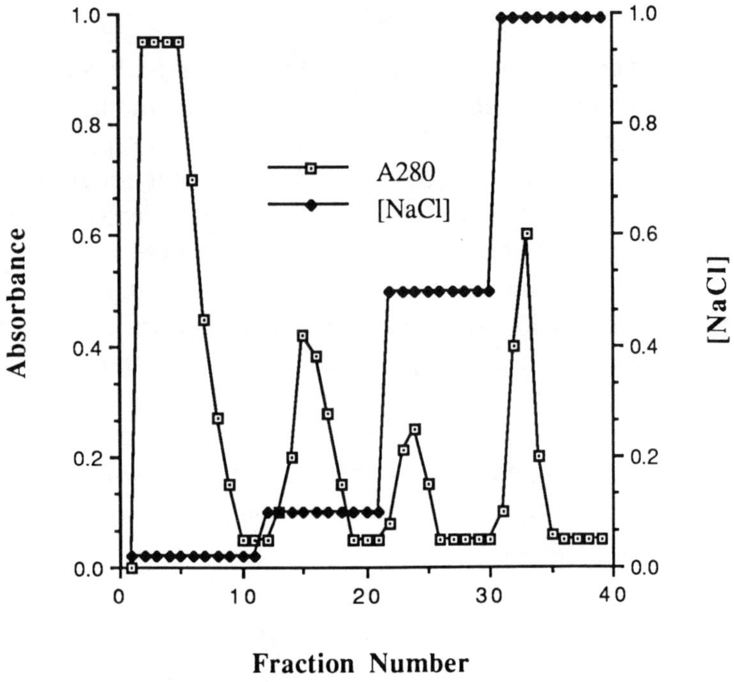

Fig. 4. Chromatogram representing ion-exchange separation with a step gradient.

centration. When the flow-through fractions contain negligible contaminants, elution may begin.

2. Two types of sample elution are most commonly used in ion-exchange chromatography: step elution and gradient elution. Step elution involves increasing the buffer ionic strength in discrete jumps (such as 0.1*M* NaCl, followed by 0.5*M* NaCl, followed by 1.0*M* NaCl), whereas gradient elution requires a steady increase in the ionic strength concentration (from 20 m*M* NaCl to 1.0*M* NaCl, for example). Step elution is a simple and rapid method, although each jump in ionic strength may elute a number of components. Gradient elution offers more discrete separation of protein or peptide peaks.
3. Step elution volumes should be calibrated by trial and error, although a good starting strategy is to use ten bed volumes for each step in order to have maximal separation of peaks from each step increase in ionic strength. The appropriate number of bed volumes of the first elution buffer (e.g., 20 m*M* Tris-HCl, 0.1*M* NaCl) is added following the column wash (*see* Fig. 4). This step is followed by addition of the second elution buffer (e.g., 20 m*M* Tris-HCl, 0.5*M* NaCl), then the third elution

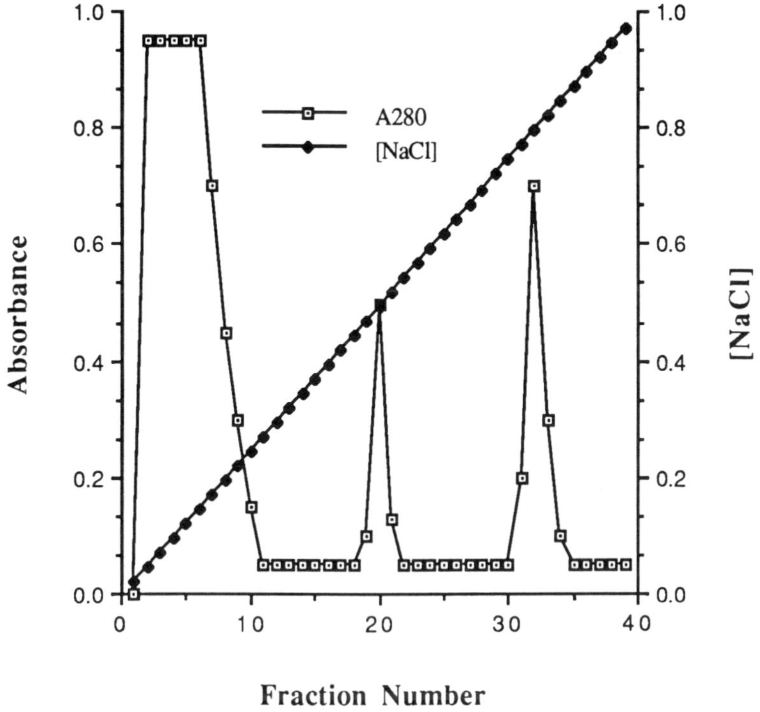

Figure 5. Chromatogram representing ion-exchange separation with a linear salt gradient.

buffer, and so on. The new buffer should be added after the previous buffer elutes to the top of the matrix in order to have a well-defined and reproducible increase in salt concentration.

4. Gradient elution parameters also must be defined in each experimental situation *(3)*. The total gradient buffer volume should equal approximately five column bed volumes. A sample gradient elution profile is shown in Fig. 5, in which the starting buffer is 20 m*M* Tris-HCl, 20 m*M* NaCl and the ending buffer is 20 m*M* Tris-HCl, 1.0*M* NaCl. A gradient maker can be constructed in the laboratory and is also commercially available.
5. The experimental conditions for protein elution should be defined so that the sample of interest emerges from the column as well resolved from other components as possible (*see* Notes 1 and 2). If the sample remains bound, it is advisable to use a counterion with a higher binding affinity for the ion exchanger during elution (e.g., switch from NaCl to Na_2PO_4; *see* Note 4). During initial trials, if increasing salt concentrations are insufficient to elute the sample, harsher eluting techniques, such as applying detergents or

denaturing agents or changing the pH, may be tried. After a preliminary characterization of the fractionated elution samples, fractions are frequently pooled to facilitate further analysis (*see* Figs. 4 and 5).

3.3. Batch Elution

1. Batch elution is a simple alternative to column chromatography when the resolution of separation is less important. In this procedure, the sample is stirred gently with the ion-exchange matrix for about 1 h or until the sample component of interest has been adsorbed to the matrix.
2. The buffer is then removed by filtration or centrifugation. A higher ionic strength buffer is added to the matrix, and the sample is again stirred. The supernatant will contain the component of interest when it is desorbed from the ion exchanger. Batch elution is frequently used for large sample volumes and for protein concentration.

3.4. Column Regeneration and Storage

1. Following sample elution, the matrix should be regenerated to remove any remaining contaminants, thus preparing the matrix for future separation or concentration procedures. Matrices that resist swelling because of changes in ionic strength (such as Sepharose, Sephacel, and G-25-based Sephadex) may be regenerated in the column; otherwise, the matrix must be removed for regeneration and repacked prior to subsequent use.
2. Sephadex, CM-Sepharose CL-6B, and DEAE-Sephacel ion exchangers can be regenerated with several column volumes of buffer containing salt of ionic strength up to 2*M* (ideally containing the appropriate counterion to the ion exchanger for the subsequent separation in order to simplify re-equilibration).
3. DEAE-Sepharose CL-6B exchangers should be regenerated with one bed volume of 1*M* sodium acetate (pH 3.0) followed by 1.5 bed volumes of 0.5*M* sodium hydroxide, which should be left in the column overnight, and then 1.5 bed volumes of 1*M* sodium acetate (pH 3.0) before re-equilibrating with the starting buffer. Consult the manufacturer's instructions for harsher treatments to remove any remaining lipids or detergents from the ion-exchange matrix.
4. Proper storage of ion-exchange matrices is critical in order to maintain column reproducibility and to reduce the frequency of preparing new columns. All matrices should be stored in buffer containing some salt and an antimicrobial agent. Antimicrobials include 0.002% hibitane (chlorohexidine) for anion exchangers, and 0.02% sodium azide or 0.005% merthiolate (Thimerosal or ethyl mercuric thiosalicylate) for cation exchangers. Certain matrices can also be autoclaved to prevent microbial growth.

4. Notes

1. A normal ion-exchange chromatography step will give a protein yield of 60–80%. Low yields may be the result of adherence of the macromolecule to glass or to the column matrix, and this interaction is generally disrupted with a high salt concentration, detergent, or organic solvent treatment. The relationship between the isoelectric point of the protein and the experimental pH should also be considered (*see* Fig. 2). Furthermore, the loss of a crucial cofactor during chromatography may destabilize or inactivate a protein, contributing to "sample loss." Remixing fractions containing the protein and its cofactor might reactivate an enzyme.
2. Poor resolution can be improved with a slower flow rate, longer column, lower applied protein concentration, different gradient slope, or different eluting counterion.
3. If the flow rate decreases significantly during chromatography, this is most frequently the result of compression of the matrix, clogging of the column support, trapped air bubbles in the tubing, or deposition of viscous material on top of the column. To remove precipitated material from the top of the ion-exchange matrix, scrape off the top layer of the matrix and remove, then gently stir the top 1–2 cm of matrix, and allow to settle before continuing with the elution. The sample should be more thoroughly filtered before application in the future. A well-maintained ion-exchange column allows efficient screening of a large number of elution conditions that may be necessary for optimization of a purification protocol.
4. Counterions remain in equilibrium with the functional group of the ion-exchange matrix, and they play a key role in determining the elution characteristics of the sample. An "activity series" defines the relative affinities of counterions for a matrix. For cation exchangers, the counterion activity series is $Ag^+ > Cs^+ > K^+ > NH_4^+ > Na^+ > H^+ > Li^+$ (where Ag^+ binds more tightly to a cation-exchange matrix than Cs^+). Likewise, the anion-exchange activity series is $I^- > NO_3^- > PO_4^- > CN^- > HSO_3^- > Cl^- > HCO_3^- > HCOO^- > CH_3COO^- > OH^- > F^-$. Therefore, if a protein is tightly attached to the column matrix, elution may be improved with a stronger counterion.
5. A valuable use for ion-exchange chromatography is protein or peptide concentration. Since a majority of proteins are negatively charged at pH 8, it is a relatively simple matter to apply a dilute, low ionic strength protein solution to an anion-exchange matrix and elute the proteins with a high salt step. The capacity of a milliliter of ion-exchange matrix may be up to 30 mg of a complex protein mixture, and most proteins are eluted with 1*M* NaCl. Significant concentration of the protein solution can be achieved rapidly in this manner. Thus, ion-exchange chromatography can be a valuable tool in the purification or concentration of proteins or peptides.

References

1. Scopes, R. K. (1987) *Protein Purification: Principles and Practice.* Springer-Verlag, New York.
2. Roe, S. (1989) *Separation Based on Structure in Protein Purification Methods: A Practical Approach* (Harris, E. L. V. and Angal, S., eds.), IRL, Oxford, pp. 175–244.
3. Pharmacia Fine Chemicals (1991) *Ion Exchange Chromatography: Principles and Methods.* Uppsala, Sweden.

CHAPTER 3

Reversed-Phase HPLC

Analytical Procedure

Udo Nirenberg

1. Introduction

In peptide chemistry, HPLC has gained importance as an analytical tool because of its exquisite sensitivity, speed, and resolving power *(1,2)*. This chapter outlines the use of HPLC for purity control and content determination of synthesized or isolated peptides. Besides different HPLC methods, like ion-exchange chromatography and size-exclusion chromatography, the reversed-phase method predominates and has become the method of choice for peptide separations.

Stationary phases typically used in reversed-phase chromatography are silica-based supports modified by chemically bonded octyl (C8) or octadecyl (C18) groups. These allow for a hydrophobic surface where the separation takes place.

To obtain a sufficient interaction of the peptide with the hydrophobic surface of the stationary phase, it is necessary to reduce the polar character of the peptide and eliminate any hydrophilic interactions between matrix and peptide. This is made possible by carrying out the chromatography with mobile phases at pH 2–3 where the carboxylic groups of aspartic acid and glutamic acid side chains are forced into the protonated form. Furthermore, the mobile phase must contain buffer anions that act as counterions to form ion pairs with the basic side chains of amino acids, like arginine or lysine. This allows the masking of the positive charges. Suitable systems that meet these requirements are, e.g., trifluoroacetate

From: *Methods in Molecular Biology, Vol. 36: Peptide Analysis Protocols*
Edited by: B. M. Dunn and M. W. Pennington Copyright ©1994 Humana Press Inc., Totowa, NJ

or phosphate. In some cases, triethylamine is added to suppress the interaction of residual free silanol groups of the matrix (stationary phase) and to reduce polar anionic groups. Such an addition often results in a better peak shape. The most favored solvent as organic modifier to elute peptides is acetonitrile.

All these eluent components meet one important requirement—they are transparent in the UV range down to 200–220 nm, where the peptides are detected because of the UV absorption of the peptide bond. In some cases, it is possible to use longer wavelengths for detection, e.g., 280 nm where Trp and Tyr absorb because of their aromatic chromophores.

2. Materials

2.1. Chemicals

1. Acetonitrile, HPLC grade (ACN).
2. Methanol, HPLC grade (MeOH).
3. Water, HPLC grade (purified and filtered) (H_2O).
4. Trifluoroacetic acid, for sequence analysis (TFA).
5. Triethylamine, p.a. (TEA).
6. Phosphoric acid (85%), p.a. (H_3PO_4).
7. Heptanel-sulfonic acid sodium salt, for ion-pair chromatography.
8. Potassium dihydrogen phosphate (KH_2PO_4).

2.2. Equipment and Supplies

1. HPLC solvent delivery system, binary gradient capability.
2. Injector, 10-µL sample loop.
3. Variable-wavelength UV detector.
4. Data capture system.
5. Reversed-phase C-18 column (4.6 mm ID × 250 mm length, 5 mm particle size, 300 Å pore size).
6. Helium purge capability.
7. Analytical balance.
8. Volumetric flasks (5, 10, 20, 50 mL).
9. Volumetric pipet (1 mL).

3. Methods

3.1. HPLC System

This part lists the chromatographic conditions for a standard HPLC system and describes the preparation of the eluents. Before use, degas all solutions by purging with helium for approx 5 min to remove oxygen and avoid formation of bubbles in the HPLC system.

3.1.1. Standard Conditions

1. Column: analytical HPLC column C18 (4.6 × 250 mm); particle size 5 mm; pore size 30 nm (300 Å).
2. Pump: eluent-flow: 1 mL/min; gradient program: linear binary (AB) gradient; or isocratic.
3. Detector: wavelength $\lambda = 220$ nm.
4. Injector: sample loop 10 µL.
5. Temperature: ambient.
6. Eluents: *see* Section 3.1.2.

3.1.2. Preparation of the Eluents

Three different eluent systems are described, the TFA and TEAP systems *(1–8)* as standard systems suitable for the most peptides, and the ion-pair (IP) system *(9–11)* for hydrophilic peptides that are not retained by the other two eluent systems. Mix each eluent in the following way.

3.1.2.1. TFA System

Eluent A: 2000 mL H_2O + 20 mL ACN + 2 mL TFA; Eluent B: 2000 mL ACN + 2 mL TFA.

3.1.2.2. TEAP System

Eluent A: 1800 mL TEAP* + 200 mL ACN; Eluent B: 800 mL TEAP* + 1200 mL ACN.

3.1.2.3. *TEAP

Add 22 mL H_3PO_4 to 1700 mL H_2O in a 2-L volumetric flask. Adjust pH to 2.3 with TEA (~20 mL) and make to volume with H_2O.

3.1.2.4. IP System

Eluent A: Add 4.0 g heptanel-sulfonic acid sodium salt and 13.6 g KH_2PO_4 to 1700 mL H_2O in a 2-L volumetric flask. Adjust pH to 3.5 with H_3PO_4. Eluent B: 600 mL H_2O + 1400 mL MeOH.

3.2. System Suitability Test

Before starting the chromatography of the sample, you have to evaluate the performance of the HPLC system by a system-suitability test (SST). The resolution of the components of a test sample (peptide mixture) should always be the same under optimized HPLC conditions. If the components are very similar and they elute very closely, it can be checked easily. The chromatogram of such a test sample is shown in Fig. 1.

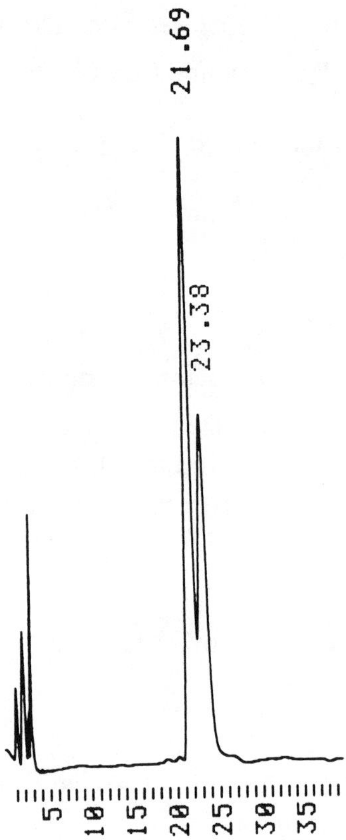

Fig. 1. System-suitability test; HPLC system: TFA; sample: diastereomers of Met(O)-enkephalin (the sulfur of Met is the chiral center); conc.: 0.5 mg/mL in H_2O; gradient: isocratic (93% A/7% B).

3.3. Purity

To determine the purity of a peptide, two methods are described.

1. The 100% method is a simple way to check the purity of a peptide in a single chromatographic run. You get the amount of the interesting peptide and impurities as area% (integral) relative to the total integral area response. With this method, you work over a range of two to four orders of magnitude of integrals for main component and impurity, respectively. Therefore, a linearity of integral vs amount is not necessarily given, and an accurate evaluation is not always possible.
2. A more accurate method for determination of peptide purity is the use of an external standard. If an impurity is known and available, it is prefer-

ably used as standard. In case of an unknown impurity, the product itself serves as external standard. This is acceptable because, in general, the impurities are of peptide origin with a comparable absorbance at the detection wavelength.

3.3.1. Chromatographic Conditions

3.3.1.1. Standard

See Section 3.1.1.

3.3.1.2. Gradient Program

After 3 min isocratic elution, start a linear gradient with an increase of 1 vol% organic modifier per minute in the eluent (increasing amount of organic modifier, e.g., ACN, forces the elution of the peptide). The retention time of the product should be 15–25 min.

3.3.2. 100% Method (Purity Check)

1. Sample preparation: Dissolve 1 mg sample in 1 mL solvent (H_2O, AcOH or another suitable solvent).
2. Sample analysis: First run a chromatogram of the solvent the sample is dissolved in (blank) and then chromatograph the sample solution.
3. Evaluation: In general, your data acquisition system (e.g., integrator) calculates the peptide purity and amount of impurities in area% automatically as follows:

$$\text{Peptide purity (area\%)} = [\text{peak area (peptide)/peak area (total)}] \times 100$$

$$\text{Impurity (area\%)} = [\text{peak area (impurity)/peak area (total)}] \times 100 \quad (1)$$

See Fig. 2 for an example of this procedure.

3.3.3. External Standard Method

1. Sample preparation: Weigh in duplicate accurately 10–20 mg of the sample in a 10-mL volumetric flask.
2. Add 7 mL solvent (H_2O, AcOH or another suitable solvent), shake until the sample is completely dissolved, and make to volume.
3. Standard preparation: Weigh accurately 10–20 mg of the standard (concerning peptide or known impurity) in a 20-mL volumetric flask. Add 15 mL solvent (same as for sample preparation), shake until the standard is completely dissolved, and make to volume. Transfer 1 mL of this solution in a 50-mL volumetric flask, and dilute to volume.

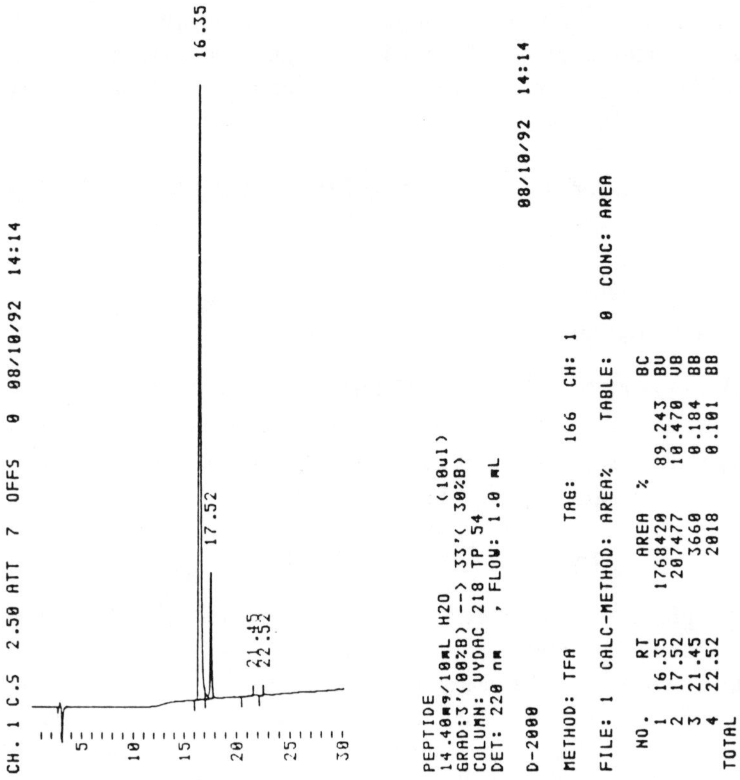

Fig. 2. Purity check of a peptide.

4. Analysis: Before filling the sample loop, flush it with the solution you want to chromatograph (three- to fivefold loop volume). First inject the solvent (blank), then your standard, and subsequently chromatograph the two sample solutions.
5. Evaluation: The amount of each impurity (wt%) is calculated in the following way:

$$\text{Impurity (\%)} = [I(imp) \times m(std) \times c(std) \times 100 / I(std) \times m(s) \times c(p) \times 2 \times 50] \quad (2)$$

where I = integral of the impurity (imp) and standard (std), m = weight of the sample (s) and standard (std), and c = content of the standard (std) and product (p).
6. Purity of the concerning peptide: purity (%) = 100% – sum of each impurity.

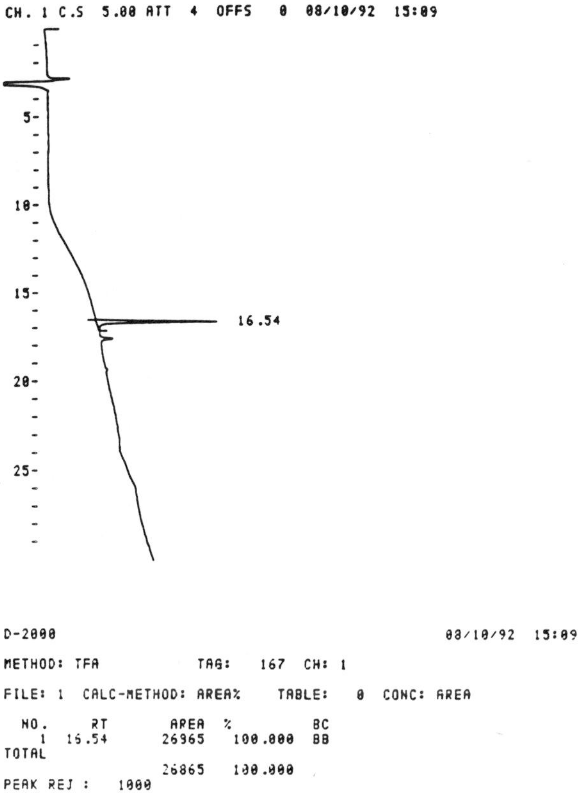

Fig. 3. Chromatogram of the standard.

3.3.3.1. Example: External Standard Method (Figs. 3 and 4)

Sample: weight = 14.40 mg
content = 73.5%
Standard: weight = 13.73 mg
content = 73.5%
Integral (standard) = 26,865

Retention, time, min	Integral, impurity	Wt%, impurity
17.52	207477	7.36%
21.45	3660	0.13%
22.52	2018	0.07%

Peptide purity = 100% – (7.36% + 0.13% + 0.07%) = 92.4%

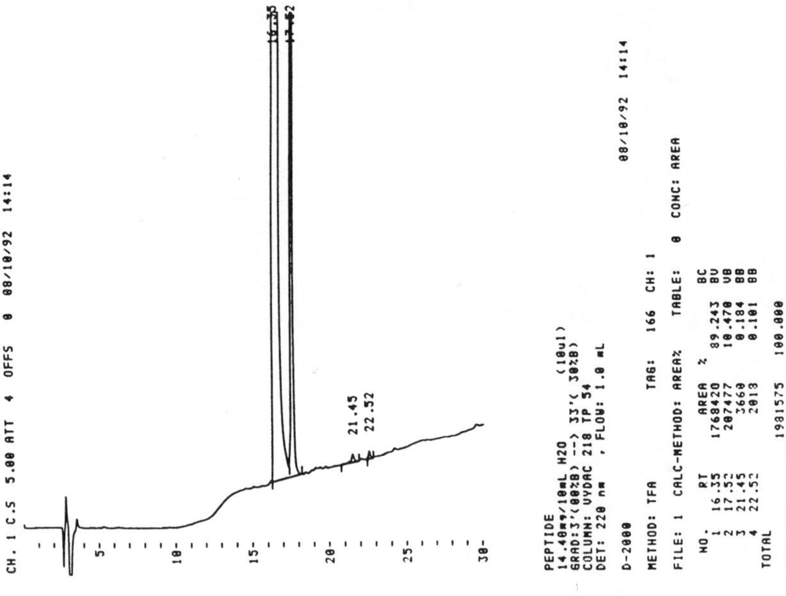

Fig. 4. Chromatogram of the sample.

3.3.4. Evaluation of the Chromatogram

To evaluate the chromatogram:

1. Have a look at the total peak (the shape should be tall and symmetric) (Fig. 5).
2. Expand the chromatogram (if you use an integrator or recorder, set the chart speed high enough to get broad peaks) (Fig. 6).
3. Print the baseline and integration marks (Fig. 6).

3.4. Determination of Peptide Content

By comparing the integrals of the peptide concerned and a standard, it is possible to determine the content of the peptide in a sample. Two requirements have to be met. The standard must be the same peptide as the product concerned, and the content of the standard has to be exactly known.

3.4.1. Chromatographic Conditions

3.4.1.1. Standard

See Section 3.1.1.

3.4.1.2. Gradient Program

Isocratic—the retention time of the peptide should be 10–20 min.

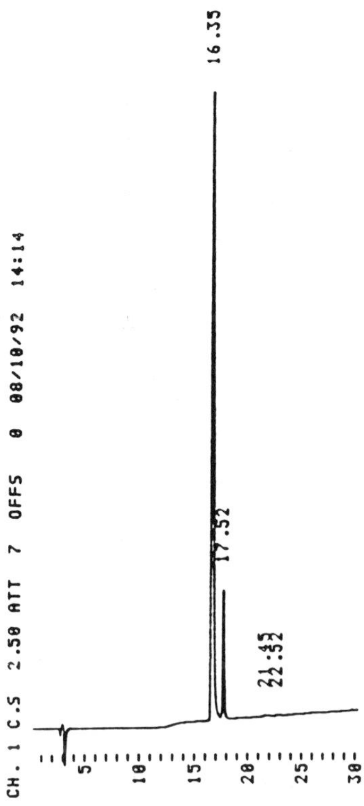

Fig. 5. Chromatogram of a peptide (total peak).

3.4.2. Sample Preparation

1. Weigh in duplicate accurately 10–20 mg of the sample in a 10-mL volumetric flask.
2. Add 7 mL solvent (H_2O, AcOH or another suitable solvent), shake until the sample is completely dissolved, and make to volume.
3. Transfer 1 mL of this solution in a 5-mL volumetric flask, and dilute to volume.

3.4.3. Standard Preparation

Same as sample preparation.

3.4.4. Analysis

Before filling the sample loop, flush it with the solution you want to chromatograph (three- to fivefold loop volume). First inject the solvent (blank), then your standard, and subsequently chromatograph the two sample solutions.

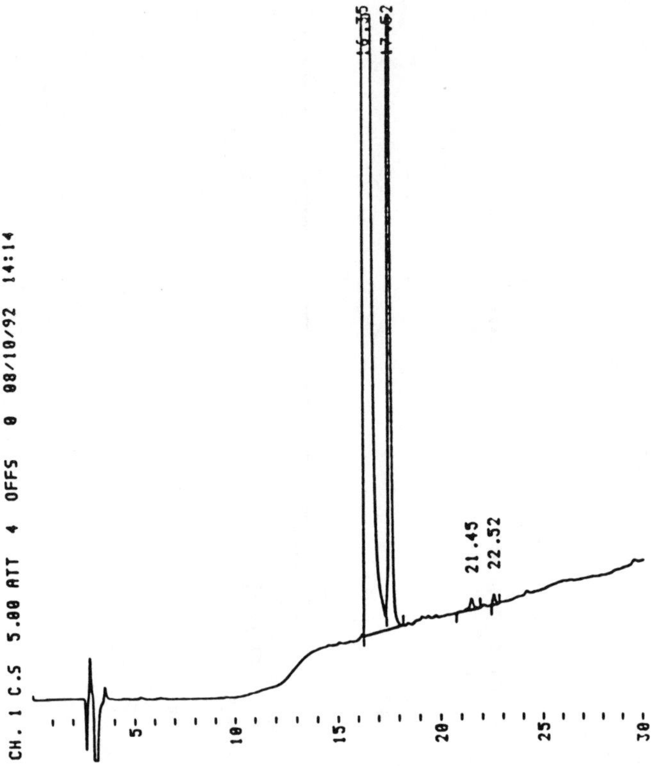

Fig. 6. Expanded chromatogram of a peptide.

3.4.5. Evaluation

The content of the peptide in the sample is calculated as follows:

$$\text{Peptide content (\%)} = [I(p) \times m(std) \times c(std)/I(std) \times m(s)] \quad (3)$$

where I = integral of the product (p) and standard (std), m = weight of the sample(s) and standard (std), and c = content of the standard. (Hygroscopic peptides: it may be necessary to determine the content of H_2O before standard preparation to correct the content of the standard.)

3.4.5.1. Example (Figs. 7 and 8)

Determination of peptide content:

Sample:	weight	=	12.32 mg
Standard:	weight	=	12.72 mg
	content	=	78.5%

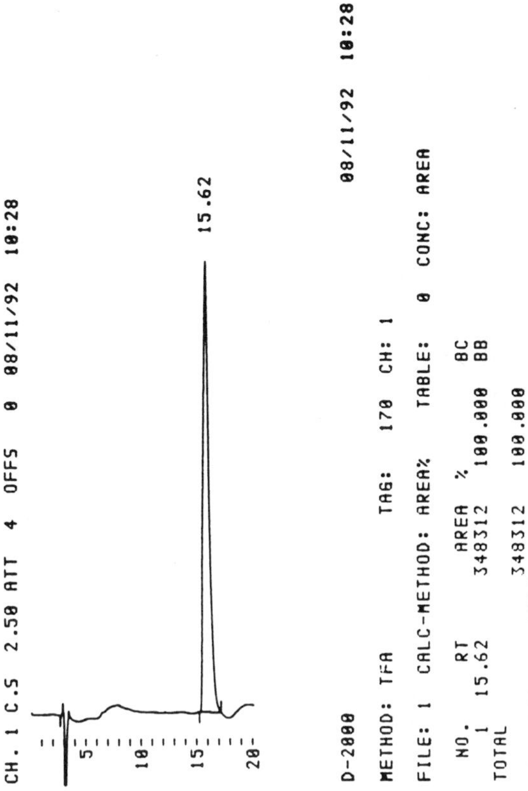

Fig. 7. Chromatogram of the standard.

Integral: standard: I(std) = 348,312
product: I (p) = 325,864

$$\text{Peptide content} = (325{,}864 \times 12.72\ \text{mg} \times 78.5\% / 348{,}312 \times 12.32\ \text{mg}) = 75.8\% \quad (4)$$

4. Notes

1. If you are working with materials that contain a complex matrix, it is recommended to filter your sample (0.45-mm membrane filter) and to protect your analytical column by using a precolumn.
2. If you do not get a sufficient resolution with the standard chromatographic conditions, the following parameters could be optimized:
   a. Gradient: Choose a gradient with an increase of organic modifier <1 vol%/min.

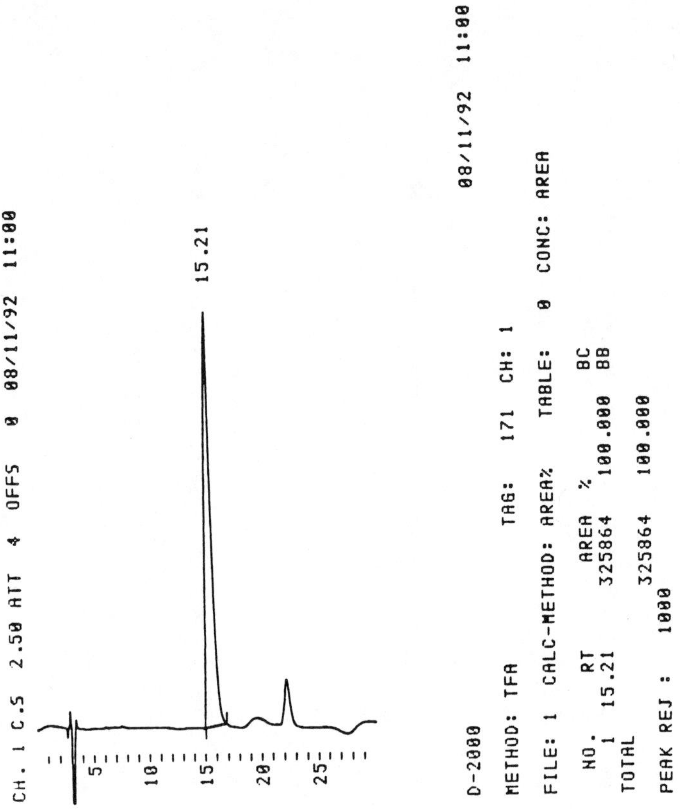

Fig. 8. Chromatogram of the sample.

b. Temperature: Increase the column temperature (column thermostat).
c. Column: Choose a column with a narrower pore volume (e.g., 10 nm [100 Å]) and/or smaller particle size (e.g., 3-mm particles).
d. Solvent composition: Choose a pH value or additives to increase the hydrophobic interactions between the peptide and stationary phase.
e. Flow rate.

References

1. Hearn, M. T. W. (ed.) (1991) *HPLC of Peptides, Proteins, and Polynucleotides.* VCH, New York.
2. Henschen, A., Hupe, K. P., Lottspeich, F., and Voelter, W. (eds.) (1985) *High Performance Liquid Chromatography in Biochemistry.* VCH, New York.
3. Bennet, H. P. J., Browne, C. A., and Solomon, S. (1980) The use of perfluorinated carboxylic acids in the reversed-phase HPLC of peptides. *J. Liquid Chromatogr.* **3**, 1353–1365.

4. Bennet, H. P. J., Browne, C. A., Goltzman, D., and Solomon, S. (1980) in *Proceedings of the 6th American Peptide Symposium* (Gross, E. and Meienhofer, J., eds.), Pierce Chemical Company, Rockford, IL, p. 121.
5. Guo, D., Mant, C. T., and Hodges, R. S. (1987) Effects of ion-pairing reagents on the prediction of peptide retention in reversed-phase high-performance liquid chromatography. *J. Chromatogr.* **386**, 205–222.
6. Guo, D., Mant, C. T., Taneja, A. K., Parker, J. M. R., and Hodges, R. S. (1986) Prediction of peptide retention times in reversed-phase high-performance liquid chromatography. *J. Chromatogr.* **359**, 499–517.
7. Mant, C. T. and Hodges, R. S. (1989) Optimization of peptide separations in high-performance liquid chromatography. *J. Liquid Chromatogr.* **12**, 139–172.
8. Rivier, J. E. (1978) Use of trialkyl ammonium phosphate buffers in reverse phase HPLC for high resolution and high recovery of peptides and proteins. *J. Liquid Chromatogr.* **1**, 343–366.
9. Tomlinson, E., Jefferies, T. M., and Riley, C. M. (1978) Ion-pair high-performance liquid chromatography. *J. Chromatogr.* **159**, 315–358.
10. Gloor, R. and Johnson, E. L. (1977) Practical aspects of reverse phase ion pair chromatography. *J. Chromatogr. Sci.* **15**, 413–423.
11. Bidlingmeyer, B. A. (1980) Separation of ionic compounds by reversed-phase liquid chromatography: an update of ion-pairing techniques. *J. Chromatogr. Sci.* **18**, 525–539.

CHAPTER 4

Reversed-Phase High-Performance Liquid Chromatography

A Semipreparative Methodology

Michael E. Byrnes

1. Introduction

As solid-phase peptide synthesis techniques improved and the rate at which a peptide could be synthesized increased, purification was identified as the new bottleneck in the production of high-quality peptides. Purification took a giant leap forward with the introduction of reversed-phase high-performance liquid chromatography (RP-HPLC; *see* Chapter 3) to the synthetic laboratory. In fact, HPLC technology has been instrumental in the purification and characterization of most biologically active peptides and proteins *(1)*.

The difficulty level of the peptides attempted by solid-phase techniques has consistently increased, creating new separation problems. These problems include closely related species caused by side-chain modification, as well as deletion or addition sequences. Elimination of these impurities is crucial in order to assess the biological properties of a given compound accurately.

Additionally, since peptide drugs have now become a reality *(2–4)*, purification of intermediate and large quantities of these compounds has created a new demand—scale-up procedures from the analytical scale to semipreparative and ultimately the large commercial-scale purification. Analytical-level purifications are routinely performed on microbore and standard analytical columns. These separations generally separate from

From: *Methods in Molecular Biology, Vol. 36: Peptide Analysis Protocols*
Edited by: B. M. Dunn and M. W. Pennington

5 μg up to 1 mg of product depending on the complexity of the mixture. Semipreparative purifications are a broader, more ill-defined area since it can include samples from 5 mg up to 100 g. The macroscale purification is utilized at levels exceeding kilogram and multikilogram amounts.

This chapter will describe several different procedures that we have found to be extremely useful in separating fairly complex mixtures of crude products. Additionally, we have included examples of oxidatively folded types of molecules that present a purification nightmare to most researchers. The purpose of this chapter is to serve as a starting point for those with little or no experience in isolating peptides by semipreparative procedures.

2. Materials

2.1. Instruments and Columns

2.1.1. Preparative HPLC System

1. Waters, DELTA PREP 3000 Pump System (Max Flow Rate = 180 mL/min).
2. Waters Lambda Max (Model 481) LC spectrophotometer.
3. Waters 1000 PrepPak column Module (Standard radial psi = 700).
4. Chart recorder.

2.1.2. Analytical HPLC System

This consists of a Beckman System Gold: Pump Model 126, Detector Model 166.

2.1.3. HPLC Columns and Packings

Standard HPLC columns contain either spherical or natural (asymmetric) silica base derivatized with a polymeric carbon chain. The most common are octadecyl silica (C_{18}) linked columns and are most commonly utilized for small- to medium-sized peptides (5–50 residues). Larger and more hydrophobic peptides are more easily eluted from a C_4 column. C_8 columns are also commercially available, as well as columns with various ion-exchange substituents.

2.1.3.1. Prep System (2.0-g Load Capacity)

1. Waters PrepPak 500 Cartridge.
2. Delta Pak™ C_{18}, 300 Å, 15 μm column (47 × 300 mm).

2.1.3.2. Small-Scale Prep (Up to 100-mg Load Capacity)

This consists of a Vydac C_{18}, 300 Å, 15–20 μm (2.2 × 25 cm) column #218TP152022.

2.1.3.3. Analytical Scale (Fraction Analysis)

This consists of a Vydac, protein and peptide C_{18}, Å 5 mm, (0.46 × 25 cm) #218TP54.

2.2. Reagents

All reagents should be of the highest chromatographic quality to ensure accurate and reproducible results.

1. Acetonitrile (MeCN): Fisher Optima™ grade. A slightly lower grade may be substituted for large-prep runs because of repetitive washings, large volume, and high flow rate (4–6 L/run at 100 mL/min).
2. Trifluoroacetic acid (TFA): Aldrich 99+% (corrosive, toxic, hygroscopic). HPLC-grade TFA is essential to maintain chromatographic integrity.
3. Triethylamine (TEA): Fisher reagent grade (flammable, causes severe burns and irritation).
4. Phosphoric acid (H_3PO_4): Aldrich (85 wt% solution/H_2O) corrosive.
5. Sodium chloride (NaCl).

2.3. Mobile Phase

All mobile-phase formulations may be extrapolated to accommodate the specific purification scale. (In Section 3.2. and figure legends, A = Aqueous buffer; B = Organic modifier).

2.3.1. *TFA System* (4) *(pH = 2.6)*

1. 0.1% TFA/distilled H_2O (v/v) (HPLC-grade H_2O for analytical scale).
2. Acetonitrile (0.1% TFA for analytical scale).

2.3.2. *Ammonium Acetate System*

1. 0.05% Acetic acid/distilled H_2O (v/v); pH as desired (4–6) with NH_4OH.
2. Acetonitrile.

2.3.3. *Phosphate System* (5,6) *(TEAP) pH as Desired*

1. Triethylammonium phosphate (TEAP) 2.3 (pH = 2.3).
2. 0.0125% TEA/distilled H_2O (v/v), (adjust pH with H_3PO_4; *see* Note 1).
3. Acetonitrile (*see* Note 2).

2.3.4. *Sodium Chloride System* (7) *(pH = 2.5)*

1. 0.15*M* NaCl/distilled H_2O (pH to 2.5 with HCl).
2. Acetonitrile (*see* Note 2).

3. Methods

Purification of natural or synthetic peptides is not the result of a single scheme. It is, however, the qualitative consequence of the "synthesis-to-product" cycle. For that reason, each step is described below and illustrated in Fig. 1.

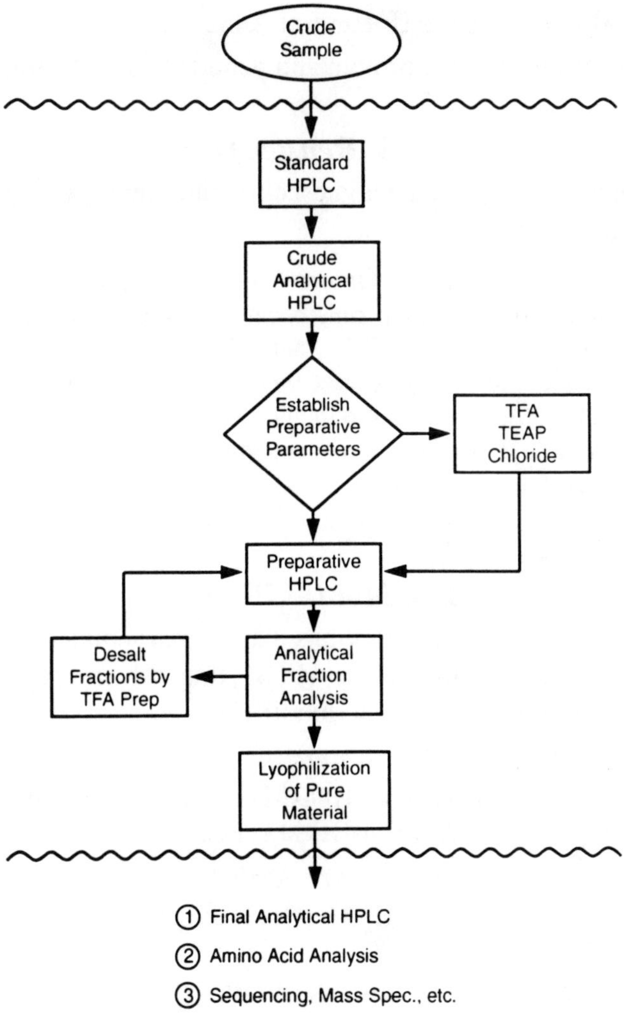

Fig. 1. Purification flowchart.

3.1. Sample Preparation

RP-HPLC analysis and purification require maximum solubility of the material in solvents most compatible with the instrument, while not hindering recovery of the peptide from the solvent. Crude products may be dissolved in a variety of aqueous-based solvent systems. Many of these may contain acetic acid, guanidine, urea, HCl, TFA, or MeOH in differing relative concentrations (*see* Note 3). These are not harmful to the instrument, because the peptide remains hydrophobically bound to the column whereas the solubilizing agent is quickly eluted.

A number of prepurification steps may be performed to facilitate purification and increase column life. For example, crude peptides received directly from cleavage usually contain substantial amounts of residual scavenger or extracting solvents, such as ether or ethyl acetate. Lyophilization of this crude material followed by filtration will eliminate particulates and most residual scavengers, as well as volatile contaminants. Lyophilized solutions are generally water-based or easily evaporated. Furthermore, solutions containing water/acetic acid mixtures must be diluted to prevent thawing while lyophilizing. Samples that are dissolved in urea mixtures because of poor solubility should not be lyophilized, but loaded on the RP-HPLC column immediately. Alternatively, a guard column or an in-line precolumn filter may be attached to the system, but these may cause pressure discrepancies if not maintained properly, thus decreasing column efficiency or damaging the instrument.

Gradient formation must be determined from an analytical chromatogram prior to sample loading. Gradient formation, solvent selection, and detection are among the most important parameters, and will be addressed later in this discussion. Solubility is a key factor in liquid chromatography because the sample must be adsorbed to the column only to be desorbed later at the critical concentration by the organic modifier *(8)*. However, the solubilizing media must lie within the pH range of the column being used. Most reversed-phase columns are effective at acidic pH values and may be irreversibly destroyed by introduction of a solution of a higher pH value *(3)*. Finally, introduce a reasonable amount of material so as to maximize column and detector abilities. Overloading the system often greatly reduces the desired resolution and separation.

3.2. HPLC System Requirements

The HPLC system is usually prepared to the optimum parameters for the specific peptide prior to introducing the completed material (*see* Note 4).

3.2.1. Stationary Phase

One characteristic of an HPLC purification that the scientist can control is the stationary phase or "column packing." Most peptides and small proteins can be purified using reversed-phase C_4, C_8, or C_{18} columns. Columns linked with a C_{18} aliphatic chain are most commonly used for smaller and less hydrophobic peptides (<40 amino acid residues). Larger and more hydrophobic peptides are more effectively purified by a C_4-substituted column *(9)*. C_8 packings work in much the same way as C_{18} and C_4, but requires intermediate values of solvent for elution of desired product. For most peptide applications, a pore size of 300 Å is recommended. However, in certain cases, such as extremely hydrophilic peptides or very difficult to resolve shoulders, a 100-Å pore size may be utilized to facilitate purifications by increasing surface interactions *(10)*.

3.2.2. HPLC Mobile Phase

The mobile phase of the HPLC system is the most important variable with respect to determining component elution. It is a combination of varying percentages of aqueous and organic solvents, utilizing a variety of gradient conditions, such as linear, hyperbolic, step, and isocratic elution. Selection of solvents and gradient conditions will determine the behavior of the target molecule.

Two of the most useful and convenient systems are those involving: (TFA system *[5,6]*) acetonitrile vs distilled H_2O (0.1% TFA) and a more ion-pairing system involving triethylammonium phosphate (TEAP *[7]*): acetonitrile vs distilled H_2O (buffered to a specific pH with TEA and H_3PO_4). Acetonitrile is preferred by this author because of the ease with which it can be removed by lyophilization to yield a "fluffy" peptide, as well as the excellent absorbance properties at wavelengths at which the peptide absorbs.

The TFA system is ideal for cases where the product peak is a large percentage of the total crude cleavage mixture or there are no closely eluting hydrophilic or hydrophobic contaminants, known as "shoulders," on the desired peak. A simple one-step HPLC purification is easily accomplished followed by lyophilization of the desired fractions collected.

However, in such cases where the synthesis or crude material is heavily contaminated, the use of one or more of the TEAP systems followed by a TFA desalting run may be necessary. This TEAP strategy offers enhanced purification capabilities in most cases *(6)*, but requires greater time and effort commitment, and is therefore only used when necessary. Generally, TEAP 2.3 is used as the initial purification, with consequent TEAP runs of increasing pH as needed (up to pH = 7.0 for column stability). When an appropriate level of purity has been achieved, the phosphate salt is removed by diluting with H_2O (2–3×) and reloading the pure material. A shorter gradient may be utilized to desalt the material on the

TFA system because of the higher degree of purity. Also, lyophilization of TEAP fractions will result in a harmful phosphoric acid syrup—hence the need for subsequent desalting.

3.2.3. System Operation

Optimal separation and resolution are achieved by gradient determination specific to each peptide and are directly comparable to the analytical profile. Therefore, the analytical profile is the major reference point for parameter determination by giving sample purity and identification of contaminant shoulders and approximate organic solvent concentration needed for elution (*see* Note 5). Consequently, one may predict a suitable solvent system as well as appropriate gradient conditions. **Note:** The TEAP system is used for situations where closely eluting shoulders appear, and material tends to elute 5–8% (B) (organic) earlier in TEAP than in an identical TFA system *(6)*. Retention time, resolution, and separation are affected slightly by the amount to be purified. Quite often, results improve at lower levels (200–700 mg of most peptides at the preparative scale). Most peptides act favorably in TEAP systems. However, there are exceptions that are difficult to recover successfully (*see* Notes 6 and 7).

3.2.3.1 System Start-Up and Gradient Determination

The standard RP-HPLC system is stored at 100% of the organic modifier for overnight and multiday periods to prevent microbacterial accumulation in aqueous solvents. The system must be re-equilibrated to the initial starting conditions prior to sample loading. We generally employ a reverse gradient to accomplish this procedure. A reverse gradient is effected by accomplishing a rate of change, from 100% of the organic modifier to 0%, resulting in maximal cleansing of the RP-HPLC column in a reasonable amount of time. A time of 20–30 min is generally sufficient, followed by approx 10 min isocratically at 0% B, or the initial starting conditions, to equilibrate the system fully. Three to five column volumes of aqueous solvent are often a sufficient volume for this purpose.

The RP-HPLC system must be checked for correct operation parameters. If the system is equipped with a radial compression chamber, the radial pressure must be checked for a steady and suitable operating pressure. The Waters Delta-Prep 3000 operates optimally at 650–700 psi. It is imperative to maintain a steady system pressure throughout the HPLC run. Therefore all leaks and excessive backpressures must be eliminated.

Clogged precolumn filters as well as filter paper applied directly to the RP-HPLC column are most commonly found to produce increased pressure readings. These should be checked and replaced regularly. Pressure variations are also a consequence of irregular flow patterns. Flow rates must be examined routinely and at different flow rate values, thereby revealing any pump system or general flow discrepancies. Irregular flow and major pressure fluctuations may also result from worn or damaged check valves.

Detection parameters are vital to each RP-HPLC run and should be adjusted accordingly prior to each sample loading. The pertinent variables to be addressed are the detection wavelength and the range of the detector.

Wavelength selection is commonly set at approx 230 nm for the semipreparative scale. The peptidyl backbone is easily detected at 220 nm. However, limiting factors, such as larger sample size or smaller flow cell dimensions, indicate a need for a slightly higher wavelength setting to reduce sensitivity of the instrument.

The range function can be viewed as a "window," and operates in a similar fashion. The range employs aufs units (absorbance units full scale) and translates absorbance data to the chart recorder. The range scale commonly exhibits settings as 0.001, 0.01, 0.05, 0.1, 0.2, 0.5, 1.0, and 2.0 increasing in magnitude. Increasing values expands the range and therefore decreases sensitivity. Likewise, decreasing range values results in increased sensitivity of detection, allowing one to see minor components more easily. Sample sizes of 100–500 mg can be effectively run at a range of 0.2–0.5 aufs. The range should be increased according to the number of aromatic residues or fluorescent groups contained in the peptide. Additionally, alternative wavelengths can be employed, such as 254, 275, 280, or others depending on the wavelength that allows one to observe only components containing aromatic residues. With these wavelengths, the range must be decreased because of the lower absorbance value at these wavelengths.

Following a check of all parameters, a gradient is constructed, and the sample loaded at either 0% of the organic modifier or the initial conditions of the gradient. Generally, this gradient is constructed to accomplish a specific rate of change of organic modifier, relative to an aqueous cosolvent. This gradient must result in a rate of change sufficient to achieve maximal separation of impurities from the desired product (*see*

Note 8). As a general rule, the gradient is designed so the elution percentage of the organic modifier is the midpoint of the gradient and the sample is loaded at the initial gradient conditions. Our laboratory typically loads samples at 0% organic modifier to ensure sample binding to the column matrix. Consequently, a peptide eluting from the column at 30% of the organic modifier analytically will be run on a gradient of approx 15–35% at the semipreparative level, with the sample being loaded at 0–15% organic concentration.

For efficiency purposes, the gradient should be designed for a run time of approx 1 h. This will allow for a maximum number of runs and column washes in a working period, while maintaining efficient separations. However, maintaining the percentage change in organic modifier while increasing run time (10–15 min) may also increase separation.

Equilibration to the initial conditions, following sample adsorption to the column matrix, allows elution of solubilizing agents, such as acetic acid and urea, as well as early eluting contaminants. The gradient program is then initiated, and the observed eluting peaks collected manually or automatically with a fraction collector into appropriately sized containers, such test tubes or flasks. Optimal separation of even moderately impure samples is achieved by limiting the size (volume) of the individual fraction collected. In our laboratory, we have employed the following standard parameters: a flow rate of 100 mL/min, gradient 20% change in organic modifier in 1 h, and the fractionation method by manual collection in 1.6×25 cm test tubes. Also, more complex mixtures may be more easily purified by collecting even smaller volumes or "half-fractions."

Alternatively, most instruments offer a "pause" feature that allows a gradient to be held at any point during the run. Use of this feature is helpful for separating closely eluting contaminants by effecting an isocratic procedure (a constant unchanging concentration of organic modifier) within the gradient, wherein the gradient is operated up to a point near the peptide eluting concentration. The gradient is halted at the particular organic percentage, allowing a gradual separation of product from contaminants.

The practice of isocratic elution may be taken a degree further by employing an entire isocratic RP-HPLC run at a constant organic modifier concentration over a predetermined time parameter. Care must be taken to ensure efficient binding of the sample to the column's stationary matrix. Consequently, the product gradually separates away from contaminants

Table 1
Purification Yields for Representative Peptides

Peptide	Crude start amt.	HPLC	Final yield
CRF	500 mg	① 0.15*M* NaCl	100 mg
	100 mg	② TEAP 4.7	(semipure)
		③ TFA desalt	24 mg
GRF	4 × 500 mg	① TEAP 2.25	1.161g
		② TEAP 6.8	
		③ TFA desalt	
Echistatin	1.8 g[a]	① TEAP 2.3	91 mg
		② TFA desalt	
Charybdotoxin	1.8 g[a]	① Drop sample pH to 2.5	80 mg
		② TEAP 2.3	
		③ TFA desalt	

[a]These numbers represent an approximation of theoretical weight based on the final resin weight. These samples were oxidized directly without lyophilization following cleavage.

at the midpoint of the run. Generally, the concentration is kept constant over a time period of 30–60 min. Also, mixtures of different organic modifiers, such as isopropanol:acetonitrile, may be utilized in an isocratic procedure to enhance separation parameters further.

Following a successful prep run, one must analyze the fractions that were collected. Fraction analysis is of vital importance and, therefore, should be analyzed by the most rigorous methods available (*see* Note 5). Analytical columns and reagents should be of the highest grade available and the gradient formation equally as rigorous. A smaller percent change in organic modifier and a lower flow rate (~1 mL/min) often provide the desired results.

Finally, the RP-HPLC semipreparative column is washed vs the organic solvent and equilibrated prior to the next run. Multiple gradient washings at lesser time intervals (0–100% B in 10 min × 3) often are more effective than one longer wash (0–100% B in 30 min × 1). Following the column wash, the system is equilibrated to 0% B or the subsequent initial conditions. For storage purposes, the system should remain equilibrated in 100% of the organic solvent.

3.3. Purification Examples

3.3.1. Growth Hormone-Releasing Factor (GRF) (11)

N-Acetyl-Tyr_1-D-Arg_2-GRF(1-29)AMIDE(human)(GRF Antag) (Table 1): The peptide was dissolved in 20% AcOH. The GRF antagonist

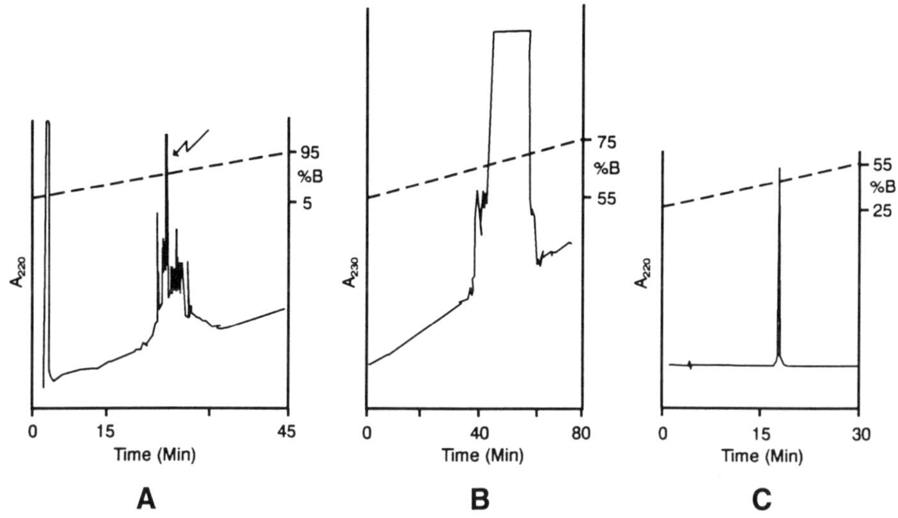

Fig. 2. GRF antagonist: (**A**) Crude mixture analysis: analytical; 5–(45 min)–95% B; flow rate = 1.5 mL/min; chart speed = 0.5 cm/min; range = 0.2 aufs. (**B**) Preparative HPLC: 55–(80 min)–75% B; flow rate = 100 mL/min; chart speed = 0.25 cm/min; atten. = 0.5 aufs. (**C**) Final pure analytical HPLC: 25–(30 min)–55% B; flow rate = 1.0 mL/min; chart speed = 0.5 cm/min.

was purified by TEAP 2.3 followed by TEAP 6.8 to eliminate closely eluting contaminants (Fig. 2A), and subsequently desalted on the TFA system prior to lyophilization. This procedure resulted in a product with a purity level of >98%. However, a later purification run demonstrated the product to oxidize partially at a Met during the higher pH procedures and without resolving the contaminants at lower pH runs. This peptide modification was eliminated and the contaminants separated by employing a unique solvent system (Fig. 2B). This specific system consisted of an aqueous solution (A) of 0.15*M* NaCl (pH 2.4 with HCl) and the organic modifier (B) of 10% MeCN in MeOH. The resulting product was desalted using a TFA prep, active run prior to lyophilization. Purity was determined to be >98% (Fig. 2C).

3.3.2. Corticotropin-Releasing Factor (CRF) Ovine (12)

CRF ovine (Table 1) was solubilized in 50% AcOH for purification. The crude RP-HPLC profile showed a broadening of the target peak (Fig. 3A) as a result of closely eluting contaminants. The TEAP system

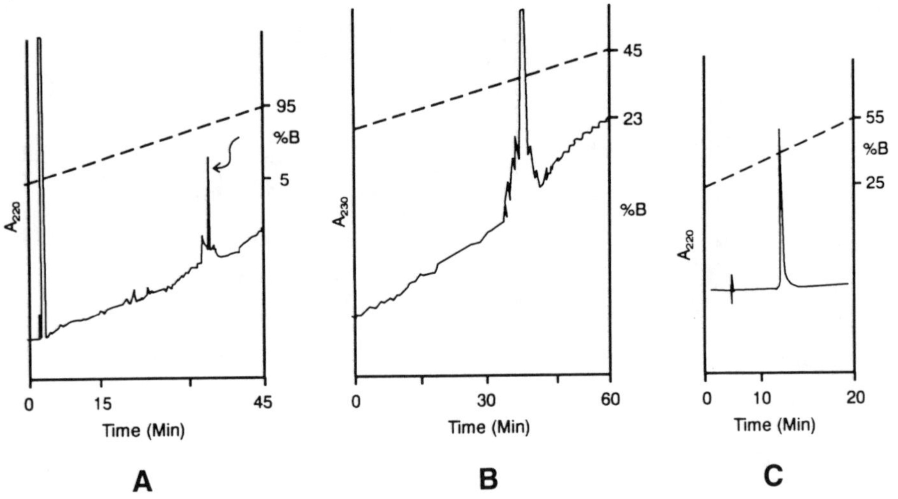

Fig. 3. CRF ovine: (**A**) crude mixture after lyophilization and filtration; linear gradient of 5–(45 min)–95% B; flow rate = 1.5 mL/min, chart speed = 0.5 cm/min; range = 0.2 aufs. (**B**) Preparative HPLC TEAP 2.3: linear gradient of 23–(60 min)–45% B; flow = 100 mL/min; chart speed = 0.25 cm/min; atten. = 0.5 aufs. (**C**) Final pure analytical HPLC: gradient = 25–(20 min)–55% B; flow 1.0 mL/min; tailing is seen routinely with CRF ovine.

was employed because of its excellent separation abilities under these circumstances (Fig. 3B). The final pure product (Fig. 3C) was obtained following a TFA desalt RP-HPLC and lyophilization.

3.3.3. Echistatin

Echistatin (Table 1) is a 49 amino acid polypeptide from the venom of the saw-scaled viper, *Echis carinatus*, and contains four disulfide bonds *(13,14)*. The free peptide was air-oxidized, and the crude RP-HPLC showed two peaks of similar intensity. The latter peak is the target peak, the earlier eluting peak represented an oxidation of methionine-to-methionine sulfoxide (Fig. 4A). Purification by TEAP 6.0 resulted in the most efficient separation of the two peaks at the semipreparative level (Fig. 4B). The relevant fractions were desalted by TFA RP-HPLC and lyophilized yielding a product of >98% purity (Fig. 4C).

3.3.4. Charybdotoxin

Charybdotoxin (Table 1) is a 37 amino acid peptide found in *Leiurus quinquestriatus*, scorpion venom, and contains three disulfide bonds *(15)*.

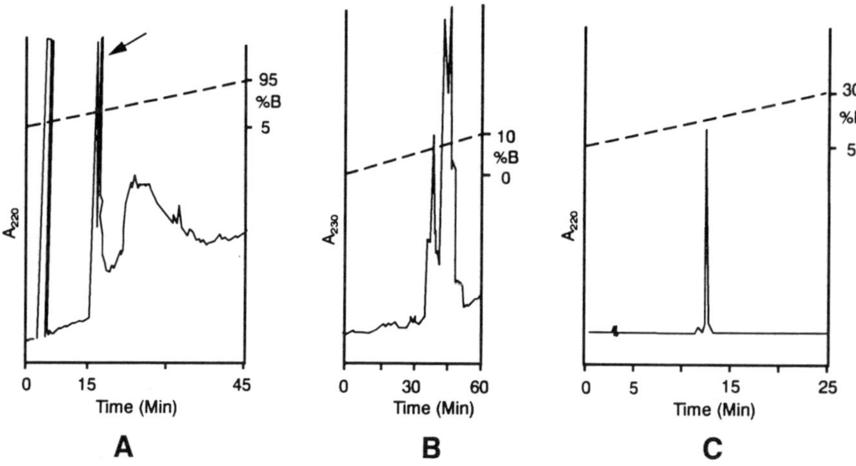

Fig. 4. Echistatin: **(A)** Crude mixture after disulfide bond formation; gradient = 5–(45 min)–95% B; flow = 1.5 mL/min; atten. = 0.2 aufs. **(B)** Preparative HPLC: gradient = 0–(60 min)–10% B; flow = 100 mL/min; atten. = 0.1 aufs. **(C)** Final pure analytical HPLC: gradient = 5–(25 min)–30% B; flow = 1.0 mL/min.

Shown in Fig. 5A is the fully reduced crude peptide amidst a mountain of peaks. On air oxidation, the biologically active molecule migrated to a retention time preceding the bolus of contaminant peaks (Fig. 5B). The pH of the solution was dropped to 2.5 prior to loading onto the RP-HPLC column. In order to ensure sufficient purity, the material was eluted from the column using the TEAP 2.3 system (Fig. 5C). The resulting fractions were desalted by the TFA RP-HPLC system and lyophilized to yield a product of >98% purity (Fig. 5D).

4. Notes

1. The TEAP system may be adjusted as desired in order to maximize the individual purification by varying the pH values between 2.3 and 6.8.
2. It is advisable to store the HPLC system in 100% acetonitrile to avoid halide corrosion of the metal parts and prevent microbacterial accumulation in aqueous solvents. The system must be washed following purifications involving halides.
3. Lyophilization of peptides solubilized in urea or guanidine is not recommended. A desalting run using the TFA system is recommended instead. This will produce a more realistic RP-HPLC profile and, combined with sample filtration, will maintain column integrity.

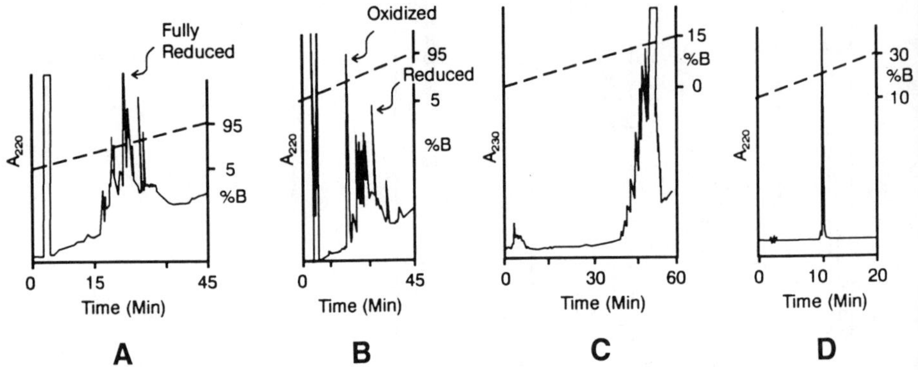

Fig. 5. Charybdotoxin: (**A**) crude sample before air oxidation; gradient = 5–(45 min)–95% B; atten. = 0.2 aufs.; flow = 1.5 mL/min; inj. vol. = 250 mL of 4-L solution. (**B**) Crude sample following disulfide bond formation; HPLC conditions same as above. (**C**) Preparative HPLC, TEAP 2.3: linear gradient = 0–(60 min)–15% B; atten. = 0.2 aufs.; flow = 100 mL/min; (**D**) Final pure analytical HPLC: gradient = 10–(20 min)–30% B; flow = 1.0 mL/min.

4. The smallest change in one of the purification parameters may greatly increase the difficulty of the immediate operation. For example, a minor change in system, such as fluctuation of room temperature, may result in differences in elution time as well as contaminant separation, thus causing irreproducible results.
5. Another common variable is the pre- or postpurification oxidation of susceptible amino acids (Cys, Met). This problem, although inconvenient, is a preventable problem. Extreme hydrophilicity or hydrophobicity of a peptide can create an ineffective purification owing to premature or incomplete elution from the column matrix. Since most alkyl chain-linked columns have an effective sample load range (100 mg–2.0 g) and pH (2–7.5) *(3)*, these parameters may be varied to affect the peptide purification. The following examples are of unique situations encountered in our lab as well as possible solutions.
6. Each HPLC run results in partial sample loss owing to nonspecific adsorption to the column matrix. Therefore, minimizing the number of runs will give a greater yield in recovered product.
7. Hydrophilic peptides that do not adsorb well to the column matrix may be more successfully purified using a smaller pore size particle (100 Å), thus promoting increased surface interaction *(10)*.
8. Never discard eluent prematurely! If sample is not observed eluting during run, one may check the loading wash as well as the 100% acetonitrile wash to find the product. Absorbance detection may misrepresent actual product yield.

Acknowledgments

I thank Dr. Michael Pennington for insight and helpful discussion. I also thank Carla DuRant for typing the text.

References

1. Merrifield, R. B. (1963) Solid phase peptide synthesis. I. The synthesis of a tetrapeptide. *J. Am. Chem. Soc.* **85,** 2149–2154.
2. Barany, G., Kneib-Cordonier, N., and Mullen, D. G. (1987) Solid phase peptide synthesis: a silver anniversary report. *Int. J. Peptide Protein Res.* **30,** 705–739.
3. Krause, E., Smettan, D., Loth, F., and Herma, H. (1990) Polyalkylenes used as stationary phases for preparative reversed-phase liquid chromatography. *J. Chromatogr.* **520,** 263–269.
4. Bennett, J. P. J., Hudson, A. M., McMartin, C., and Purdon, G. E. (1977) Use of octadecasilyl-silica for the extraction and purification of peptides in biological samples. *Biochem. J.* **168,** 9–13.
5. Rivier, J. (1978) Use of trialkylammonium phosphate (TAAP) buffers in reverse phase HPLC for high resolution and high recovery of peptides and proteins. *J. Liquid Chromatogr.* **1,** 343–367.
6. Hoeger, C., Galyean, R., Boublik, J., McClintock, R., and Rivier, J. (1987) Preparative reversed phase high performance liquid chromatography: effects of buffer pH on the purification of synthetic peptides. *Biochromatography* **2(3),** 134–142.
7. Zanelli, J. M., O'Hare, M. J., Nice, E. C., and Corran, P. H. (1981) Purification and assay of bovine parathyroid hormone by reversed-phase HPLC. *J. Chromatog.* **223,** 59–67.
8. Kamp, R. M., Bosserhoff, A., Kamp, D., and Wittman-Leibold, B. (1984) Application of high performance liquid chromatographic techniques to the separation of ribosomal proteins of different organisms. *J. Chromatogr.* **317,** 181–192.
9. Rivier, J., McClintock, R., Galyean, R., and Anderson, H. (1984) Reversed phase HPLC: preparative purification of synthetic peptides. *J. Chromatogr.* **288,** 303–324.
10. Fallick, G. J. and Waters, J. L. (1972) Making maximum use of high speed L.C. *Am. Lab.* **4(8),** 21–32.
11. Robberecht, P., Coy, D. H., Waelbroeck, M., Heiman, M. L., de Neef, P., Camus, J. C., and Christophe, J. (1985) Structural requirements for the activation of rat anterior pituitary adenylate cyclase by growth hormone-releasing factor (GRF): discovery of (N-Ac-Tyr$_1$-D-Arg$_2$)-GRF(1-29)-NH_2 as a GRF antagonist on membranes. *Endocrinology* **117(5),** 1759–1764.
12. Morell, J. L. and Brown, J. H. (1985) Solid phase synthesis of ovine corticotropin releasing factor. *Int. J. Peptide Protein Res.* **26,** 49–54.
13. Gan, Z., Gould, R. J., Jacobs, J. W., Friedman, P., and Polokoff, M. A. (1988) Echistatin: a potent platelet aggregation inhibitor from the venom of the viper, *Echis Carinatus. J. Biol. Chem.* **263(36),** 19,827–19,832.
14. Garsky, V. M., Lumma, P. K., Freidinger, R. M., Pitzenberger, S. M., Randall, W. C., Veber, D. F., Gould, R. J., and Freidman, P. A. (1989) Chemical synthesis of echistatin, a potent inhibitor of platelet aggregation from *Echis Carinatus*: synthesis and biological activity of selected analogs. *Proc. Natl. Acad. Sci. USA.* **86,** 4022–4026.

15. Sugg, E. E., Garcia, M. L., Reuben, J. P., Patchett, A. A., and Kaczorowski, G. J. (1990) Synthesis and structural characterization of charybdotoxin, a potent peptidyl inhibitor of the high conductance Ca^{2+}-activated K^+ channel. *J. Biol. Chem.* **265(31),** 18,745–18,748.

CHAPTER 5

Applications of Strong Cation-Exchange (SCX)-HPLC in Synthetic Peptide Analysis

Dan L. Crimmins

1. Introduction

The ability to produce a wide array of synthetic peptides routinely for diverse biomedical research applications has increased dramatically over the past 10 years. This is primarily a result of improvements in automated solid-phase peptide synthesis, impressive developments in orthogonal synthesis strategies, and an expanding supply of protected amino acids. Analytical characterization of the synthesized product is an absolute necessity because of the myriad of potential synthesis problems *(1)* comprising sequence deletions, incomplete deprotection, sequence fragmentation, and unexpected adducts as a consequence of improper resin cleavage to name a few. No single analytical technique will consistently provide sufficient information to assess sample purity, and for this reason, a combination of high-performance liquid chromatography (HPLC), N-terminal sequence analysis, amino acid composition, capillary electrophoresis, and mass spectrometry is used.

The most popular chromatographic method to analyze synthetic peptide components is reversed-phase (RP)-HPLC *(2–4)*. In some instances, however, alternative modes of chromatography may prove beneficial. Strong cation-exchange (SCX)-HPLC is one such example and is particularly well suited for characterization of peptides *(5–8)*. Generally, SCX-HPLC is performed near pH 3.0, where the peptide carboxylates are predominantly in a protonated form so separation mainly results from

From: *Methods in Molecular Biology, Vol. 36: Peptide Analysis Protocols*
Edited by: B. M. Dunn and M. W. Pennington Copyright ©1994 Humana Press Inc., Totowa, NJ

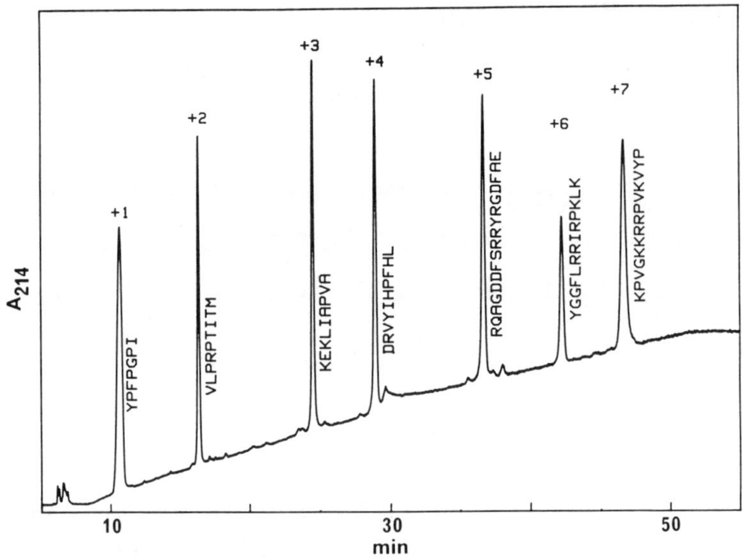

Fig. 1. SCX-HPLC for a peptide mixture ranging in net nominal positive charge of +1 to +7. The standard single-letter codes for the amino acids are used, and the chromatogram was monitored at 214 nm (A_{214}) and displayed at 80 mV Full Scale (mVFS). The "free" amino terminus, arginine (R), lysine (K), and histidine (H) are each assigned a value of +1 at pH 3.0.

differences in positive charge. A sulfoethyl aspartamide SCX column (*see* Note 1) has been shown to be extremely effective for the general analysis of synthetic peptides *(5–8)*, N-terminally blocked peptides *(5,9)*, and peptide fragments derived from proteolytic digests *(10–14)*. It also complements information from capillary electrophoresis *(15)*, and was used successfully to analyze and isolate disulfide-linked peptides *(16,17)*.

The utility of SCX-HPLC to analyze several types of peptides resides in the monotonic elution of these peptides as a function of their nominal net charge at pH 3.0 *(5,6)*, is illustrated in Fig. 1. This experimental observation allows one to predict elution characteristics rationally for groups of peptides. In this chapter, I use this information to analyze disulfide-linked dimer synthetic peptides. Both homo-peptide and hetero-peptide dimers were produced and their chromatographic behavior investigated by SCX-HPLC. For comparative purposes, these dimers were also analyzed by standard C_{18}RP-HPLC, since this mode of chromatography is likely to be more familiar to most readers.

2. Materials

2.1. Reagents, Chemicals, and Solvents

Peptides were purchased from Peninsula (Belmont, CA) and used without further purification. The lyophilized peptides were reconstituted to a nominal concentration of 1 mg/mL with 0.1% (v/v) acetic acid, the resulting solutions blanketed with N_2, and then stored at 4°C until use. Acetonitrile (MeCN) and methanol (MeOH) were high-performance liquid chromatography (HPLC)-grade Burdick and Jackson products obtained from Baxter (McGaw Park, IL); Fisher (St. Louis, MO) supplied 85% (w/v) HPLC-grade phosphoric acid (H_3PO_4) and HPLC-grade sodium acetate; the reversed-phase (RP) ion-pairing agent trifluoroacetic acid (TFA) and amino acid analysis reagents phenylisothiocyanate (PITC), 6*N* HCl, triethylamine (TEA), and standard H were purchased from Pierce (Rockford, IL). Sigma (St. Louis, MO) was the source for Ellman's reagent (DTNB), 5,5'-dithiobis(2-nitrobenzoic acid) and dithiothreitol (DTT). High-quality water was obtained by sequential passage through ion-exchange and carbon canisters and a Milli-Q (Millipore, Boston, MA) apparatus with 0.2-µm final filter *(5,9)*. All other materials were of the highest quality available and obtained from local suppliers.

2.2. HPLC Instrumentation and Columns

A Waters (Milford, MA) gradient HPLC system consisting of two 510 pumps, a 680 gradient controller, a 710 WISP autosampler, a 490 four-channel detector, and a 1122 column temperature controller was used. Data were acquired and analyzed with Nelson Analytical (now PE-Nelson, Cupertino, CA) 700 series A/D boxes and 4400 series software *(5,9)*.

The NEST group (Southborough, MA) supplied the 300-Å, 5-µ strong cation-exchange (SCX) sulfoethyl aspartamide column, 200 mm × 4.6 mm I.D. (*see* Note 1) and the 300-Å, 5-µ C_{18}RP Vydac column, 250 mm × 4.6 mm I.D. Derivatized amino acid hydrolysates were analyzed on a 300-Å, 5-µ Altex (Beckman, San Ramon, CA) ODS C18 PTH column, 250 mm × 4.6 mm id.

2.3. HPLC Mobile-Phase Preparation

2.3.1. SCX-HPLC

A 2X stock solution, without MeCN, of [A] and [B] is prepared as follows.

2X [A]: 10 m*M* sodium phosphate, pH 3.0. Add 1.36 mL of 85% (w/v) H_3PO_4 to ~1800 mL high-quality water, and then titrate to pH 3.0 with 10*N* NaOH, q.s. to 2 L with high-quality water. Recheck pH before filtering through a Rainin (Woburn, MA) 0.2-µm Nylon 66 filter.

2X [B]: 10 m*M* sodium phosphate, 1*M* NaCl, pH 3.0. Add 1.36 mL of 85% (w/v) H_3PO_4 and 116.9 g of NaCl to ~1800 mL high-quality water. Titrate to pH 3.0 with 10*N* NaOH, q.s. to 2 L with high-quality water. Recheck pH before filtering through a Nylon 66 filter as above.

Working [A]: 5 m*M* Sodium phosphate pH 3.0; 25% (v/v) MeCN. Mix 500 mL of 2X [A], 250 mL of MeCN, and 250 mL of high-quality water.

Working [B]: 5 m*M* sodium phosphate, 500 m*M* NaCl pH 3.0; 25% (v/v) MeCN. Mix 500 mL of 2X [B], 250 mL of MeCN, and 250 mL of high-quality water.

2.3.2. Standard C_{18}RP-HPLC

[A]: 0.1% TFA. Mix 1 L of high-quality water and 1 mL of neat TFA.

[B]: 90% MeCN, 0.095% TFA. Mix 900 mL of MeCN, 100 mL of high-quality water, and 0.95 mL of neat TFA.

2.3.3. Amino Acid Analysis C_{18}RP-HPLC

Buffer [A]: 150 m*M* sodium acetate, 0.05% (v/v) TEA pH 6.35. Mix 40.8 g of sodium acetate trihydrate and 1 mL of neat TEA to ~1900 mL high-quality water. Titrate to pH 6.35 with glacial acetic acid, q.s. to 2 L with high-quality water. Recheck pH prior to filtering through a Nylon 66 filter.

Buffer [B]: 60% (v/v) MeCN. Mix 1200 mL of MeCN with 800 mL of high-quality water.

2.3.4. Reaction Buffer

[A]: 100 m*M* sodium phosphate, 5 m*M* EDTA, pH 7.4. Add 4.32 g of Na_2HPO_4 with 2.70 g of NaH_2PO_4 monohydrate and 0.93 g Na_2EDTA dihydrate to ~450 mL with high-quality water. Titrate to pH 7.4 with 10*N* NaOH, q.s. to 500 mL with high-quality water. Recheck pH before filtering with a Nylon 66 filter, and store at 4°C.

[B]: 50 m*M* sodium phosphate, 2.5 m*M* EDTA, 0.25 m*M* DTNB, pH 7.4. Add 9.9 mg of DTNB, 50 mL of 100 m*M* sodium phosphate, 5 m*M* EDTA, pH 7.4, to ~95 mL with high-quality water. Titrate to pH 7.4 with 1*N* NaOH, q.s. to 100 mL with high-quality water. Recheck pH, and store at 4°C.

3. Methods

3.1. Preparation of Disulfide-Linked Peptides (16)

3.1.1. Preparation of Homo-Peptide Disulfide-Linked Dimers

Prepare homo-peptide dimers by reacting 10 mL of stock peptide (0.67–1.45 m*M*) with 80 µL of 100 m*M* sodium phosphate, 5 µ*M* disodium ethylenediaminetraacetic acid (EDTA), pH 7.4, and 10 µL of freshly prepared 0.25 m*M* DTNB in 50 m*M* sodium phosphate and 2.5 m*M* EDTA pH 7.4. Blanket the reaction mixture with N_2 and incubate for 2 h at room temperature. Also prepare a mixed-disulfide peptide:2-nitro-5-thiobenzoic acid (TNB) adduct, and use as a chromatographic marker by increasing the concentration of DTNB 10-fold to 2.5 m*M*. Prepare control samples by omitting DTNB from the 50 m*M* sodium phosphate and 2.5 m*M* EDTA, pH 7.4 solution.

3.1.2. Preparation of Hetero-Peptide Disulfide-Linked Dimers

A few modifications to the above protocol are made to produce the desired hetero-peptide dimers (*see* Note 2). React 10 µL of the "first" peptide plus 80 µL of the 100-m*M* sodium phosphate buffer with 10 µL of a 0.75-m*M* (threefold increase from above) DTNB solution to produce a mixed-disulfide, peptide:TNB adduct. After a 2-h incubation at room temperature under N_2 blanket, add 10 µL of the "second" peptide, and allow the thiol-disulfide interchange reaction to proceed for an additional 2 h at room temperature.

3.1.3. Reduction of Disulfide-Linked Peptide Dimers

To reduce disulfide bonds, add 10 µL of 0.5*M* aqueous DTT (77.2 mg of DTT plus 1 mL of high-quality water) to either the 100 µL of the homo-peptide dimer solution or the 110 µL of the hetero-peptide dimer solution, and allow to react for 1 h at 45°C under N_2 blanket.

3.2. HPLC Operation

3.2.1. SCX-HPLC (See *Note 3)*

1. Use a standard linear gradient at a flow rate of 1 mL/min with the column at 28°C. 0% [B] to 100% [B] over 60 min, 100% [B] isocratic for 10 min, 100% [B] to 0% [B] over 5 min, 0% [B] isocratic for 25 min.
2. Next sample injection may be at 100 min.
3. Wash the column with high-quality water, and store in MeOH if it will not be used the next day.

3.2.2. Standard C_{18}RP-HPLC (see *Note 4)*

1. Use a standard linear gradient at a flow rate of 1 mL/min with the column at 37°C. 0% [B] to 60% [B] in 60 min, 60% [B] to 100% [B] in 5 min, 100% [B] isocratic for 5 min, 100% [B] to 0% [B] in 5 min, and 0% [B] isocratic for 25 min.
2. Next sample injection may be at 100 min. Wash the column, and store in 100% [B] indefinitely.

3.2.3. Amino Acid Analysis C_{18}RP-HPLC (see *Note 5*)

1. Use a combination of a linear and a convex gradient at a flow rate of 1.2 mL/min with the column at 38°C.
2. 10% [B] isocratic for 5 min, 10% [B] to 50% [B] over 30 min with convex gradient #5, 50% [B] to 100% [B] over 2.5 min with a linear gradient, 100% [B] isocratic for 5 min, 100% [B] to 10% [B] over 2.5 min with a linear gradient, and 10% [B] isocratic for 15 min. Next sample injection at 60 min.
3. Store the column in 100% [B] for several days. Otherwise, wash the column with high-quality water, and store in MeOH.

3.3. Analysis of HPLC Fractions

Collect fractions (0.5 mL) from the SCX chromatogram at 0.5-min intervals. It is possible to analyze these samples directly without resorting to desalting if an aliquot of 50 µL or less is taken for analysis. The peptide content of the hetero-peptide dimers from SCX-HPLC is assessed by amino acid composition (*see* Note 6). A standard manual procedure comprising vapor-phase hydrolysis with 6*N* HCl + 1% (w/v) phenol for 1 h at 150°C in a Waters work station and precolumn PITC derivatization of the resulting hydrolysate may be used *(18–20)*.

3.4. Results

3.4.1. SCX- and C_{18}RP-HPLC of Homo-Peptide Disulfide-Linked Dimers

Monomeric synthetic peptides containing a cysteine residue may become oxidized to a disulfide-linked dimer during resin cleavage or sample work-up and manipulation. For some biochemical applications, the dimer may be the desired species, whereas for others, a free sulfhydryl residue is required (e.g., conjugation to carrier proteins for antibody production). In either case, a reliable analytical procedure is required to assay the molecular composition of the peptide solution accurately. I have taken a chromatographic approach to address this issue, although other techniques are available *(16)*.

The homo-peptide dimer of <ESGLGCNSFRY (<E = pyroglutamic acid) was produced as described in Section 3.1.1. via a set of thiol-disulfide interchange reactions. First, the free sulfhydryl internal cysteine peptide was reacted with DTNB to yield a mixed-disulfide peptide:TNB adduct. Because only ~ 0.5 mEq of DTNB was used in

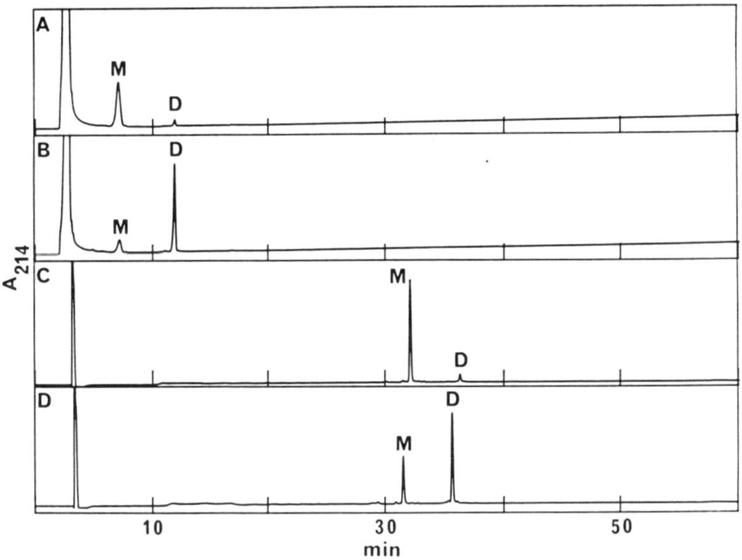

Fig. 2. SCX- and C_{18}RP-HPLC of <ESGLGCNSFRY. The homo-peptide disulfide-linked dimer was produced as described in Section 3.1.1. **(A)** SCX-HPLC of the control (–DTNB) peptide, A_{214} = 300 mVFS; **(B)** SCX-HPLC of the disulfide-linked dimer (+DTNB), A_{214} = 300 mVFS; **(C)** C_{18}RP-HPLC of the control (–DTNB) peptide, A_{214} = 400 mVFS; and **(D)** C_{18}RP-HPLC of the disulfide-linked dimer (+DTNB), A_{214} = 400 mVFS. For all, M = monomer and D = dimer. The homo-peptide dimer was reduced with DTT (*see* Section 3.1.3.), and the chromatogram of the resulting sample was similar to A or C (data not shown). <E = pyroglutamic acid.

the reaction, an appreciable amount of peptide with an unmodified cysteine remained. A second thiol-disulfide interchange reaction thus occurred with the peptide sulfhydryl group displacing the TNB moiety, the net result being a dimeric peptide species with a nominal positive charge at pH 3 of +2 compared to the monomeric peptide of charge +1.

Figure 2 displays the SCX- and C_{18}RP-HPLC profiles for both the reaction mixture (panels B and D, respectively) and a control (–DTNB) sample (panels A and C, respectively). As expected, the dimer species elutes later on the SCX column because of an increase in net positive charge. In this example, the resolution of the monomer (M) and dimer (D) pair on either column is about the same. Area integration of the relevant A_{214} UV peaks for the +DTNB chromatograms gave 80% dimer for SCX-HPLC (panel B) and 82.6% dimer for C_{18}-HPLC (Panel D). Four additional distinct homo-dimer peptides were produced *(16)*, and the average ratio of the percent dimer from SCX-HPLC to percent dimer from C_{18}RP-HPLC was 1.03. This further emphasizes that HPLC can reliably assess the distribution of monomeric and disulfide-linked dimeric species of synthetic peptides.

3.4.2. SCX- and C_{18}RP-HPLC of Hetero-Peptide Disulfide-Linked Dimers

3.4.2.1. SCX and C_{18}RP Chromatograms

Preparation of the hetero-peptide dimer followed the protocol in Section 3.1.2., and the resulting solution was analyzed by SCX and C_{18}RP chromatography. Peptide 1 was an 11-mer, SLRRSSCFGGR, with an internal cysteine at position 7 and a net charge of +4, whereas peptide 2 was a +2 charge, N-terminal cysteine 10-mer, CQDSETRTFY. Peptide 1 is first incubated with DTNB to produce the peptide 1:TNB adduct after which peptide 2 is added, which produces the peptide 1: peptide 2 = D_{1+2} disulfide-linked hetero-peptide dimer.

All possible reaction products were represented in the A_{214} SCX (panel A) and C_{18}RP (panel C) chromatograms of Fig. 3. The peaks were assigned from chromatographic analysis of individual peptide reactions. Importantly, the disulfide-linked hetero-peptide dimer (labeled D_{1+2}) was widely separated from other components on the SCX column (panel A), but much less so when analyzed by C_{18}RP-HPLC (panel C). This observation was also true for five other peptide pairs used for the production of hetero-peptide dimer *(16)*. Furthermore, quantitation of all relevant peaks in each 214-nm chromatogram displayed in Fig. 3 gave 52% hetero-dimer from SCX-HPLC (panel A) and 49% hetero-dimer from C_{18}RP-HPLC (panel C). Simultaneous detection at 325 nm was useful for identification of TNB adducts *(21)*. SCX analysis in this case (panel B) shows two well-resolved peptide:TNB adducts with each individual peptide:TNB species eluting before the corresponding monomeric unmodified peptide. This is likely a result of an increased negative charge contributed by the TNB group. Note also that these two peptide:TNB adducts were not resolved by standard C_{18}RP-HPLC (panel D).

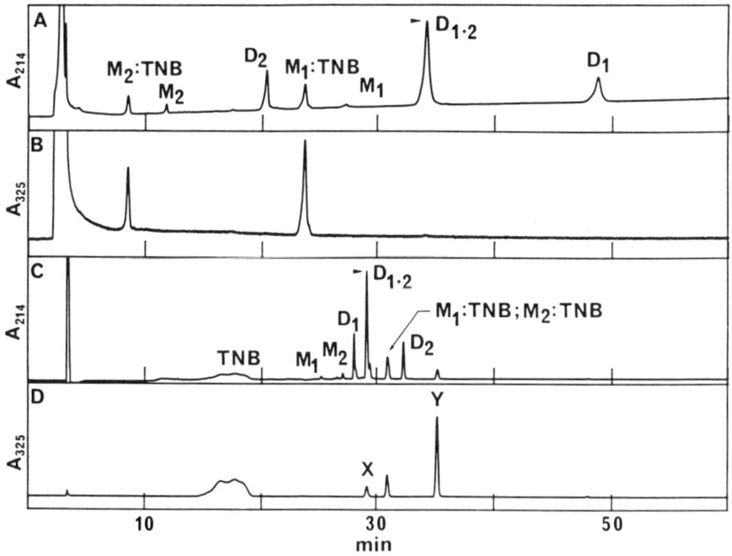

Fig. 3. SCX- and C_{18}RP-HPLC of SLRRSSCFGGR (peptide 1) plus CQDSETRTFY (peptide 2). The hetero-peptide disulfide-linked dimer was produced as described in Section 3.1.2. **(A)** SCX-HPLC of the reaction mixture, A_{214} = 150 mVFS; **(B)** SCX-HPLC of the reaction mixture, A_{325} = 25 mVFS; **(C)** C_{18}RP-HPLC of the reaction mixture, A_{214} = 365 mVFS; and **(D)** C_{18}RP-HPLC of the reaction mixture, A_{325} = 90 mVFS. In D, the peaks labeled X and Y are unidentified nonpeptide species as determined from amino acid analysis. For A and C, M1(2) = monomer of peptide 1(2); D1(2) = dimer of peptide 1(2); M1(2):TNB = TNB adduct of the monomer of peptide 1(2), and D1 + 2 = the hetero-peptide disulfide-linked dimer indicated by an arrowhead.

3.4.2.2. Amino Acid Analysis of SCX-HPLC Fractions

The hetero-dimer peak (D_{1+2}) in Fig. 3A was subjected to standard HCl hydrolysis and PITC derivatization (*see* Sections 3.2.3. and 3.3.). The observed residue values for the hetero-dimer were close to the theoretical expected values, which are listed in parentheses: D/N = 0.92 (1); E/Q = 1.8 (2); S = 3.6 (4); G = 2.0 (2); R = 4.3 (4); T = 1.9 (2); Y = 1.1 (1); C = not determined (2); L = 1.1 (1); and F = 2.2 (2). An excellent correlation was also evident between the experimentally observed values for five other hetero-peptide dimers with the expected residue composition *(16)*. Thus, the overall method that consists of the production and subsequent chromatographic analysis of the peptide species can successfully identify and isolate the hetero-peptide dimer.

4. Notes

1. This column is manufactured by PolyLC (Columbia, MD) under the trade name PolySULFOETHYL Aspartamide. Two ligand densities are available at either a high load for separation of low to moderately charged peptides (+1 to +3/+4) or a low load for separation of peptides with charges of +4 and greater at pH 3.
2. To obtain appreciable amounts of hetero-peptide disulfide-linked dimers, it is important to produce near-quantitative amounts of the peptide 1:TNB adduct. This was accomplished by reacting peptides with roughly 1.5 Eq of DTNB. No separation of the desired adduct from excess reagent was performed in these studies, although it would be possible to do so.
3. Each distinct set of peptides to be separated by SCX-HPLC will require some optimization of the gradient depending on the nominal net positive charges of the peptide species.
4. Different peptides will require HPLC gradient optimization. The one listed is a good starting point, however.
5. Use of a different C_{18} column and changes in HPLC hardware, e.g., pumps, autosampler, and tubing lengths, will require alterations in the HPLC gradient.
6. Many pre- and postcolumn amino acid hydrolysate derivatization procedures exist. The choice of a particular method will depend on several factors, including sensitivity, speed of analysis, interferences from sample matrices, and robustness. PITC derivatization is a good compromise and has been successfully used in this laboratory for a variety of samples over the last 7 yr.
7. Note added in proof: the homo- and heterodimer peptides described in this chapter and in ref. *16* have since been analyzed by MALDI-MS and ESI-MS. The correct molecular weight was observed for each of the disulfide-linked peptides.

Acknowledgments

The excellent secretarial assistance of Pat Parvin is sincerely appreciated. I thank Richard S. Thoma for careful reading of and suggestions about the manuscript. This chapter is dedicated to the loving memory of my mother, Jacqueline Lillian Donnelley Crimmins.

References

1. Smith, A., Young, J. D., Carr, S. A., Marshak, D. R., Williams, L. C., and Williams, K. R. (1992) State-of-the-art peptide synthesis: comparative characterization of a 16-mer synthesized in 31 different laboratories, in *Techniques in Protein Chemistry III* (Angeletti, R., ed.), Academic, San Diego, CA, pp. 219–229.

2. Szokon, G., Torok, A., and Penke, B. (1987) High-performance liquid chromatographic detection of side reactions in peptide synthesis. *J. Chromatogr.* **387,** 267–280.
3. Pedroso, E., Grandas, A., Amor, J.-C., and Giralt, E. (1987) Reversed-phase high-performance liquid chromatography of protected peptide segments. *J. Chromatogr.* **409,** 281–290.
4. Mant, C. T. and Hodges, R. S. (1990) HPLC of peptides, in *HPLC of Biological Macromolecules: Methods and Applications* (Gooding, K. and Regnier, F., eds.), Marcel Dekker, New York, pp. 315–325.
5. Crimmins, D. L., Gorka, J., Thoma, R. S., and Schwartz, B. D. (1988) Peptide characterization with a sulfoethyl aspartamide column. *J. Chromatogr.* **443,** 63–71.
6. Alpert, A. J. and Andrews, P. C. (1988) Cation-exchange chromatography of peptides on poly (2-sulfoethyl aspartamide)-silica. *J. Chromatogr.* **443,** 85–96.
7. Burke, T. W. L., Mant, C. T., Black, J. A., and Hodges, R. S. (1989) Strong cation-exchange high-performance liquid chromatograpy of peptides. Effect of non-specific hydrophobic interactions and linearization of peptide retention behavior. *J. Chromatogr.* **476,** 377–389.
8. Alpert, A. J. (1991) Ion-exchange high-performance liquid chromatography of peptides, in *High-Performance Liquid Chromatography of Peptides and Proteins: Separation, Analysis, and Conformation* (Mant, C. and Hodges, R., eds.), CRC, Boca Raton, FL, pp. 187–194.
9. Crimmins, D. L., McCourt, D. W., and Schwartz, B. D. (1988) Facile analysis and purification of deblocked N-terminal pyroglutamyl peptides with a strong cation-exchange sulfoethyl aspartamide column. *Biochem. Biophys. Res. Commun.* **156,** 910–916.
10. Andrews, P. C. (1988) Ion-exchange HPLC for peptide purification. *Pept. Res.* **1,** 93–99.
11. Crimmins, D. L., Thoma, R. S., McCourt, D. W., and Schwartz, B. D. (1989) Strong cation-exchange sulfoethyl aspartamide chromatography for peptide mapping of *Staphylococcus aureus* V8 protein digests. *Anal. Biochem.* **176,** 255–260.
12. Schlabach, T. D., Colburn, J. C., Mattaliano, R. J., and Yuen, S. (1989) Meeting the challenge in peptide fragment purification for protein sequencing, in *Techniques in Protein Chemistry* (Hugli, T., ed.), Academic, San Diego, CA, pp. 497–505.
13. Iadorola, P., Zapponi, M. C., Minchiotti, L., Meloni, M. D., Galliano, M., and Ferri, G. (1990) Separation of fragments from human serum albumin and its charged variants by reversed-phase and cation-exchange high-performance liquid chromatography. *J. Chromatogr.* **512,** 165–176.
14. Swiderek, K., Jaquet, K., Meyer, H. E., Schachtele, C., Hofmann, F., and Heilmeyer, L. M. G. (1990) Sites phosphorylated in bovine cardiac troponin T and I. Characterization by ^{31}P-NMR spectroscopy and phosphorylation by protein kinases. *Eur. J. Biochem.* **190,** 575–582.
15. Crimmins, D. L. (1992) Comparative analysis of synthetic peptides by free-solution capillary electrophoresis (FSCE) and strong cation-exchange (SCX)-HPLC, in *Techniques in Protein Chemistry III* (Angeletti, R., ed.), Academic, San Diego, CA, pp. 171–181.

16. Crimmins, D. L. (1989) Analysis of disulfide-linked homo- and hetero-peptide dimers with a strong cation-exchange sulfoethyl aspartamide column. *Pept. Res* **2,** 395–401.
17. Andrews, P. C. (1990) Selective isolation of disulfide-containing peptides from trypsin digests using strong cation exchange HPLC, in *Current Research in Protein Chemistry: Techniques, Structure, and Function* (Villafranca, J., ed.), Academic, San Diego, CA, pp. 95–102.
18. Bidlingmeyer, B. A., Cohen, S. A., and Tarwin, T. L. (1984) Rapid analysis of amino acids using pre-column derivatization. *J. Chromatogr.* **336,** 93–104.
19. Heinrikson, R. L. and Meridith, S. C. (1984) Amino acid analysis by reverse-phase high-performance liquid chromatography: precolumn derivatization with phenylisothiocyanate. *Anal. Biochem.* **136,** 65–74.
20. Cohen, S. A. and Strydom, D. J. (1988) Amino acid analysis utilizing phenylisothiocyanate derivatives. *Anal. Biochem.* **174,** 1–16.
21. Sliwkowski, M. X. and Levine, R. L. (1985) Labeling of cysteine-containing peptides with 2-nitro-5-thiobenzoic acid. *Anal. Biochem.* **147,** 369–373.

CHAPTER 6

Principles and Practice of Peptide Analysis with Capillary Zone Electrophoresis

Thomas E. Wheat

1. Introduction

1.1. Application of Capillary Electrophoresis in Peptide Analysis

Capillary electrophoresis (CE) of peptides is emerging as a complement to established separation techniques *(1,2)*. Although reversed-phase high-performance liquid chromatography (HPLC) has proven to be the most generally useful and accepted method for peptide analysis and purification, no single separation technique can resolve all possible peptides *(3)*. Because the selectivities of HPLC and CE are derived from different chemical and physical properties, it is likely that peptides that are difficult or impossible to resolve in one technique will be much easier to separate in the other. CE is also attractive because the separations are relatively rapid and consume very little sample. Further, the instruments are easy to operate and to maintain, since there are so few moving parts. The technique is economical, since little or no organic solvent is used and since there is no column. Many of the early concerns about reproducibility in CE have been addressed in the newer commercial instruments. One can expect the technique to be similar to HPLC in quantitative precision and accuracy, as well as separation reproducibility. However, the greatest value of CE is that its selectivity is orthogonal to HPLC, helping to ensure that each observed peak in a separation is in fact a

From: *Methods in Molecular Biology, Vol. 36: Peptide Analysis Protocols*
Edited by: B. M. Dunn and M. W. Pennington Copyright ©1994 Humana Press Inc., Totowa, NJ

single peptide *(3,4)*. Some general considerations of the parameters controlling this selectivity are useful in obtaining the best possible results for a given sample.

1.2. Principles of Electrophoresis

Electrophoresis is the movement of charged molecules in an electric field *(1,2)*. Its separation selectivity is derived from differences in electrophoretic mobility among peptides. The direction of migration is determined by the net charge of the peptide, and the rate of migration is a function of the magnitude of the charge, the mass of the peptide, its Stoke's radius, and its intrinsic viscosity. These parameters, of course, apply in any form of electrophoresis. CE is different in that the separation occurs in a fused silica capillary, typically 25–100 μ in diameter and 50–100 cm long. This format enhances resolution by reducing band broadening, that is, increasing efficiency. A number of factors contribute to this reduced dispersion. First, in most forms of CE, there is no solid matrix. There is, therefore, no multipath band broadening as observed with gel or paper electrophoresis, and no dispersive transport owing to the resistance to mass transfer, as in HPLC. Further, there is neither turbulent nor parabolic flow of liquid. Band broadening in CE results primarily from diffusion and from interaction of the sample with the wall of the capillary. The latter effect is discussed below. Diffusion is related to time, temperature, viscosity, and the diffusion coefficients of the sample molecules. In CE, the time and temperature terms are minimized because heat dissipation is better. In classical electrophoretic separations, resolution is limited by the inefficient dissipation of Joule heating in the relatively thick electrophoretic path. With a capillary, the temperature remains lower because the surface-to-volume ratio is larger. This permits the use of higher field strengths that yield more rapid separations, minimizing the time for diffusion. Lower temperatures also reduce the rate of diffusion. Heat dissipation is, therefore, a major source of increased resolution in CE.

Resolution is also affected by applied voltage and resultant current, capillary length and diameter, buffer, and sample introduction. Resolution increases with higher voltages or, more exactly, higher field strength (V/cm). However, current increases with voltage, reducing resolution because increased Joule heating increases band broadening. With adequate heat dissipation, current and voltage are linearly related, so it is common to select the highest voltage where this relationship (Ohm's Law) holds. This point is empirically determined, as described below, for a particular combination of buffer and capillary dimensions.

In general, resolution increases with capillary length at the expense of increased run time. Although applied voltage is increased with length, resolution is not further enhanced because the separation is sensitive to field strength.

Resolution is increased with smaller capillary diameter where higher field strengths can be used and because heat is dissipated more efficiently. Field strength is limited by current, which is proportional to the square of capillary radius. Although heat dissipation is usually considered from the wall of the capillary to the environment, resolution is most sensitive to the temperature gradient from the center of the capillary to the wall. This gradient produces a parabolic migration profile, since mobility is higher and viscosity lower at the relatively warmer center of the capillary. This radial temperature gradient is also proportional to the square of the radius. Although smaller-diameter capillaries give better resolution, they also reduce sensitivity because the detector path length is the diameter of the capillary. In addition, small-diameter capillaries are more prone to plugging.

Band broadening can be affected by the pH, ionic strength, and composition of the buffer, in addition to the effects on selectivity. As pH increases or ionic strength decreases, electroosmotic flow increases, reducing run time and band broadening, as described below. Increasing ionic strength may sharpen peaks by reducing interaction with the wall of the capillary and, sometimes, by minimizing aggregation. However, current and Joule heating increase, and electroosmotic flow decreases with higher electrolyte concentrations. In addition, electrolyte components that affect viscosity alter the separation, because both electroosmotic flow and electrophoretic mobility decrease with increasing viscosity.

Resolution is affected by sample concentration and volume, as well as diluent composition. In general, the best separations are observed when both the sample volume and concentration are as low as possible. If the sample concentration is too high, the peptides locally displace buffer ions. This distorts the field strength and can create local discontinuities in pH. If the sample volume is too high, it will exceed the effects of band broadening during the separation. Further, resolution will be lost as closely spaced bands begin to overlap. The requirements of detector sensitivity define minimum concentration and volume. To some extent, proper selection of sample diluent can help to meet this requirement. Peptides should be dissolved at the same pH as the running buffer. In addition, the ionic strength of the sample should be equal to or lower than the electrolyte. Substantial resolution will be lost if the sample ionic strength is too high. A lower ionic strength diluent can give both increased resolution and sensitivity. In this case, the stacking effects of a discontinuous buffer, as familiar from polyacrylamide gel electrophoresis, serve to sharpen the bands *(5)*. Very low ionic strength diluents can give as much as 10-fold enhanced sensitivity without loss of resolution.

CE differs from other electrophoretic techniques in the contribution of electroendosmosis, or electroosmotic flow, to the separation *(1)*. This is the bulk flow of electrolyte as a consequence of the surface charge on the wall of the capillary. It exists in all forms of electrophoresis, but the magnitude is small where the supporting matrix has a relatively low charge density and also resists the flow of liquid. In an open capillary, there is no such resistance to flow, so the electroosmotic flow can be relatively large. The motive force originates with the attraction of electrolyte cations to form a double layer at the negatively charged wall of the capillary. These cations are driven toward the negative electrode, or cathode, and, since they are hydrated by hydrogen bonding, the bulk liquid will be dragged along. Since this driving force is generated uniformly along the length of the capillary, there is no pressure drop to produce parabolic or laminar flow that contributes to band broadening. The magnitude of the flow is a function of the surface charge, or ζ potential, of the ionic strength, specific cations, viscosity, and the field strength. The surface charge is a function of buffer pH, so that the flow is maximal above pH 8–9 and falls off as the pH is reduced, reaching essentially imperceptible levels below pH 3.0. Flow is also reduced as the ionic strength is increased.

The manipulation of electroosmotic flow can enhance the utility of CE for peptide separations. The observed migration of a peptide is the vector sum of its electrophoretic mobility and the electroosmotic flow. Since the electroosmotic flow rate will usually be greater than the electrophoretic mobility, the net migration of both positively and negatively charged, as well as neutral, species will be toward the cathode. It is, therefore, practical to analyze both acidic and basic peptides in a single separation.

1.3. Control of Peptide Electrophoresis

Within the framework of these general principles, the various parameters that can be manipulated to resolve peptides can be described. It is useful to consider separately those factors that influence the electrophoretic mobility of the peptides and those that affect the electrophoretic process, particularly band broadening.

Of the several properties that control the mobility of peptides *(6)*, only charge is amenable to systematic experimental manipulation. This property can be easily changed over a wide range since the charge of a peptide is determined by the pH of the buffer. At relatively low pH, all the ionizable groups are protonated, ensuring a full positive charge on all basic groups while neutralizing the acidic side chains. Essentially all the peptides are, therefore, positively charged, and the primary selectivity arises from differences in the number of basic groups on each peptide. Selectivity differences based on mass and size in solution are superimposed on this charge discrimination. At the other extreme of high pH, functional groups are largely unprotonated, so most peptides have a net negative charge and selectivity is primarily based on the number of acidic functionalities. Between these extremes, each peptide has a titration curve for charge as a function of pH that depends on its sequence. In principle, the electrolyte pH should be selected to give the greatest difference in charge among the peptides to be separated, but in practice, pH is empirically chosen to give the greatest differences in migration times, reflecting the additional effects of mass and size in solution. However, since these latter properties are essentially fixed, the selectivity is optimized by manipulating charge with electrolyte pH.

Mobility can, to some extent, be directly affected by the composition of the buffer. Changes in ionic strength may alter the shape of the peptide or influence aggregation, as well as interaction with the wall as discussed below. Certain buffer ions may form complexes with particular peptides, altering electrophoretic mobility. Organic solvents and other buffer additives may change solubility.

The interaction of peptides with the wall of the capillary produces tailing peaks, poor recovery, and alterations in electroosmotic flow as the peptides coat the capillary. This binding most often results from the attraction of cationic groups on the peptides to anionic silanol groups on the capillary. There are several ways to eliminate this interaction. First, the negative charge on the capillary can be neutralized by selection of an electrolyte, usually near pH 2.0, that protonates the silanol groups. Alternatively, a high-pH buffer, above 10, can be used to titrate most of the positive charges on the peptide. Such manipulation of pH may alter the selectivity of the separation or the solubility of the peptides. Further, removal of silanol charge at low pH also eliminates electroosmotic flow. Alternatively, peptides can be displaced from the wall by increasing the electrolyte ionic strength, particularly the cation concentration. This approach increases the conductivity of the buffer, limiting the field strength. This effect may be obviated by the use of a zwitterionic reagent to increase the cation concentration effectively without increasing the conductivity. A zwitterionic reagent that maintains zero net charge over the full useful pH range is commercially available (AccuPure™ Z-1_{methyl}; trimethylaminopropylsulfonic acid; Waters, Milford, MA). Finally, the wall of the capillary can be chemically coated to shield the silanol groups. This approach has not proven generally useful for peptides. It is difficult to prepare a surface that is stable for a long series of runs, and, as the bonded phase is gradually hydrolyzed, both the degree of peptide interaction and the electroosmotic flow change. Further, some peptides will directly interact with the coating by, for example, hydrophobic mechanisms. It should be noted that the problem of wall interaction is much less significant with small peptides, <50 or 60 residues, than for intact proteins. Even very basic peptides give good separations at modest ionic strengths on the order 100–200 m*M* with respect to sodium, where the current is reasonable.

1.4. Development of Peptide Separations

Within these general principles, a series of experiments can be defined to find useful separation conditions for a peptide sample. The operational details are given below, but, first, the rationale for this approach should be described *(4)*. For maximum resolution and speed, the separation should be developed on the smallest-diameter capillary compatible with the sensitivity requirements of the sample and at the highest field strength allowed by the buffer. The first run should be made at high pH, usually pH 10.5. These conditions are selected because most peptides tend to be more soluble at high pH and because the electroosmotic flow will be high, ensuring that all sample components will pass through the detector in a reasonable time. This run serves both as a useful separation and as a reference for the total peak area that can be used to estimate whether all sample components have reached the detector with other separation conditions. Even if the resolution at high pH appears good, the sample should be run a second time at low pH, usually near pH 2.0. Because the charge selectivity is different at the two pH extremes, this pair of runs is likely to reveal comigrations, even if neither set of conditions provides complete resolution of all sample components. In some applications, for example, assessing the purity of fractions collected from an HPLC separation, these two runs may prove sufficient. That is, a single HPLC peak that also shows a single CE peak at two extreme pH values is very likely a single peptide. If the sample is more complex or more complete resolution is required, it is necessary to examine intermediate pH buffers. The most useful values are those near the pKs of ionizable amino acid side chains. This includes pH 3.0–4.0 for aspartate and glutamate, pH 6.0–7.0 for histidine, and pH 9.0–11.0 for tyrosine and lysine. The pK of arginine is too high to provide useful titration. The charge of the peptide will be most sensitive to small pH adjustments in these ranges because the side chains will behave as though they have fractional charges. Since the pK for a given functional group is sensitive to the surrounding sequence and microenvironment, there is potential for subtle selectivity changes. The useful pH range can be estimated for a given sample. For example, tryptic peptides should all have a single basic residue at the carboxyl terminal, so the greatest charge selectivity is expected from differences in the number of acidic residues and from variations in the pKs of these side chains. Electrolytes near pH 3.5–4.5 should, therefore, be tested. On the other hand, the best selectivity for a chymotryptic digest of a basic protein should be in the range of pH 9.0–11.0. The charge distribution for a synthetic peptide is easily estimated from the intended sequence.

After identifying the best electrolyte pH for the particular sample, other parameters can be modified to optimize the separation. It is usually desirable to compare two different electrolyte concentrations, differing by twofold. If solubility is in question, the inclusion of an organic solvent, 10–20% acetonitrile or an alcohol, can be tested. Alternative buffer systems may affect some samples. Additives, such as zwitterions, ion-pairing reagents, chaotropic agents, and submicellar concentrations of detergents, are useful on occasion. Finally, it may prove necessary to test another electrophoretic mode, such as micellar electrokinetic capillary chromatography (MECC). Although this technique is outside the scope of this discussion, some general comments are appropriate.

In MECC, the electrolyte includes micelle-forming surfactants, and peptides partition between these micelles and the electrolyte *(7)*. Both the micelles and the peptides are moved through the capillary by the combination of electrophoretic and electroosmotic forces. Selectivity is a combination of differences in peptide electrophoretic mobility and hydrophobicity, tending to combine the advantages of reversed-phase HPLC and CE. Further, the wide range of electrolyte compositions permits novel combinations of ion suppression and ion pairing that are not possible with silica-based HPLC columns. Although sodium dodecyl sulfate has been the most common MECC reagent, a wide range of other anionic, cationic, and neutral surfactants have shown promise as well as unique selectivities. The utility of this form of CE will increase as it is further developed. However, almost all peptide samples have given useful separations in simple, free-solution CE, as outlined above and described in detail below.

2. Materials

2.1. Required Solutions

It has proven useful to condition and clean capillaries with both the acidic and alkaline solutions described below. Buffers are selected based on their pKs, which should be within about 1 pH U of the desired pH. In all cases, reagents are prepared from the highest-grade salts commercially available and dissolved in MilliQ™ water (Millipore Corporation, Bedford, MA). Electrolytes are filtered through a 0.45-μ filter, such as Millipore HA, and thoroughly degassed before use.

2.1.1. Cleaning Solutions

2.1.1.1. Alkaline Cleaning Solution

Sodium hydroxide (0.5*M*): Dissolve 20 g of sodium hydroxide with water to a final volume of 1 L.

2.1.1.2. Acidic Cleaning Solution

Phosphoric acid (1.0*M*): Dilute 69 mL of concentrated phosphoric acid with water to a final volume of 1 L.

2.1.2. Electrolytes

2.1.2.1. Phosphoric Acid, pH 2.0

Dilute 2.3 mL of concentrated phosphoric acid with water to a final volume of 1 L (100 m*N* or 0.033*M*). This electrolyte can be stored indefinitely at 4°C.

2.1.2.2. Sodium Phosphate, pH 2.0–3.0

Dissolve 13.8 g of monobasic sodium phosphate monohydrate ($NaH_2PO_4 \cdot H_2O$) in 900 mL of water. Titrate to the desired pH with phosphoric acid, and bring to a final volume of 1 L. Alternatively, dilute 6.9 mL of concentrated phosphoric acid to 900 mL, and titrate to the desired pH with sodium hydroxide. Bring to a final volume of 1 L. The first formulation gives a constant 0.1*M* sodium concentration, and the phosphate concentration will vary with pH. The latter electrolyte will be constant 0.1*M* phosphate, and sodium will vary with pH. The two formulations may give different results, particularly for basic peptides. Store either buffer at 4°C for <1 wk.

2.1.2.3. Sodium Citrate, pH 3.0–5.0

Prepare 0.25*M* stocks of citric acid and sodium citrate. Dissolve 52.5 g of citric acid, monohydrate, or 48.0 g of citric acid, anhydrous, to 1 L with water. Dissolve 73.5 g of trisodium citrate, dihydrate, to 1 L with water. These stocks may be stored at 4°C for several weeks. To prepare 0.025*M* working electrolyte, dilute 100 mL of stock to 1 L, and blend the dilute citric acid and sodium citrate to the desired pH using a pH meter. For 0.05*M* electrolyte, dilute 200 mL of each stock to 1 L and blend. Working electrolytes should be stored at 4°C for <1 wk.

2.1.2.4. Sodium Phosphate, pH 6.0–8.0

For pH 6.0, 0.175*M* with respect to sodium, dissolve 18.9 g monobasic sodium phosphate, monohydrate ($NaH_2PO_4 \cdot H_2O$) and 5.1 gm dibasic sodium phosphate, heptahydrate ($Na_2HPO_4 \cdot 7H_2O$) to 1 L with water. For pH 6.8, 0.175*M* with respect to sodium, dissolve 8.3 g monobasic sodium phosphate, monohydrate ($NaH_2PO_4 \cdot H_2O$) and 15.4 g dibasic sodium phosphate, heptahydrate ($Na_2HPO_4 \cdot 7H_2O$) to 1 L with water. For pH 7.6, 0.175*M* with respect to sodium, dissolve 1.7 g monobasic sodium phosphate, monohydrate ($NaH_2PO_4 \cdot H_2O$), and 21.8 g dibasic sodium phosphate, hepthydrate ($Na_2HPO_4 \cdot 7H_2O$) to 1 L with water. For electrolytes that are 0.35*M* with respect to sodium, use twice the weight of each salt. If other hydration states of the salts are more readily available or are preferred, the weights must be recalculated to give the same molarity. Store at 4°C for <1 wk.

2.1.2.5. Sodium Borate, pH 9.5–10.5

Prepare a stock of sodium borate, 0.2*M* with respect to borate by dissolving 19.07 g sodium borate, decahydrate ($Na_2B_4O_7 \cdot 10H_2O$) or 10.06 g sodium borate, anhydrous ($Na_2B_4O_7$) to 1 L with water. To make

0.025*M* working electrolyte, dilute 125 mL of stock to 1 L with water, and titrate to the desired pH with sodium hydroxide. For a 0.05*M* buffer, dilute 250 mL of stock to 1 L and titrate. The stock can be stored for several weeks, and the working electrolyte for several days at 4°C.

2.2. Capillary

2.2.1. Preparation of Capillary

The polyimide-coated, fused silica capillary used for the separation can be purchased ready to use, installed in the appropriate holder, from the instrument manufacturer. Although this is convenient and ensures a properly assembled device, it is more economical to prepare the working capillary from rolls purchased in bulk. The desired length is cut from the spool by scoring the capillary with a diamond cutter or silicon wafer, and breaking by gently bending. The sample introduction end must be clean and square. It is sometimes useful to examine this end under a microscope. A detector window must also be formed by removing a short length of the polyimide coating. The required length of this window and its distance from the end of the capillary are dependent on the specific instrument. Although the polyimide can be removed by gentle scraping, it is more convenient to burn it away by briefly passing the capillary through the flame of a butane lighter. The residue is then removed by carefully wiping with a lint-free tissue moistened with methanol. This window must be handled carefully since it is more fragile than the polyimide-coated capillary. For this reason, the window should not be longer than necessary, and overheating should be avoided. This capillary is then installed in the appropriate holder and placed in the instrument following the manufacturer's instructions.

2.2.2. Conditioning and Testing the Capillary

A new capillary, or a capillary judged in need of cleaning and conditioning, is alternately purged with base and acid, typically 10 min with 0.5*M* sodium hydroxide, 2 min with water, 10 min with 1.0*M* phosphoric acid, and 2 min with water. This cycle is repeated three times before a final flush with base and purging with electrolyte until a stable detector baseline, current, and electroosmotic flow are observed. This may require as much as 1–2 h, depending on the electrolyte.

The capillary should be tested for physical occlusions and for the chemical condition of the wall. Capillary occlusions can be detected during the conditioning and cleaning by monitoring at or below 214 nm. Any plug in the capillary will increase the time required for a change in solution to cause a detector response. With some instruments, it is easy to introduce a small air bubble during the purge and to measure the time interval to the detector. This is the preferred method, since it can be used at any time and does not change the chemical equilibrium of the capillary. This measurement should be made with each new capillary, and recorded for comparison with other capillaries and as an aid in troubleshooting.

The most useful and convenient indicator of the chemical condition of the capillary surface is electroosmotic flow, measured as the migration time of a neutral marker, such as 0.5% formamide, under defined buffer conditions. The electrolyte for sample analysis should be used if it provides appreciable electroosmotic flow. In addition to recording migration, the shape of this marker peak should be quantitated by calculating the number of theoretical plates as a measure of system band broadening. Again, this test should be performed on each new capillary as a basis for troubleshooting.

2.3. Instrument Setup

Most details of instrument configuration and operation are outside the scope of this discussion, because they are specific to a particular manufacturer. Some general principles are, however, common, including electrolyte, detection, control of sample introduction, and voltage selection.

2.3.1. Electrolytes

The electrolytes used in CE must be free of particles and dissolved gases. The former is accomplished by passing them through a 0.45-μ filter, whereas the latter requirement is best met by sonication under a vacuum. Particles can occlude the capillary while gases form bubbles with Joule heating. Such bubbles increase baseline noise, add detector spikes, and in the worst case, interrupt the electric field. Both the anode and the cathode reservoirs should contain the same electrolyte and should be relatively large to minimize any effects of buffer depletion during electrophoresis. The liquid levels in the two reservoirs must be the same to prevent siphoning, since such a flow contributes to band broadening and irreproducible migration times.

2.3.2. Detection

The detection wavelength is chosen for maximum sensitivity. It should in general be as low as possible. This is primarily constrained by instrument design, but it is also sensitive to operating conditions, such as capillary diameter and UV absorbance of the electrolyte. Although the absorbance of a peptide will always be higher at a lower wavelength, the optical noise also tends to increase, particularly with electrolytes that absorb strongly in the UV and with variable-wavelength detectors. There is, therefore, for each combination of optics, capillary diameter, and electrolyte, a point where the noise increase is so large that sensitivity and the signal-to-noise ratio are compromised. This defines the wavelength selected for that run.

2.3.3. Sample Introduction

The sample may be introduced into the capillary in two ways: electrokinetic and hydrodynamic injection. Most commercial instruments provide both options, and each has advantages.

For electrokinetic injection, the end of the capillary and one electrode are placed in the sample, and voltage is applied. The sample migrates into the capillary electrophoretically. The amount injected depends on the electroosmotic flow, as well as the electrophoretic mobility and the concentration of the sample components. For a given sample, electrokinetic injection is usually reproducible. However, it is unsatisfactory for comparing different samples because the peptides with the highest electrophoretic mobility enter the capillary in greater proportion. For the same reasons, it is inappropriate for comparing different electrophoretic conditions where both flow and mobility are different.

In hydrodynamic injection, a volume of liquid is forced into the capillary by a pressure differential between the sample and detector ends of the capillary. This volume depends on the magnitude of the pressure differential, the dimensions of the capillary, and the viscosity of the liquid in the capillary. The pressure difference can be generated by applying a gas pressure to the inlet side, a vacuum to the outlet side, or by raising the sample vial above the outlet to fill by siphoning. The last technique is the oldest and remains among the most reliable, because the motive force is dependent on gravity, a constant parameter. The accuracy and precision of this approach depend on how well the height differential, usually about 10 cm, is established and how exactly the time, often 10 s, is controlled. The autosampler in a well-designed instrument should execute and control this movement more than adequately for useful results. The accuracy and precision of gas- or vacuum-driven systems depend on how well the instrument controls these forces. In operating such a system, care must be taken to ensure a gas-tight seal at the vial. In the case of a vacuum-driven system, this is easily established at the start of a series of runs since the seal does not move. In contrast, the seal in a pressure-driven system moves from vial to vial, so the manufacturer's recommendations should be followed exactly to maintain accurate and precise injections.

All modes of injection are sensitive to differences in viscosity among samples, and to differences in height between the liquid level in the sample vial and the receiver vial. The impact of sample viscosity is effectively quite small in hydrodynamic injection, even though viscosity obviously affects flow rate. Since the relevant term is viscosity through the length of the capillary and since the sample "plug" is <1% of this length, sample viscosity has little impact. Height differences can introduce larger errors. Since a 10-cm pressure head for 10 s is a typical hydrostatic injection, each 1 mm difference in height will contribute about a 1% change in injected volume for all modes of injection.

2.3.4. Voltage Selection

The voltage selected for the separation should provide the highest possible field strength to ensure the best resolution. Field strength is limited by the dissipation of Joule heating, so in practice, the maximum applied voltage is a function of capillary dimensions, buffer, environment, and instrument design. A variety of guidelines for voltage selection have been suggested, including maintaining current below 100 μA or keeping total power below 1 W (A × V). However, it is good practice to prepare an Ohm's Law plot empirically for each combination. Measure the current at a series of increasing voltages, usually at 2-kV intervals. Over the useful range, current will be slightly above the predicted linear relationship. That is, doubling the voltage will give a few percent more than twice the current. At some point, Joule heating will exceed the thermal dissipation capacity of the system, and the current will increase much more than expected. The voltage selected for an analysis should be just below this breakpoint. If that field strength is too low, either a smaller-diameter capillary or a lower-conductivity electrolyte should be selected. Useful field strengths are usually in the range of 250–500 V/cm, and values as low as 100 V/cm can give good separations.

2.3.5. Sample Preparation

The peptide sample must be dissolved in a buffer compatible with the electrolyte and must be free of particulates. The sample volume must be above the minimum required for the particular instrument. The sample container must be compatible with the instrument and must also be of a material to which peptides do not adsorb. Adsorption becomes more significant at lower concentrations, and the instrument manufacturer should be consulted for solutions to this problem that are compatible with the design. As discussed above, the sample buffer should be the same pH as the electrolyte and 2- to 10-fold lower in ionic strength. Particulates can be removed by passing it through a small-volume filter, such as a Millex HV_4 (Millipore Corp., Bedford, MA). If the available sample volume is too small to permit the loss of a few microliters in filtration, the sample should be centrifuged at 10,000*g* or more. The concentration required is determined by detection sensitivity. The limit of detection is determined by many factors, including instrument design and operating parameters, such as capillary diameter, UV absorbance of the electrolyte, wavelength, and so on. In favorable cases, concentrations as low as 1 µg/mL can give useful analyses, but it is often better or necessary to work at 100 µg/mL or more.

Injection parameters are ultimately related to sensitivity and resolution requirements as discussed above. In general, hydrodynamic injection is preferred to avoid differential introduction of sample components, and the sample volume should be on the order of 1% of the capillary volume. The conditions to effect this injection can often be found in the operator's manual. It can be easily measured by positioning the capillary in a sample vial containing a UV-absorbing material, such as 0.5% formamide in electrolyte, and starting the injection mode. Measure the time to the detector response to obtain the injection flow rate, or simply use 1% of this measured time for the injection.

3. Method

3.1. Sequence of Experiments

Using the conditions defined as above, run the sample in 0.025*M* sodium borate, pH 10.5, 100 m*N* phosphoric acid, and at least one intermediate pH chosen from among 0.025*M* sodium citrate, pH 4.0; sodium phosphate, 1.75 m*M* sodium, pH 6.8, and 0.025*M* sodium borate, pH 9.5. For each experiment, use a sufficiently long run time that all components can pass the detector. As with any analytical separation, it is difficult to prove complete recovery without determining a mass balance or identifying each component with an independent technique. It is usually prudent to set the initial run time for an unknown sample to at least three to five times the migration time of the neutral marker.

3.2. Evaluation of Results

3.2.1. Resolution

Resolution is judged by the number of peaks and the baseline between peaks. In judging the number of peaks, it is essential to ensure that each component originates from the sample by making blank runs. In addition, under conditions where some peptides are expected to have no net charge, they will comigrate with the artifactual detector response at the migration time of the neutral marker. This baseline upset is related to refractive index differences between the sample diluent and the electrolyte. It usually does not obscure peptides at high pH, above pH 10.5, where almost all peptides have a net negative charge. It is not observed at pH 2.0, where there is no electroosmotic flow to move the diluent through the detector and where very few peptides do not have a net positive charge.

3.2.2. Sensitivity

Sensitivity is usually defined as the signal-to-noise ratio and is evaluated for the particular application. If, for example, CE is used to assess the homogeneity of an HPLC fraction before applying the peptide to a sequencer, it would be sufficient to establish that the major component represents 90–95% of the sample. Since a peptide can be detected at a signal to noise of about 3–5, the height of the major component should be at least 30 (10% minor component; S/N = 3) to 100 (5% minor component; S/N = 5) times greater than the baseline noise. For applications requiring more stringent detection, the main component must be correspondingly larger.

3.2.3. Separation Reproducibility

Separation reproducibility is usually judged primarily by consistency of migration time, but changes in resolution can also arise from increased band broadening or deterioration of peak shape. Again, the reproducibility requirements are dependent on the application. Within broad limits, there is little impact of separation consistency where samples are simply being checked for homogeneity. Reproducibility must be much better where two samples are being compared or where a peak is to be tentatively identified by its migration time. One useful guideline is that peaks must be separated by 2–4 SD of their migration time.

3.2.4. Quantitation

Quantitation is much the same as in HPLC with the important exception that CE measurements are sensitive to migration time in ways that HPLC is not sensitive to retention time. In both methods, peak area and height are used as measures of amount. Height is the detector response at the center of a peak, whereas area is the integrated sum of detector response over time. The underlying assumptions are that detector response is linear with concentration and that time is proportional to volume. The latter is valid in HPLC where a constant flow is maintained. Essentially, detector response integrated over time is equivalent to concentration integrated over volume, i.e., amount. In CE, however, the rate at which a peak moves through the detector is a function of electrophoretic mobility and electroosmotic flow. The volume of liquid over which a detector response is observed is, therefore, not directly related to time. The volume occupied by a peak is a function of its migration time. To illustrate, two peaks can be compared. Assume that both are 15-s wide, but one has a migration time of 5 min and the other 10 min. If it is 100 cm from the injection end to the detector, the first peak is moving at 20 cm/min, whereas the second migrates 10 cm/min. The first peak, therefore, occupies 5 cm of the capillary, whereas the second is only 2.5 cm. If the capillary has a diameter of 50 μ, the volume occupied by the first peak ($\pi r^2 l$) is about 10 nL ([3.1416][.0025 cm]^2[5 cm] = 0.0000098 cm^3]), whereas the second peak is about 5 nL. The areas reported by an integrator will not, therefore, permit comparisons of absolute amount between peaks or between different runs. These areas may be corrected or normalized by dividing the reported area by migration time. Several data devices are commercially available with CE software that performs this calculation automatically. With this correction, peak areas are related to amount in the same way that they are in HPLC, and comparisons can be made within the limits of variation in extinction coefficients among peptides.

4. Troubleshooting

The basis of successful troubleshooting is established benchmarks for good performance, as described above. These include determining flow rate during a purge, measuring the voltage–current relationship for the electrolyte, measuring the electroosmotic flow with a neutral marker, and quantitating peak shape for the neutral marker by plate count. Any available instrument diagnostics, such as lamp energy, should also be regularly recorded in a logbook. The problem should then be defined as specifically as possible, but the various possible symptoms should be examined in the order described below.

4.1. Migration Time

4.1.1. Erratic Migration Times

In analyzing migration time problems, a distinction must be made between those showing a consistent drift in one direction over a series of runs and those that vary back and forth. The latter may include drift in one direction for a few runs and then in the opposite. Erratic migration times arise from inconsistencies in electrophoresis. If, for example, the electrolyte is outgassing to give microbubbles, field strength will be inconsistent. The same is true if the sample contains particulate material. Such problems can be recognized by erratic current as well as spikes in the detector signal. Monitoring current can also reveal malfunctions in the instrument power supply. Migration times will also vary if the wall of the capillary is changing. This can result from chemical equilibrium with the electrolyte or from the adsorption of sample components. Fluctuating temperatures can also affect migration. Finally, large variations in concentration can create local distortions in field strength that change mobility.

4.1.2. Drifting Migration Times

Migration changes consistently in one direction imply a systematic change in the electrophoretic process. They may result from modification of the capillary wall by the adsorption of sample components. There may also be long-term drifts in temperature. The electrolyte buffers may also be depleted by the migration of buffer ions. The last effect depends on voltage and time, as well as the concentration and capacity of the buffer. Depletion is characterized by stable migration for several runs followed by a significant shift that is reversed when a fresh buffer vial is used.

4.1.3. Incorrect Migration Times

With an established method, any change in the specified operating parameters can cause uncharacteristic migration times. These can include electrolyte composition and pH, voltage setting, temperature, capillary diameter, and length.

4.2. Loss of Resolution

4.2.1. Selectivity Changes

Peptides are most likely to move relative to one another if the composition of the electrolyte is not as specified. This is most common for operating pH near the pKs of a particular class of amino acids, but it may also be sensitive to ionic strength. Although such conditions are often selected for maximum resolution, extra care must be taken in preparation of electrolyte.

4.2.2. Changes in Peak Shape and Losses of Efficiency

Changes that affect some peaks more than others most often result from chemical interactions with the wall of the capillary. If, on the other hand, all sample components are broadened, a physical loss of efficiency is implied. Both effects are discussed in detail above. When either phenomenon is suspected, it is most useful to repeat the capillary test with a neutral marker.

4.3. Problems of Quantitation

Precision and accuracy should be distinguished in problems related to measurement of amount and should only be addressed when migration time is constant. Precision should first be tested with multiple runs from the same sample vial. If excessive variation is observed, there is either a mechanical problem with the sample introduction or there are short-term changes in the sample. The latter might include diluent evaporation or gradual adsorption of peptides to the sample vial. Evaporation gives larger peaks for all sample components on successive runs, whereas adsorption reduces peak size and usually affects some peaks more than others. Identifying mechanical or physical problems can best be accomplished using injections of the neutral marker in the test procedure described above.

Inaccuracy can result from errors in sample or standard preparation. If there are large differences between the sample and the standard, the linear range of the detector may be exceeded. In addition, sample adsorption to the vial and evaporation may give inaccurate quantitation as can differences that affect sample introduction, such as pH and ionic strength in electromigration or viscosity and liquid level in any mode.

References

1. Jorgensen, J. W. and Lukacs, K. D. (1983) Capillary zone electrophoresis. *Science* **222,** 266–272.
2. Issaq, H. J., Janini, G. M., Atamna, I. Z., and Muschik, G. M. (1991) Separations by high performance liquid chromatography and capillary zone electrophoresis: a comparative study. *J. Liq. Chromatogr.* **14,** 817–845.
3. Young, P. M., Astephen, N. E., and Wheat, T. E. (1992) Effects of pH and buffer composition on peptide separations by high performance liquid chromatography and capillary electrophoresis. *LC · GC* **10,** 26–32.
4. Wheat, T. E., Young, P. M., and Astephen, N. E. (1991) Use of capillary electrophoresis for detection of single-residue substitutions in peptide mapping. *J. Liq. Chromatogr.* **14,** 987–996.
5. Vinther, A. and Soeberg, H. (1991) Mathematical model describing dispersion in free solution capillary electrophoresis. *J. Chromatogr.* **559,** 3–26.
6. Florance, J. R., Konteatis, Z. D., Macielag, M. J., Lessor, R. A., and Galdes, A. (1991) Capillary zone electrophoresis of motilin peptides. Effects of charge, hydrophobicity, secondary structure, and length. *J. Chromatogr.* **559,** 391–400.
7. Terabe, S., Otsuka, K., and Ando, T. (1985) Electrokinetic chromatography with micellar solution and open tubular capillary. *Anal. Chem.* **57,** 834–841.

CHAPTER 7

Fast Atom Bombardment Mass Spectrometric Characterization of Peptides

P. R. Das and B. N. Pramanik

1. Introduction

The structural characterization of peptides and proteins is an integral part of biotechnology/pharmaceutical industry research in order to develop novel therapeutic agents through the production of these compounds by synthesis and/or recombinant DNA techniques. The most commonly used methods are the indirect approach through the translation of the DNA sequence of the gene coding for the protein and the automated stepwise Edman degradation. Nevertheless, the former can introduce inaccuracies either from mistakes in reading DNA sequencing gels or from posttranslational modifications, and the latter cannot be used for N-terminally blocked proteins/peptides and certain modified amino acids. Recent developments in desorption ionization techniques and instrumentation have made mass spectrometry a complementary alternative, and in several cases, an effective method of choice over conventional analytical methods employed in the evaluation of a solid-phase synthetic peptide or a recombinant product.

The use of mass spectrometry (MS) for peptide/protein structural analysis dates back over 30 years. Peptides/proteins being polar, labile, and involatile posed a problem for MS in the past, since the production of ions from these molecules in the gas phase is a fundamental requirement for this technique. Earlier approaches included predominantly electron ionization (EI), chemical ionization (CI), and desorption chemical

From: *Methods in Molecular Biology, Vol. 36: Peptide Analysis Protocols*
Edited by: B. M. Dunn and M. W. Pennington

ionization (DCI) MS, and generally involved derivatization of the polypeptides by acetylation or permethylation *(1–3)*. This procedure required relatively large amounts of samples, was time-consuming, and the upper mass range of application was limited to ca. 1000 Dalton of the derivatized oligopeptides. Another approach, especially suitable for mixtures, involved GC/MS of reduced, silylated peptides *(4–6)*.

With the advent of the newer desorption techniques in MS, the peptide/protein molecules can now be ionized in the mass spectrometer; without prior derivatization. The most frequently used methods include:

1. Fast atom bombardment (FAB) *(7)* MS/Cs^+ ion liquid secondary ion mass spectrometry (SIMS);
2. Plasma desorption (PD) *(8,9)* MS;
3. Electrospray ionization (ESI) *(10–12)* MS; and
4. Laser desorption (LD) *(13–15)* MS.

The much newer applications of laser desorption and electrospray ionization to peptide/protein characterization have made analysis of the intact molecule at the picomole/femtomole level possible. The electrospray technique, which is now commercialized, has extended the mass range of applicability to ca. 150,000 Dalton and has on-line LC/MS capability. Laser desorption MS, which has been shown to be capable of detecting mol wt of 300,000 Dalton using 1 pmol of sample, is becoming increasingly important in biotechnology and can be used to analyze mixtures.

The application of MS using current methods in peptide/protein analysis is well documented and reviewed *(16,17)*. This chapter focuses on the FAB MS/Cs^+ ion liquid SIMS method, which is perhaps the most widely used and has made a large impact in this field. This methodology is readily adaptable to both sector and quadrupole systems. Moreover, high-resolution accurate mass measurements using sector equipment can be performed to determine elemental compositions of specific ion peaks for their structural characterization. This can, for instance, differentiate between O from oxidation of methionine or the presence of an NH_2 group. The mass spectrometric methodologies comprise:

1. Peptide/protein structure characterization through molecular-weight determination and sequence information by MS/MS and or linked scanning techniques;
2. Peptide/protein mapping techniques, including amino and carboxy terminal sequence confirmation;
3. Disulfide bond determination;

4. Spot checks for microheterogeneities; and
5. Posttranslational modifications and their structural characterization.

Most of the examples cited in this chapter are from work performed in the authors' laboratory.

2. Methods

2.1. FAB/LSIMS Analysis

In the FAB method, peptides and proteins introduced into the mass spectrometer in the condensed phase are desorbed and ionized by bombardment with fast atoms (Xe or Ar) with 8 kV of energy. The mass limits of detection of small proteins have been increased by using Cs^+ liquid SIMS *(18–20)*. In this method, which uses the same sample preparation as in FAB, high-energy Cs^+ ion at 35–40 kV is used as the bombarding particle. This increase in energy of the bombarding particle as compared to FAB significantly increases the sensitivity of detection for larger molecules.

Typically, 1 nmol of the peptide sample, dissolved in 2 µL of 0.1% aqueous TFA, is deposited onto a stainless-steel probe tip containing 1–2 µL of matrix (thioglycerol or 5:1 DTT/DTE mixture), which is then introduced into the mass spectrometer. The FAB/LSIMS analysis was carried out on double-focusing mass spectrometers operating at 3–8 kV acceleration voltage.

2.2. Enzymatic Digestion

The protein was dialyzed against several changes of 0.1*M* ammonium bicarbonate buffer (pH 7.8) followed by a single dialysis against deionized water and lyophilization. A 200-µg aliquot of rh-IFN α-2b was dissolved in 1% ammonium bicarbonate buffer (pH 8.4 adjusted with 10% ammonium hydroxide). TPCK-treated trypsin or *Staphylococcus aureus* V8 protease was added at an enzyme-to-substrate ratio of 1:50 (w/w) and 1:25 (w/w), respectively, and the solution incubated at 37°C for 18–24 h. The reaction was stopped by flash-freezing followed by lyophilization, and the crude digest mixture was subjected to FAB and PD MS analysis. Reduction of the disulfide bonds was carried out by adding 30-fold molar excess DTT over the cysteine present and reacting under nitrogen at 40°C for 1 h.

2.3. Cysteine Modification

Complete reductive pyridylethylation of cysteine was accomplished by reacting 100 µg of rh-IFN α-2b with 4-vinylpyridine in the presence of tri-*n*-butylphosphine *(21,22)*.

2.4. Chromatography

The reductive pyridylethylation reaction product of rh-IFN α-2b was dissolved in 0.1% aqueous TFA containing 27% (v/v) acetonitrile, applied onto a reversed-phase Vydac C_4 guard column (25 × 4.6 mm, 5 μm; The Separations Group, Hesperia, CA), and eluted with a linear gradient of 0.1% aqueous TFA and 0.1% TFA in acetonitrile (B) with UV detection at 215 nm. The gradient for the separation was 27–63% B in 15 min at a flow rate of 1.0 mL/min. The modified protein appeared as a single, sharp peak with a retention time of 10.1 min, eluting at an acetonitrile composition of ca. 50%. Fractions containing the protein product were collected, pooled, and evaporated to dryness in a Savant SpeedVac. The sample was resuspended in 1% ammonium bicarbonate buffer and digested with trypsin under conditions similar to those described for rh-IFN α-2b. After digestion, the sample was concentrated to 20 μL and analyzed by fast atom bombardment mass spectrometry (FAB MS).

3. Molecular-Weight and Amino Acid Sequence Determination

The first and most important single piece of information in characterizing any peptide by MS is its molecular weight. Once a purified sample of a peptide is obtained, FAB MS or Cs^+ ion liquid SIMS can be used to determine its molecular weight. A general strategy for peptide characterization is summarized in Table 1. The usual upper mass ranges of applicability are ca. 4000 Dalton for FAB and ca. 10,000 Dalton for liquid SIMS as observed in our laboratory. Currently, MS can provide molecular-weight information from picomole/femtomole quantities of purified peptide. The ion signal characterizing the molecular weight is observed as $(M + H)^+$ and/or $(M + Na)^+$, $(M + K)^+$ if any salts are present. Depending on the nature of the peptide, the presence of salts can interfere with the observation of molecular-weight-related ion signals or enhance ionization, and produce strong signals corresponding to the pseudomolecular ion *(23)*. Peptides containing a greater proportion of acidic residues may also be analyzed in the negative ion mode. The most commonly used matrices are glycerol, thioglycerol, *m*-nitrobenzyl alcohol, or a mixture of dithiothreitol/dithioerythritol, and their proper selection can determine the success of the analysis. TFA is usually added to enhance protonation of the sample molecule. We have observed that the right combination of matrix and cosolvent (TFA/water, DMSO, methanol) significantly

Table 1
Strategy For Peptide Characterization

- Step one
  - Calculate $(M + H)^+$ from amino acid analysis
  - Determine $(M + H)^+$ by FAB MS
- Step two
  - If (M + H)+ FAB < $(M + H)^+$ AAA
    - Presence of ASN, GLN, PCA, a-$CONH_2$
    - Loss of neutral species
    - Cyclic peptide
  - If $(M + H)^+$ FAB = $(M + H)^+$ AAA
    - Absence of ASN, GLN, PCA, a-$CONH_2$
    - Free amino terminus
    - Absence of TRP
  - If $(M + H)^+$ FAB > $(M + H)^+$ AAA
    - Prosthetic groups
    - Protecting groups
    - Oxidized TRP, MET, CYS
      - Undetected or unusual amino acids
      - Reduced disulfide
- Additional considerations
  - If $(M + H)^+$ FAB appears as multiplet
    - Polyisotopic elements
  - If $(M + H)^+$ FAB changes with time
    - Reducible functionalities
    - Chemical reactions
  - If $(M + H)^+$ FAB displays satellite intensities
    - Cationated molecular ions
    - Adduct ions
    - Multiply charged ions
    - Polymeric species

*Reproduced with permission from Fraser, B. A. (1987) Fast atom bombardment mass spectrometry: application to peptide structural analysis, in *Proteins, Structure and Function* (L'Italien, J. J., ed.), Plenum, New York and London, pp. 241.

enhances the quality of the FAB/SIM spectra and considerably suppresses the relatively high background ion signals in the lower mass region (<200 Dalton). Any differences between the molecular weights calculated from amino acid composition and that obtained by MS may reveal useful structural information about the peptide. This knowledge is very useful to the synthetic peptide chemist, since such disagreement indicates the occurrence of a synthetic error or modification *(24,25)*.

Furthermore, the presence of impurities copurifying with the desired product is detected as ion signals in addition to the expected one *(24)*.

The sequencing of peptides by mass spectrometry has made an important contribution to the study of peptides and proteins. In an early study, FAB MS was successfully utilized *(7)* for structural characterization of Met-Lys-bradykinin without prior chemical modification. The determination of amino acid sequences of polypeptides containing up to 21 amino acids by FAB MS had also been reported in the early 1980s *(26,27)*. Similar FAB MS work conducted in our laboratories *(28)* on several synthetic peptides revealed comparable fragmentation patterns and is described below (Schemes I–IV).

The systematic nomenclature for designating the sequence ions from the fragmentation observed in the FAB MS of a peptide was proposed in 1984 by Roepstorff and Fohlman *(29)*. This allows the identification of cleavage points on the peptide backbone and distinction between N- and C-terminal fragment ions (Fig. 1). However, not all ions shown in this scheme are present in the MS of a peptide; the formation of the fragment ions and their relative abundances are dependent on several factors, all of which are not fully understood.

Kemptamide is a synthetic peptide with 13 amino acid residues that shows several of the sequence-related ions *(28)* depicted in the scheme by FAB MS (Fig. 2). Only major fragment ions observed in the spectrum are indicated in the figure.

A peptide containing the C-terminal 15 amino acids of recombinant human IFN γ (GIF-C15) was synthesized with the predicted structure H_2N-R-K-R-S-Q-M-L-F-R-G-R-R-A-S-Q-COOH. Figure 3 describes the FAB MS for GIF-C15 *(30)* (mol wt 1875). Along with the protonated molecular ion, several key fragment ions indicative of the amino acid residues present in its primary structure are observed in the FAB mass spectrum. These ions permitted confirmation of the sequence noted above. The presence of basic residues in the peptide molecule facilitates the production of a strong molecular ion signal and fragments. Although the singly charged protonated molecular ion was the strongest signal observed in the FAB mass spectrum of GIF-C15, doubly and triply charged ions were also detected in the spectrum. The extent of cleavage between the bonds and the relative intensities of the fragment ions could indicate the presence of specific amino acid residues in the peptide molecule. The cleavage between nitrogen and carbon bonds with a

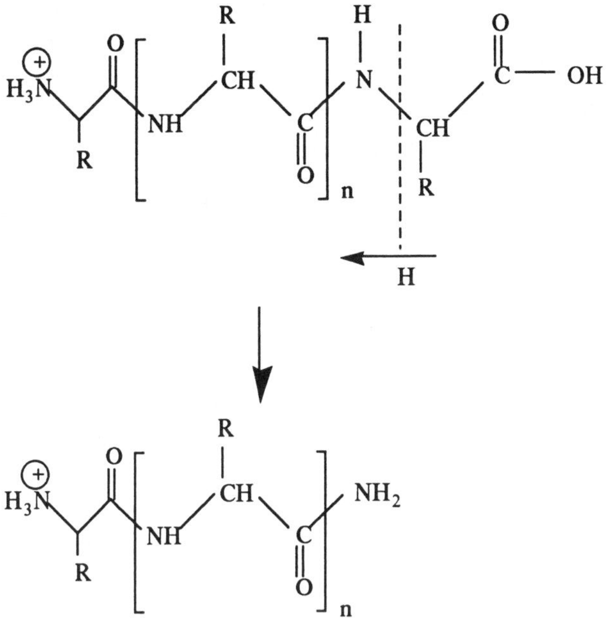

Scheme I. N—C bond cleavages (formation of amide fragmentation)

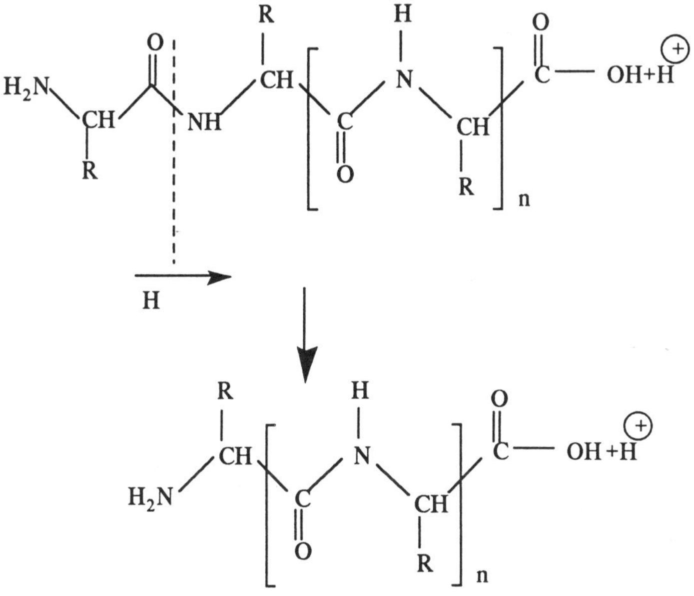

Scheme II. Peptide bond cleavage (C-terminal amnio sequence ion).

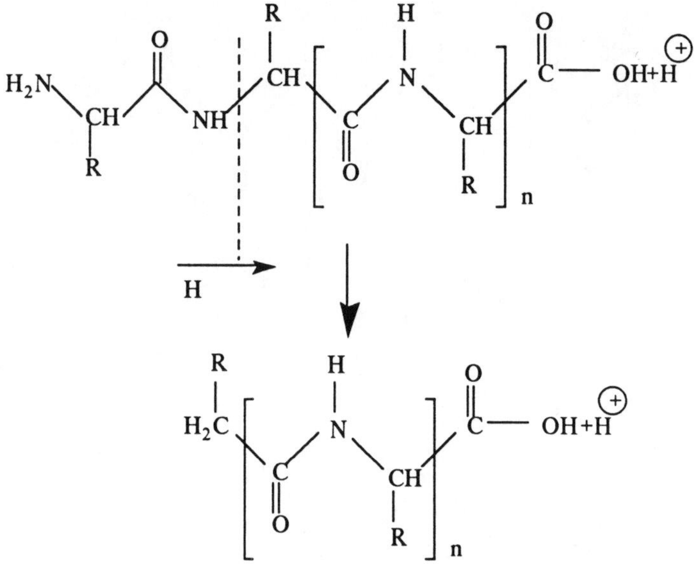

Scheme III. N—C bond cleavage (C-terminal alkyl sequence).

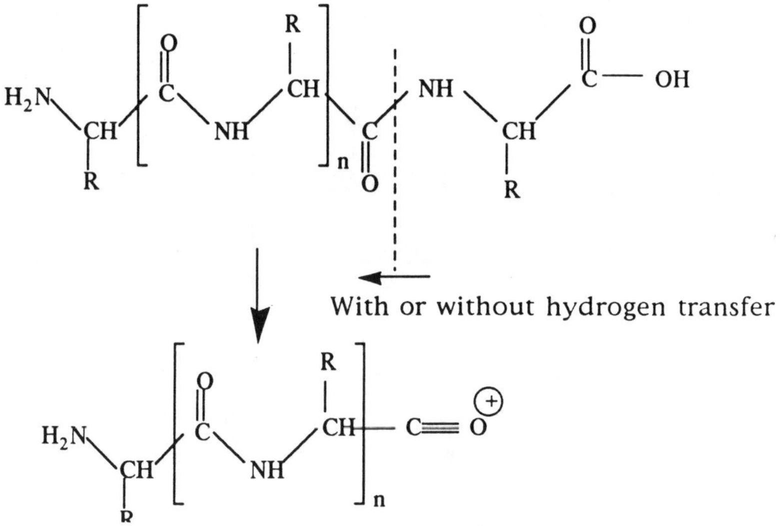

Scheme IV. Peptide bond cleavage (N-terminal oxonium ion).

hydrogen transfer to the N-terminal is favored when Arg or Lys is proximal to the cleavage site. Similar fragment ions are observed for the Leu-Leu residue at the N-terminal.

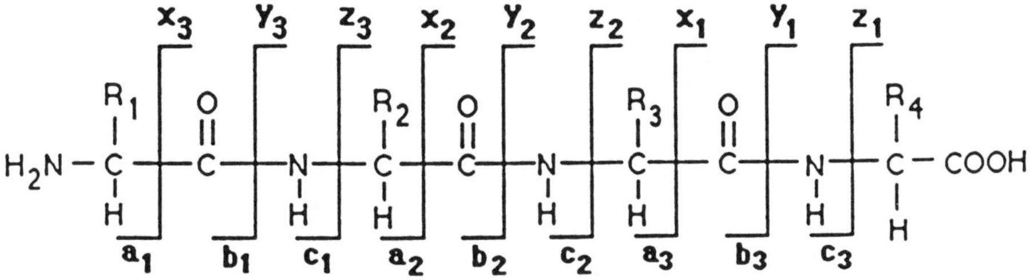

Fig. 1. Nomenclature for designating sequence ions from the fragmentation observed in FAB MS of a peptide *(29)*.

During an FAB MS analysis of a pentapeptide with its N-terminal blocked by a long-chain acyl group, significant enhancement of fragment ions formed by cleavages of the peptidic backbone (Fig. 4) was observed *(28)*. However, C—N bond cleavages were weak. The spectra are much simpler to interpret for structure information, and hence, this type of approach may be adopted as a strategy for peptides that otherwise generate weak fragment ions in their FAB mass spectrum.

In another example *(30)* (Fig. 5), a pentapeptide with a homoserine group showed a strong loss of water (m/z 545 to m/z 527), presumably through the formation of a lactone moiety. Other fragment ions, indicative of structure, are also shown in the figure.

Obtaining sequence information from fragment ions can often be hampered by incomplete or weak fragmentation, and interference from matrix or impurity-related ions. The technique of tandem MS overcomes these limitations and involves three sequential steps

1. Selection and separation of the ion of interest;
2. Collision-induced fragmentation of this ion with neutral gas molecules, such as He or Ar; and
3. Analysis of the resulting product ions.

The successful utilization of tandem MS for elucidating peptide structure is well documented *(31–33)*, and beyond the scope of discussion of this chapter.

4. Peptide Mapping

This involves chemical or enzymatic degradation of the protein, followed by MS determination of the resulting peptides *(34–36)*. In order to maximize the chances of cleavage, the protein molecule is unfolded by

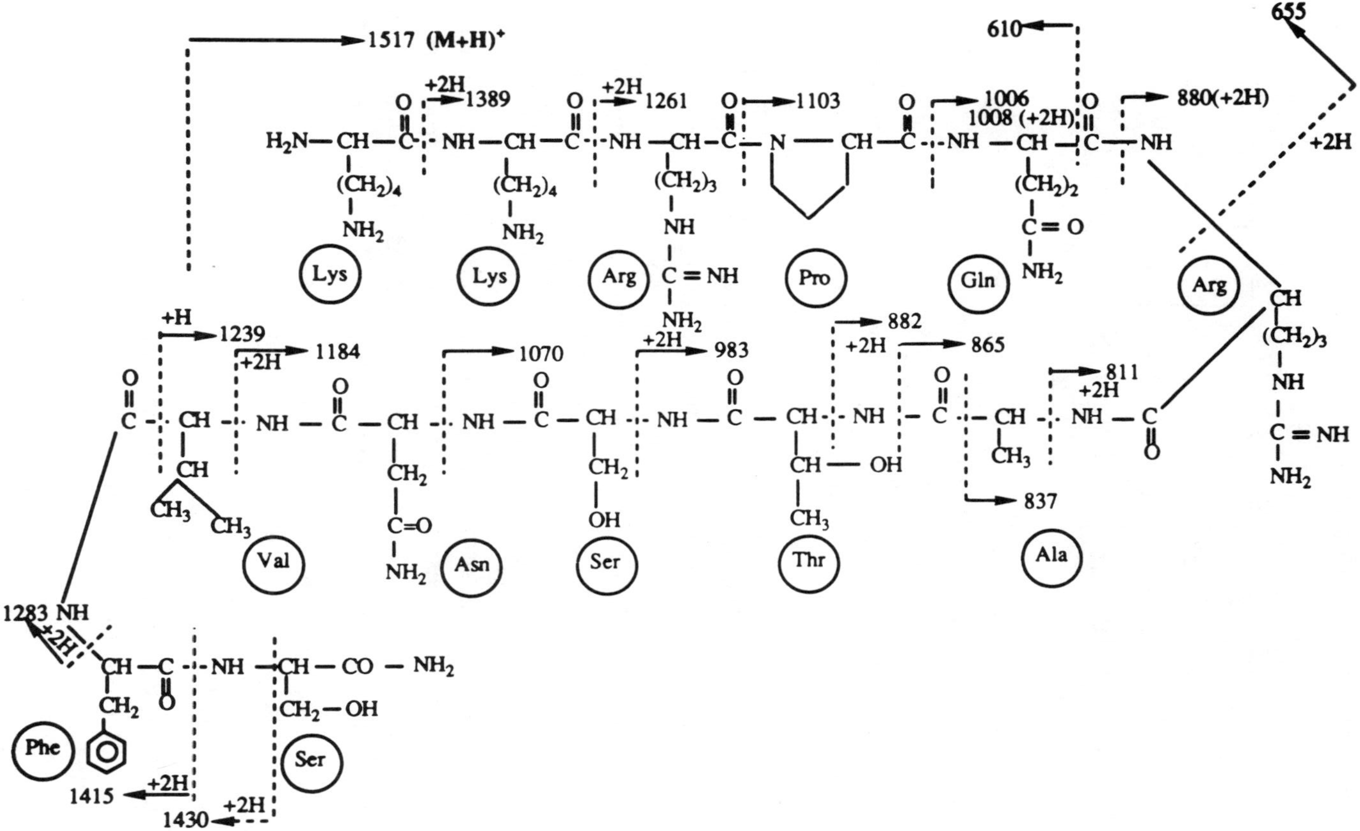

Fig. 2. FAB MS of Kemptamide (mol wt 1516). Lys-Lys-Arg-Pro-Gln-Arg-Ala-Thr-Ser-Asn-Val-Phe-Ser.NH2.

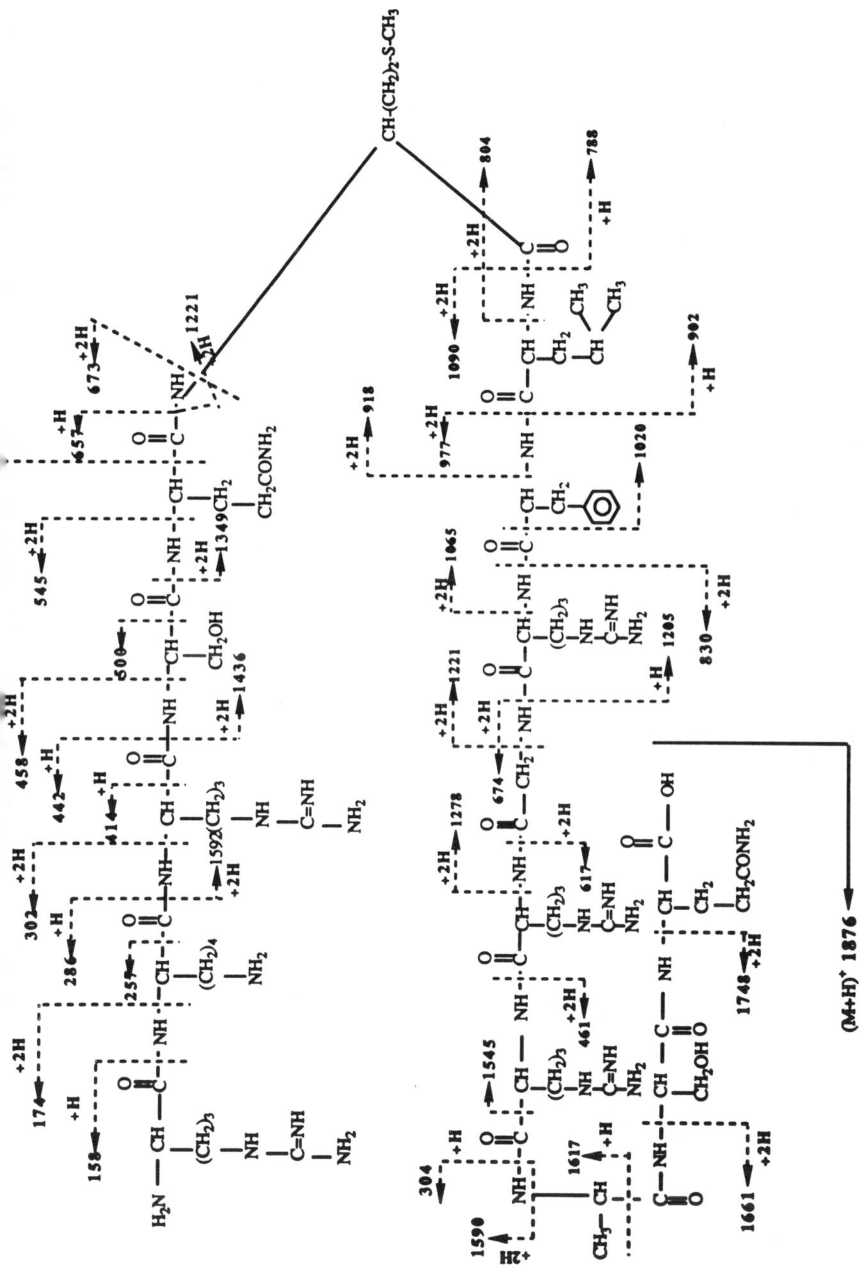

Fig. 3. FAB MS of GIF-C15 (mol wt 1875).

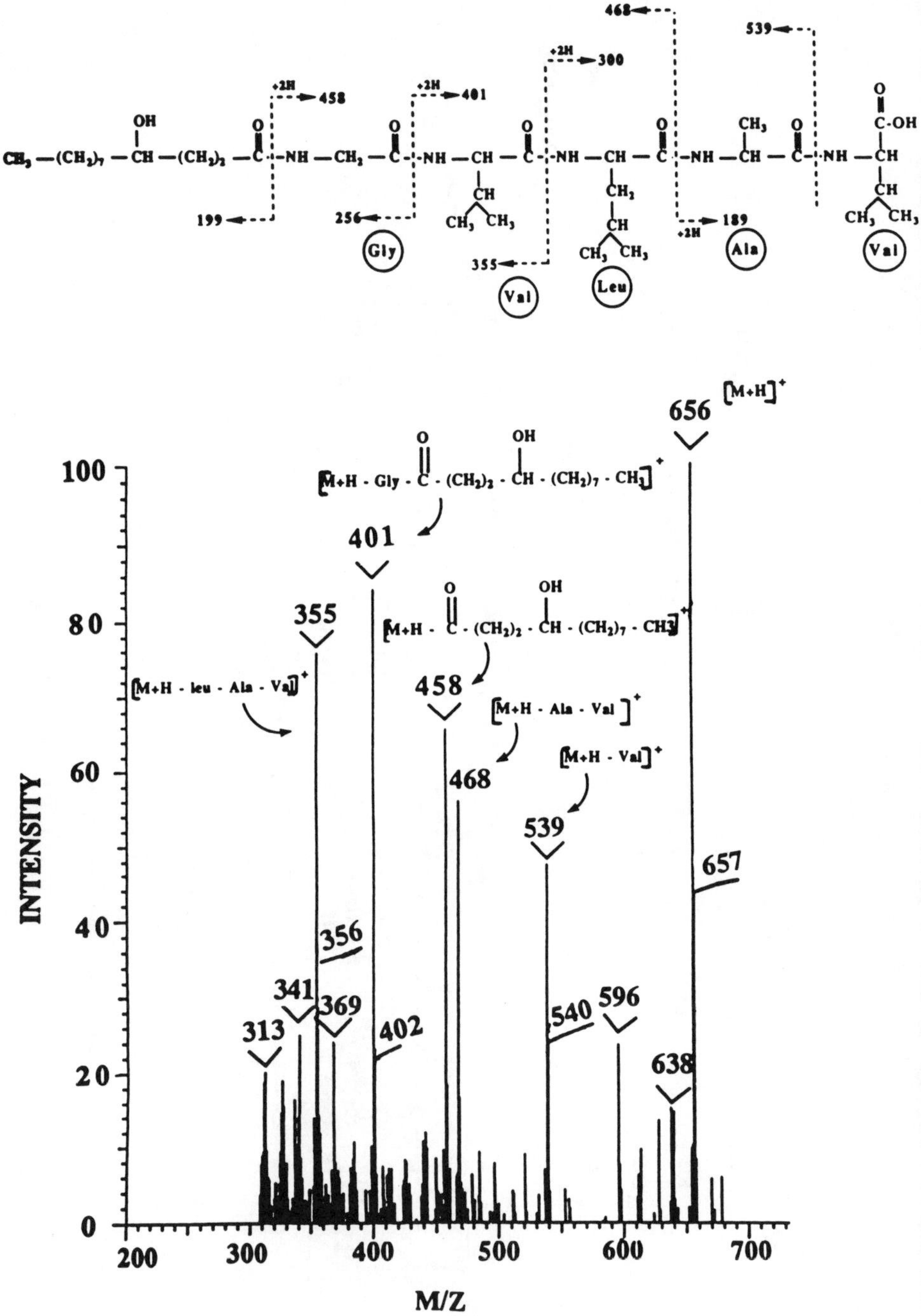

Fig. 4. FAB MS analysis of a pentapeptide with long-chain acyl group at N terminal.

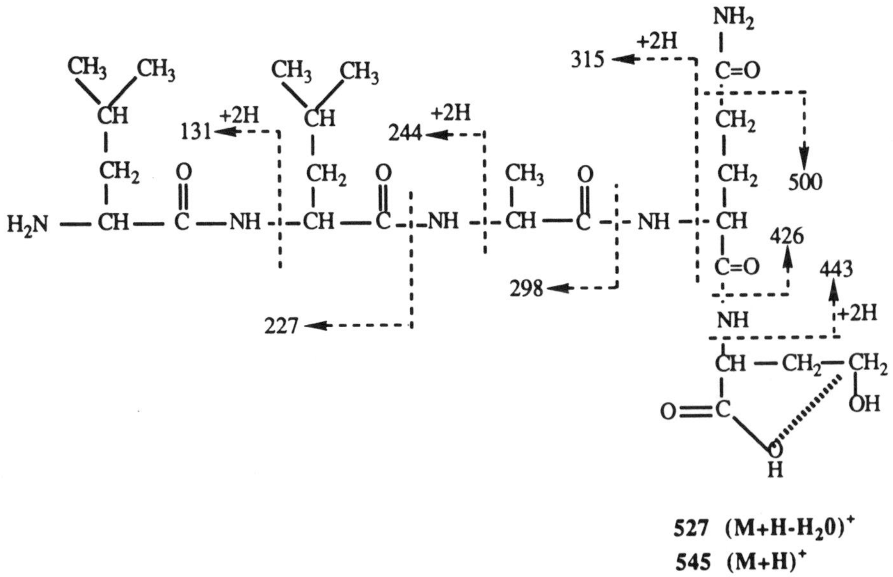

Fig. 5. FAB MS of Leu-Leu-Ala-Gln-Ser(homo) (a pentapeptide of α-2 IFN residue 17–21 generated by enzymatic digestion).

reduction and alkylation prior to enzyme treatment. MS may then be used to analyze the complex peptide mixture. However, to maximize detection limits for minor components, HPLC is used to separate the digests, and the fractions analyzed by either on-line or off-line MS. The ion signals from the peptides are matched to the sequence with computer programs *(37)* that calculate the expected masses of the peptides from the user input of the particular cleavage condition employed and the presumed sequence of the protein. Like Edman degradation, complete coverage is not possible, which necessitates the administration of multiple complementary cleavage strategies. Frequently, ambiguities arising from ion signals that can be mapped to more than one possible location in the peptide chain can be resolved by carrying out multiple, manual Edman degradation *(35)*.

The N-termini of the shortened peptides in the mixture are defined by the mass shifts, which are the MS equivalent of subtractive Edman degradation of a mixture. Several possibilities may give rise to cases where the MS ion signals cannot be fitted to any predicted sequence. Contaminating proteins, unexpected side reactions during reduction, alkylation, cyanogen bromide treatment, deamidation, oxidation, or posttranslational modification are some of the more common causes for such an observation *(16)*.

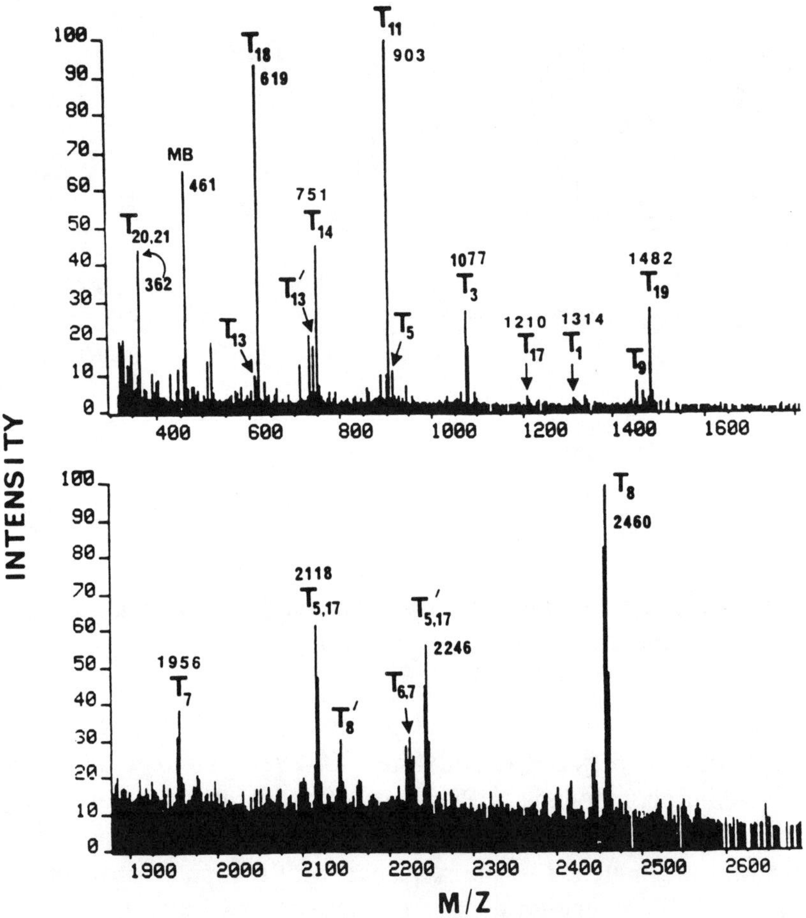

Fig. 6. FAB MS of the tryptic digest of rh-IFN α-2b. (Reproduced with permission from ref. *38*).

In the case of rh-IFN α-2b mapping procedure *(38)*, the protein was cleaved with trypsin at the C-terminal side of the lysine and arginine residues. FAB MS analysis of the resulting crude digest mixture produced ion signals (T_1........T_{21}) from all the predicted tryptic peptides in the mass range scanned, viz., 200–4800 Dalton (Fig. 6). The ion signals originating from the relatively low-mol-wt tryptic mono- and dipeptides were submerged in the high background usually present in the lower mass region of the FAB mass spectrum. In addition, several peptide ion signals arising from either incomplete or unexpected cleavages were also noted. It was observed that incomplete cleavage by trypsin occurs when there are two contiguous cleavage sites in the polypeptide chain, i.e., R-R, K-K, R-K, and K-R. Edman degradation-derived sequences also confirmed the presence of these products in the rh-IFN α-2b tryptic digest.

5. Posttranslational Modifications

FAB MS and MS in general have made extremely useful contributions to the study of posttranslational modifications providing significant advantages over many classical biochemical methods. Examples of such modifications include blocked or modified amino and carboxy termini, phosphorylated and sulfated peptides, oxidation of methionine sites, *N*- and *O*- glycosylation of specific residues, and others *(39)*.

Enzymatic acylation of the N-termini is fairly common, and most intracellular proteins in eukaryotes are N^{α}-acetylated *(40)*. A significant number have long-chain fatty acids, viz., myristic acid, attached to their N-termini *(40,41)*. In such cases, straightforward use of the Edman sequencing procedure is impossible. However, the difference in molecular weight observed by MS and that calculated from amino acid composition gives a clue regarding the nature of the blocking group. Also, the inadvertent insertion or deletion of glycine during solid-phase synthesis can be examined *(42)* by MS.

Isoprenylation of the C-termini with polyisoprenoid lipids has been determined by MS *(16)*. A nonenzymatic posttranslational modification involving deamidation of α-ASP to β-ASP, which may result in loss of biological activity *(43)*, has been studied by mass spectrometric methods *(42)*.

Enzymes also control phosphorylation and sulfation of Ser, Thr, and Tyr residues. Although such residues are not readily detectable by conventional techniques, MS has been used to study such modifications *(44,45)*. Strong molecular ions were observed for phosphorylated Ser and Tyr in both positive and negative ion FAB MS, and somewhat greater intensities of ion signals of fragments were observed from phosphotyrosine-containing peptides as compared to phosphoserine-containing ones *(44)*. FAB MS methods are often employed for locating disulfide bonds and glycosylation sites in peptides/proteins as the following examples illustrate.

6. Disulfide Bond Location

An important task in the determination of protein primary structure is disulfide bond location *(46)*. Traditionally, the possibility of disulfide "scrambling" or "reshuffling" by reduction and reoxidation of S–S bridges causes difficulty. The strategy adopted for the FAB MS peptide mapping approach aims at obtaining disulfide-linked peptides, each containing only a single S–S bridge, by chemical or enzymatic cleavage of the native protein. Usual procedures involve proteolysis by pepsin, partial acid hydrolysis, or cyanogen bromide cleavage at methionine. To minimize the chances of disulfide rearrangements, low pH is used for these reactions. The resulting mass spectra of the peptide mixtures give molecular-weight-indicative ion signals for inter- and intramolecular disulfide-bonded peptides. To enhance the ion signals of the reduced forms of the peptide, reduction in dithiothreitol/dithioerythritol can be supplemented by addition of triethylamine or ammonium hydroxide. The assignment of the cysteinyl peptides and their linkages can be made by comparing the spectra before and after the reduction process.

The disulfide bonds in *E. coli* expressed human interferon α-2b (rh-IFN α-2b) were determined employing FAB and PD mapping procedures *(38)*. This protein has four cysteines at positions 1, 29, 98, and 138 of the primary sequence in the tryptic peptides T_1, T_5, T_{10} and T_{17}. Six disulfide pairings are possible of which the only peptide signals detected in the FAB mass spectrum of the unreduced tryptic mixtures were at m/z 4616 and m/z 2118. This matches with the disulfide-linked peptides T_1-S-S-T_{10} and T_5-S-S-T_{17}, respectively (Fig. 7). On treatment of the tryptic digest mixture with DTT, the signals at m/z 4616 and m/z 2118 were diminished, whereas those corresponding to T_1, T_5, T_{10}, and T_{17} increased significantly on reduction of the disulfide bonds. Thus, in this protein molecule, disulfide bonds link Cys(1) to Cys(98) and Cys(29) to Cys(138) agreeing with previously reported HPLC results *(47)*.

In a slightly modified strategy, the cystine residues of the protein are *S*-alkylated with 4-vinylpyridine in the presence of tri-*n*-butylphosphine *(21,22)*. This strategy was exploited for detecting tryptic peptides from rh-IFN α-2b *(38)*. The tryptic peptides containing one *S*-pyridylethyl cysteine (PEC) residue produced strong FAB ion signals at 106 mass units higher than their non-PEC counterparts. The advantage of the PEC derivatization procedure is mainly twofold. *S*-alkylation with 4-vinylpyridine is highly specific for cystine residues and is not known to modify other

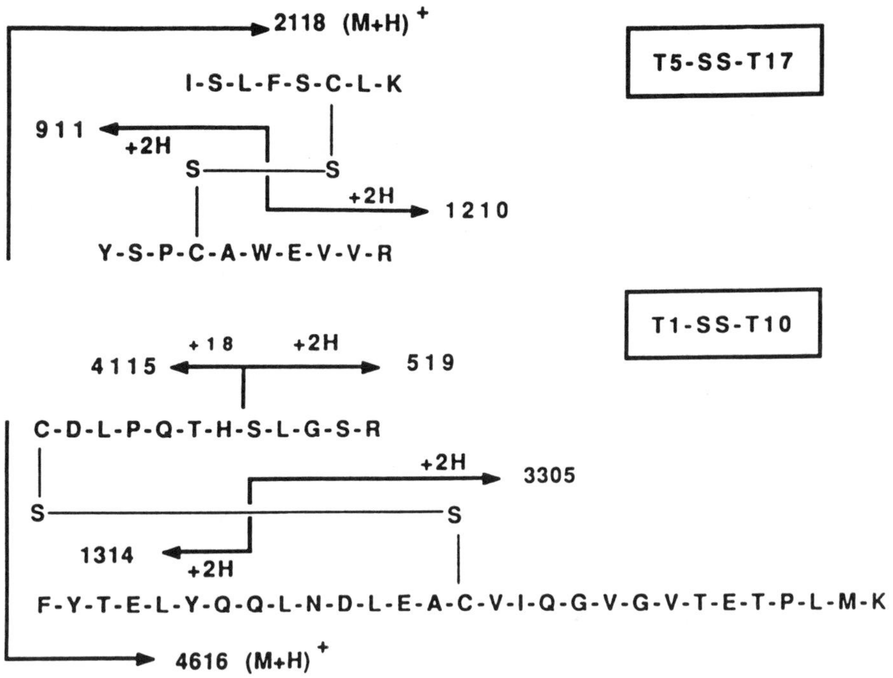

Fig. 7. Assignment of the disulfide bonds in rh-IFN α-2b observed from the FAB MS signals in the analysis of the unreduced and reduced tryptic digest mixture. (Reproduced with permission from ref. *38*).

side chains, unlike alkylation with iodoacetic acid or iodoacetamide, which reacts with histidine. Second, the PEC-containing peptides generate stronger FAB ion signals possibly because of the participation of the pyridine ring in charge stabilization. This overcomes the problem of detecting some cysteine-containing peptides that give very weak FAB ion intensities.

7. Glycosylation Sites

Human interleukin-4 (IL-4) is a naturally occurring glycoprotein that mediates the proliferation and differentiation of B-lymphocyte mast cells. The carbohydrate chains can play a significant role in the biological properties of a glycoprotein and, hence, the determination of the structure, as well as the attachment sites for the carbohydrate moieties, is important for therapeutic approval of a glycoprotein. The possible attachment site(s) of the N-linked carbohydrate of biologically active recombinant human IL-4 expressed in Chinese hamster ovary cells was determined

HUMAN INTERLEUKIN-4

```
        .         10        .         20        .         30
H K C D I T L Q E I I K T L N S L T E Q K T L C T E L T V T

        .         40        .         50        .         60
D I F A A S K N T T E K E T F C R A A T V L R Q F Y S H H E

        .         70        .         80        .         90
K D T R C L G A T A Q Q F H R H K Q L I R F L K R L D R N L

        .        100        .        110        .        120
W G L A G L N S C P V K E A N Q S T L E N F L E R L K T I M

        .        129
R E K Y S K C S S
```

Fig. 8. Human interleukin-4 sequence *(48)*.

(48). To define which of the two potential Asn glycosylation sites (Asn-X-Ser/Thr sequence) is glycosylated, *S. aureus* V8 protease cleavage of glycosylated and deglycosylated human IL-4 was carried out following reduction and carboxymethylation of the glycoprotein *(49)*. For the digest mixture, the FAB mass spectral signal of the V8 peptide fragment containing the first potential site (Asn-38) could only be observed for the *N*-glycosidase F-treated glycoprotein, whereas that for the second potential site (Asn-105) containing peptide was observed with and without enzyme treatment. In conclusion, the N-linked oligosaccharide is located at position 38 of human IL-4 (Fig. 8).

8. Conclusion

FAB MS and Cs^+ ion liquid SIMS have made enormous contributions to peptide/protein characterization and are complementary to classical approaches, such as Edman degradation and gene sequencing. In some cases, MS offers an alternative to these biochemical methods. FAB MS and liquid SIMS have been shown in the chapter to be useful for determining mol wt of polypeptides up to ca. 4000 Dalton by FAB and even higher by liquid SIMS. The fragment ions observed in FAB MS/liquid SIMS greatly aid in the sequence determination of these peptides, and peptide mapping techniques confirm their cDNA sequence. FAB MS also helps in quick detection of impuries and/or modifications during peptide synthesis when other approaches may fail. In addition to this methodology, the two techniques of laser desorption and electrospray ionization MS are currently having a great impact on biological MS. These two techniques have opened up the possibilities of detecting large biomolecules up to 300,000 Dalton with a higher level of sensitivity. Electrospray is being applied to the analysis of ionic transition metal complexes and porphyrins. A particularly exciting feature is the ability to study protein–ligand and protein–protein noncovalent interactions *(50–53)* generally at physiological pHs. With the current trend of improvements in MS instrumentation and computer algorithms, this technique is emerging as an indispensable tool for peptide/protein characterization that an even greater number of synthetic peptide chemists will turn to for solving their problems.

References

1. Morris, H. R., Williams, D. H., and Ambler, R. P. (1971) Determination of the sequences of protein-derived peptides and peptide mixtures by mass spectrometry. *Biochem. J.* **125,** 189–201.
2. Rose, K., Priddle, J. D., Offord, R. E., and Esnouf, M. P. (1980) A mass-spectrometric method for the estimation of the ratio of g-carboxyglutamic acid to glutamic acid at specific sites in proteins. *Biochem. J.* **187,** 239–243.
3. Rose, K., Simona, M., and Offord, R. (1983) Amino acid sequence determination by g.l.c.-mass spectrometry of permethylated peptides. *Biochem J.* **215,** 261–272.
4. Biemann, K., Gapp, F., and Seibl, J. (1959) Application of mass spectrometry to structure problems. *J. Am. Chem. Soc.* **81,** 2274,2275.
5. Carr, S. A., Herlihy, W. C., and Biemann, K. (1981) Advances in gas chromatographic mass spectrometric protein sequencing. *Biomed. Mass Spectrom.* **8,** 51–61.
6. Khorana, H. G., Gerber, G. E., Herlihy, W. C., Gray, C. P., Anderegg, R. J., Nihei, K., and Biemann, K. (1979) Amino acid sequence of bacteriorhodopsin. *Proc. Natl. Acad. Sci. USA* **76,** 5046–5050.
7. Barber, M., Bordoli, R. S., Sedgewick, R. D., and Tyler, A. N. (1981) Fast atom bombardment of solids (F.A.B.): a new ion source for mass spectrometry. *J. Chem. Soc. Chem. Comm.* **7,** 325–327.
8. McFarlane, R. D. and Torgerson, T. F. (1976) Californium-252 plasma desorption mass spectroscopy. *Science* **191,** 920–925.
9. Sundqvist, B. and McFarlane, R. D. (1985) ^{252}Cf-plasma desorption mass spectrometry. *Mass Spectrom. Rev.* **4,** 421–460.
10. Whitehouse, C. M., Dreyer, R. N., Yamashita, M., and Fenn, J. B. (1985) Electrospray interface for liquid chromatography and mass spectrometers. *Anal. Chem.* **57,** 675–679.
11. Loo, J. A., Udseth, H. R., and Smith, R. D. (1988) Collisional effects on the charge distribution of ions from large molecules, formed by electrospray-ionization mass spectrometry. *Rapid Comm. Mass Spectrom.* **2,** 207–210.

12. Bruins, A. P., Covey, T. R., and Henion, J. D. (1987) Ion spray interface for combined liquid chromatography/atmospheric pressure ionization mass spectrometry. *Anal. Chem.* **59,** 2642–2646.
13. von Wyssenhoff, H., Selzle, H. L., and Schlag, E. W. (1985) Laser-desorbed large molecules in a supersonic jet. *Z. Naturforsch.* **40a,** 674–676.
14. Li, L. and Lubman, D. (1989) Resonant two-photon ionization for the identification of thermal decomposition products in the laser desorption of small peptides. *Rapid Comm. Mass Spectrom.* **3,** 12–16.
15. Karas, M. and Hillenkamp, F. (1988) Laser desorption ionization of proteins with molecular masses exceeding 10000 Daltons. *Anal. Chem.* **60,** 2299–2301.
16. Carr, S. A., Helmling, M. E., Bean, M. F., and Roberts, G. D. (1991) Integretation of mass spectrometry in analytical biotechnology. *Anal. Chem.* **63,** 2802–2824.
17. Chowdhury, S. K. and Chait, B. T. (1989) Recent developments in the mass spectrometry of peptides and proteins, in *Annual Reports in Medicinal Chemistry—24,* Ch. 27, Academic, San Diego, CA, pp. 253–263.
18. Barber, M. and Green, B. N. (1987) The analysis of small proteins in the molecular weight range 10–24 kDa by magnetic sector mass spectrometry. *Rapid Comm. Mass Spectrom.* **1,** 80–83.
19. Pramanik, B., Tsarbopoulos, A., Siegel, M., Tsao, R., Reichert, P., Bartner, P., Das, P., Her, G., Doelling, V., Nagabhushan, T. L., and Trotta, P. P. (1989) Californium-252 plasma desorption and cesium ion liquid secondary ion mass spectrometry studies of some natural and recombinant proteins, in *Proceedings of the 37th ASMS Conference on Mass Spectrometry and Allied Topics,* Miami Beach, FL, pp. 893,894.
20. Tsarbopoulos, A., Pramanik, B. N., Reichert, P., Siegel, M. M., Nagabhushan, T. L., and Trotta, P. P. (1991) 252 Cf-Plasma desorption and cesium-ion liquid secondary-ion mass spectrometric analysis of recombinant proteins. *Rapid Comm. Mass Spectrom.* **5(2),** 81–85.
21. Friedman, M., Krull, L. H., and Cavins, J. F. (1970) The chromatographic determination of cystine and cysteine residues in proteins as S-β-(4-pyridylethyl) cysteine. *J. Biol. Chem.* **245,** 3868–3871.
22. Andrews, P. C. and Dixon, J. E. (1987) A procedure for in situ alkylation of cystine residues on glass fiber prior to protein microsequence analysis. *Anal. Biochem.* **161,** 524–528.
23. Pramanik, B. N., Das, P. R., and Bose, A. K. (1989) Molecular ion enhancement using salts in FAB matrices for studies on complex natural products. *J. Natl. Prod.* **52(3),** 534–546.
24. Chait, B. T. (1988) The use of ^{252}Cf plasma desorption mass spectrometry for the analysis of synthetic peptides and proteins, in *The Analysis of Peptides and Proteins by Mass Spectrometry* (McNeal, C. J., ed.), John Wiley, New York, p. 21.
25. Biemann, K. and Scoble, H. A. (1987) Characterization by tandem mass spectrometry of structural modifications in proteins. *Science* **237,** 992–998.
26. Williams, D. H., Bojesen, G., Auffret, A. D., and Taylor, L. (1981) Study of difficult peptides from paracoccus cytochrome c-550 and a dolphin cytochrome c. *FEBS Lett.* **128,** 37–39.

27. Williams, D. H., Bradley, C. V., Santikarn, S., and Bojesen, G. (1982) Fast-atom-bombardment mass spectrometry. *Biochem. J.* **201,** 105–117.
28. Pramanik, B. N., Schering-Plough Internal Memo, Dec. 21, 1984.
29. Roepstorff, P. and Fohlman, J. (1984) Biomedical mass spectrometry, in *Proposal for a Common Nomenclature for Sequence Ions in Mass Spectra of Peptides.* **11,** 601.
30. Nagabhushan, T. L., Kosecki, R., Pramanik, B., Labdon, J., and Trotta, P. P. (1989) Purification and sequencing of interferons and other biologically active proteins and polypeptides, in *Frontiers in Bioprocessing* (Sikdar, S. K., Bier, M., and Todd, P., eds.), CRC, Boca Raton, FL, Chapter 5, pp. 51–62.
31. McLafferty, F. W. (ed.) (1983) *Tandem Mass Spectrometry.* John Wiley, New York.
32. Hunt, D. F., Yates, J. R., Shabanowitz, J., Winston, S., and Hauer, C. (1986) Protein sequencing by tandem mass spectrometry. *Proc. Natl. Acad. Sci. USA* **83,** 6233–6237.
33. Biemann, K. and Martin, S. (1987) Mass spectrometric determination of the amino acid sequence of peptides and proteins. *Mass Spectrom Rev.* **6,** 1–76.
34. Gibson, B. W. and Biemann, K. (1984) Strategy for the mass spectrometric verification and correction of the primary structures of proteins deduced from their DNA sequences. *Proc. Natl. Acad. Sci. USA* **81,** 1956–1960.
35. Morris, H. R., Panico, M., and Taylor, G. W. (1983) FAB-mapping of recombinant-DNA protein products. *Biochem. Biophys. Res. Commun.* **117,** 299–305.
36. Tsarbopoulos, A., Becker, G. W., Occolowitz, J. L., and Jardine, I. (1988) Peptide and protein mapping by 252Cf-plasma desorption mass spectrometry. *Anal. Biochem.* **171,** 113–123.
37. Lee, T. D. and Vemuri, S. (1990) MacPro mass: a computer program to correlate mass spectral data to peptide and protein structures. *Biomed. Environ. Mass Spectrom.* **19,** 639–645.
38. Pramanik, B. N., Tsarbopoulos, A., Labdon, J. E., Trotta, P. P., and Nagabhushan, T. L. (1991) Structural analysis of biologically active peptides and recombinant proteins and their modified counterparts by mass spectrometry. *J. Chromatogr.* **562,** 377–389.
39. Dixon, J. E., Yazdanparast, R., Smith, D., and Andrews, P. C. (1987) Identification of posttranslational modifications in neuropeptides, in *Methods in Protein Sequence Analysis, 1986.* (Walsh, K. A., ed.), Humana, Clifton, NJ, p. 493.
40. Tsunasawa, S. and Sakiyama, F. (1984) Amino-terminal acetylation of proteins: An overview. *Methods Enzymol.* **106,** 165–170.
41. Carr, S. A. and Biemann, K. (1984) Identification of posttranslationally modified amino acids in proteins by mass spectrometry. *Methods Enzymol.* **106,** 29–58.
42. Carr, S. A., Bean, M. F., Helmling, M. E., and Roberts, G. D. (1990) Integration of mass spectrometry in biopharmaceutical research, in *Biological Mass Spectrometry* (Burlingame, A. L. and McCloskey, J. A., eds.), Elsevier, Amsterdam, p. 621.
43. Geiger, T. and Clarke, S. (1987) Deamidation, isomerization and racemization at asparaginyl and aspartyl residues in peptides. *J. Biol. Chem.* **262,** 785–794.
44. Gibson, B. W. (1990) The identification and sequence analysis of phosphorylated and sulfated peptides by liquid secondary ion mass spectrometry, in *Biological Mass Spectrometry* (Burlingame, A. L. and McCloskey, J. A., eds.), Elsevier, Amsterdam, p. 315.

45. Bateman, A., Solomon, S., and Bennett, H. P. J. (1990) Post-translational modification of bovine pro-opiomelanocortin. *J. Biol. Chem.* **265,** 22,130–22,136.
46. Morris, H. R. and Pucci, P. (1985) A new method for rapid assignment of S–S bridges in proteins. *Biochem. Biophys. Res. Commun.* **126,** 1122–1128.
47. Lydon, N. B., Favre, C., Bove, S., Neyret, O., Benureau, S., Levine, A. M., Seelig, G. F., Nagabhushan, T. L., and Trotta, P. P. (1985) Immunochemical mapping of α-2 interferon. *Biochemistry* **24,** 4131–4141.
48. Her, G. R., Pramanik, B. N., Kumarasamy, R., Bartner, P., Das, P., Tindall, S. H., Nagabhushan, T. L., Trotta, P. P., and Tsarbopoulos, A. (1990) Structural characterization of the recombinant human interleukin-4 N-linked carbohydrate chains by mass spectrometry, in *Proceedings of the 38th ASMS Conference on Mass Spectrometry and Allied Topics,* Tucson, AZ., pp. 1341,1342.
49. Carr, S. A. and Roberts, G. D. (1986) Carbohydrate mapping by mass spectrometry: a novel method for identifying attachment sites of Asn-linked sugars in glycoproteins. *Anal. Biochem.* **157,** 396–406.
50. Ganem, B., Li, Y. T., and Henion, J. D. (1991) Detection of non-covalent receptor-ligand complexes by mass spectrometry. *J. Am. Chem. Soc.* **113,** 6294–6296.
51. Ganem, B., Li, Y. T., and Henion, J. D. (1991) Observation of non-covalent enzyme-substrate and enzyme-product complexes by ion-spray mass spectrometry. *J. Am. Chem. Soc.* **113,** 7818,7819.
52. Baca, M. and Kent, S. B. H. (1992) Direct observation of a ternary complex between the dimeric enzyme HIV-1 protease and a substrate-based inhibitor. *J. Am. Chem. Soc.* **114,** 3992,3993.
53. Ganguly, A. K., Pramanik, B. N., Tsarbopoulos, A., Covey, T. R., Huang, E. C., and Fuhrman, S. A. (1992) Mass-spectrometric detection of the noncovalent GDP-bound conformational state of the human H-ras protein. *J. Am. Chem. Soc.* **114,** 6559–6560.

CHAPTER 8

Sequence Analysis of Peptide Resins from Boc/Benzyl Solid-Phase Synthesis

Jan Pohl

1. Introduction

There are numerous methods available to monitor the course of solid-phase peptide synthesis (SPPS) and to characterize the peptide intermediates synthesized therefrom. No single technique, however, is universally suitable for identifying all of the varied problems that may occur during SPPS *(1–3)* because the intermediates of SPPS, unlike those in homogeneous solution peptide synthesis, remain attached to the resin and are, therefore, not amenable to classical methods of peptide characterization.

Solid-phase sequencing (Edman degradation, *4*) of peptide resins *(5)* is a very powerful technique that determines, retrospectively or in real time and at any stage of synthesis, the sequence of the peptide being assembled, and measures the acylation efficiencies achieved in each synthetic cycle *(5–15)*. This chapter will describe the current high-sensitivity mode of automated sequencing *(16,17)* as applied to peptides synthesized using the Boc-benzyl strategy *(11,13–15)*. The major characteristics of this method are summarized in Table 1.

1.1. Basic Principles

Edman degradation *(4,18,19)* of peptide resins proceeds through three discrete steps (Fig. 1) during which the peptide remains covalently attached to the resin via its carboxy terminus. During step 1, coupling, the free amino terminus of the peptide is thiocarbamylated at pH 8–9 with phenylisothiocyanate (PITC), yielding phenylthiocarbamyl (PTC)

From: *Methods in Molecular Biology, Vol. 36: Peptide Analysis Protocols*
Edited by: B. M. Dunn and M. W. Pennington Copyright ©1994 Humana Press Inc., Totowa, NJ

Table 1
Characteristics of the Solid-Phase Sequencing Technique as Applied to Peptides Synthesized by the Boc-Benzyl SPPS Strategy

Sensitivity
~1 pmol (absolute limit); detects at least 0.1% of deletions assuming an initial load of ~5000–10,000 pmol peptide resin
Practical initial load
~1–30 Resin beads
Typical sequencing yields
Initial yield >70%; ~50–150 pmol of PTH amino acid/resin bead (assuming resin substitution of 0.4–0.75 mmol/g); repetitive yield >94–95%
Advantages
Sensitive; minimum amount of resin consumed
Provides direct sequence and preview information
Detects incomplete Boc group deprotections (unlike, e.g., the ninhydrin assay)
Stable covalent linkage of peptide to resin prevents washout of peptides; sequencing through the C-terminus is therefore routine
Analysis is not complicated by artifacts from cleavage, deprotection, and postsynthetic handling (unlike, e.g., sequencing of the cleaved peptides)
Is invaluable for analysis and rescuing of resins from "failed" syntheses
Disadvantages
Is slow and usually retrospective
Exhibits general limitations inherent to Edman chemistry
Carryover and internal cleavages obscure the accuracy of preview analysis
Specialized equipment is required, and expensive reagents are used

peptide. During step 2, cleavage, the amino-terminal residue is removed from the peptide via acidic cyclization, yielding the anilinothiazolinone (ATZ) derivative of the N-terminal amino acid and the truncated peptide resin. Finally, during step 3, conversion, the ATZ derivative, which is rather unstable, is extracted from the truncated peptide resin. The anilinothiazolinone ring of the ATZ amino acid is then opened to form the PTC-amino acid intermediate, which is cyclized (in the rate-limiting step) to yield the phenylthiohydantoin (PTH) derivative (PTH amino acid). The PTH amino acids are then separated by reversed-phase HPLC, and identified and quantitated by comparing their retention time and peak heights with known standards stored in the calibration file *(20–23)*. The truncated peptide resin is taken to the next coupling step of Edman degradation.

The chromatograms of the PTH amino acids provide the following information *(5–15)*. First, the amino acid sequence of the peptide is identified as the predominant PTH signal that is present in each sequencing

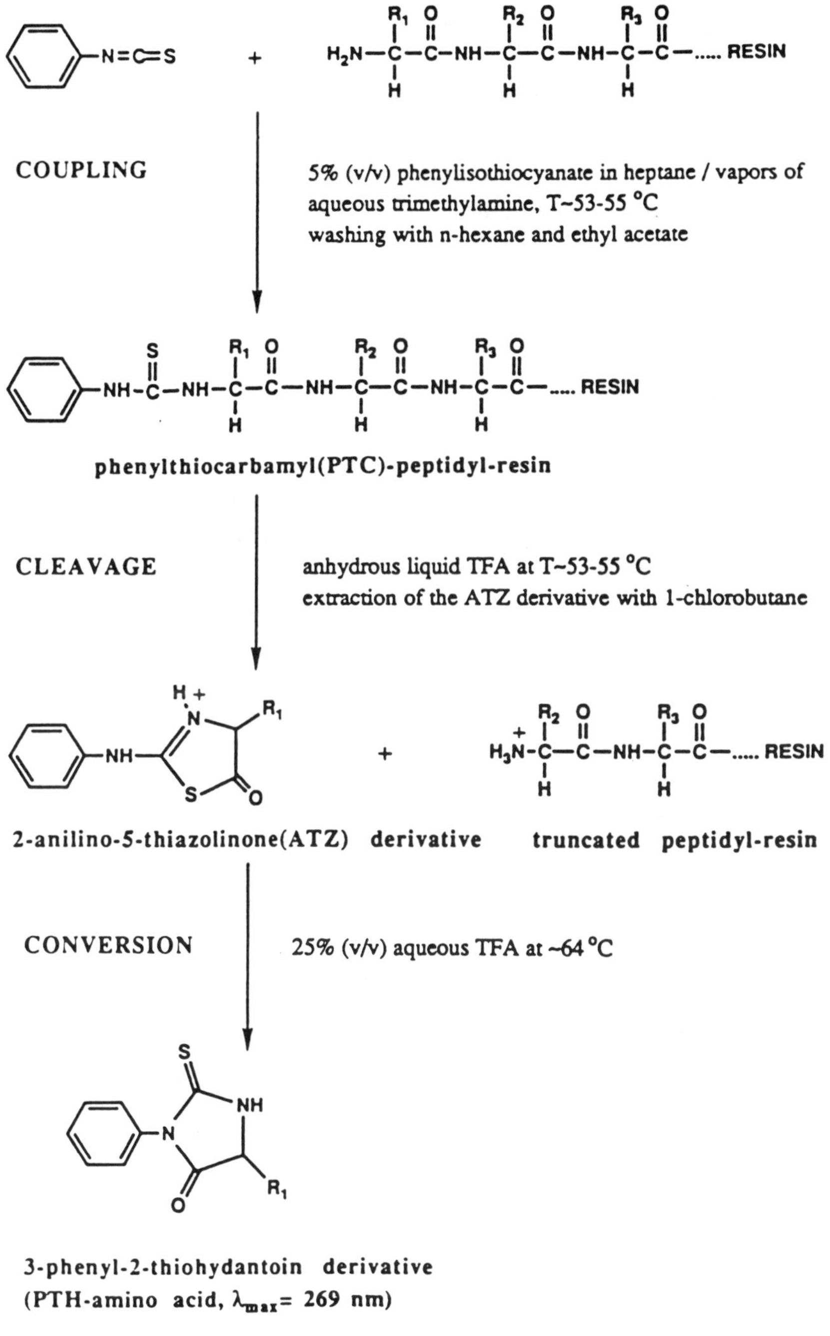

Fig. 1. The chemistry of Edman degradation.

cycle throughout the run. Structural integrity and fidelity of the peptide can therefore be directly evaluated. Second, the efficiency of acylations in each synthesis cycle can be determined: If acylations were not totally quantitative, i.e., if deletion peptides are present on the resin, the PTH derivative of each amino acid residue of the major sequence would also be found in its previous sequencing cycle(s). The relative quantity with respect to that of the major amino acid is called the sequencing preview (P) and is a measure of the inefficiency of acylation in each synthetic step. Thus, the preview for amino acid *X*, which is the jth residue from the N-terminus of the *n*-mer peptide, is *(5,11)*:

$$(P_X)j \text{ (in\%)} = 100 \times A/(A + B) \quad (1)$$

where A = sum of yield of PTH-*X* in cycles 1 to j – 1, and B = yield of PTH-*X* in cycle j. Third, since the preview accumulates from cycle to cycle in proportion to the amount of deletions, its final value (the cumulative preview) approximates the total amount of deletion peptides in the analyzed sequence *(5)*.

In practice, the majority of SPPSs do not require resin sequencing because of high coupling efficiencies, or because monitoring is done by other methods and/or the analysis is left for the cleaved product. Resin sequencing is instead being used in specialized applications, such as characterizing intermediates of long-chain peptide syntheses *(8,24,25)*, monitoring the progress of reactions that involve side-chain functionalities and cannot therefore be monitored by other methods (e.g., quantitative ninhydrin assay, ref. *26*), or establishing the integrity of resins from failed, unattended syntheses *(6–8)*.

1.2. Complications and Limitations

Major limitations of the Edman technique arise from its failure to degrade certain synthetic peptides *per se*, from its inefficiencies in stepwise degradation, and from the harshness of conditions to which the peptide and its amino acid derivatives are repeatedly exposed. As a result, the number of residues in which the sequence can be obtained is limited by such factors as carryover, internal peptide cleavages, and physical losses.

1.2.1. Nonsequenceable Peptides

Some synthetic peptides are not sequenceable *per se* or are only partially sequenceable because they cannot be either thiocarbamylated or cleaved. Examples of such peptides include peptides containing other than α-amino acid residues, and peptides modified at the N^{α}-terminus

Table 2
Examples of Peptide Modifications That Are Incompatible with Edman Sequencing

Modification	Cause/occurrence	Residues involved
N^{α}-terminus		
Acyl	SPPS, capping protocols	Nonspecific
Trifluoroacetyl	O→N shift (SPPS, PSQ[a])	Ser(Tfa),Thr(Tfa)
Aspartimide, β-aspartyl	Cyclization to aspartimide/ α,β-rearrangement (SPPS, PSQ)	Asn, Asp(OBzl), Asp(OcHex)
Pyroglutamyl-X	Cyclization (SPPS, PSQ)	Gln, Glu(OBzl), Glu(OcHex)
Pyruvyl-X	β-elimination to dehydro-Ala/ N—C bond cleavage (PSQ, SPPS)	Ser, Ser(Bzl), Cys, Cys(4-MeOBzl)
Alkyl or aryl	SPPS	Nonspecific
Side-chain		
Branching	Intentional, SPPS	Trifunctional
N→O acyl shift	SPPS, PSQ	Ser, Thr
Peptide bond surrogates		
—CH_2NH—	Intentional, SPPS	Nonspecific

[a]PSQ, terminations also occur during Edman sequencing *(27,28)*.

or at the peptide backbone (*see* examples in Table 2). These modifications can originate from the synthetic protocol or from side reactions during Boc-Bzl SPPS *(1)*, but may also occur during sequencing *(27,28)*.

1.2.2. Carryover

The coupling and cleavage steps of Edman degradation are never entirely quantitative. For this reason, removal of the N-terminal amino acid in each cycle is never complete, and a population of uncleaved peptides arises. As sequencing progresses, this fraction of the PTH signal, called carryover or lag, accumulates *(4,22,29)* and remains out of phase with the major sequence signal. For example, assuming a very low 1%/step carryover, half of the PTH signal will be out of phase in 70 cycles. The amount of carryover influences the number of residues that can be sequenced. In addition, calculations of the preview (Eq. [1]) need to be corrected for carryover *(11)*:

$$(P_X)_j \text{ (in\%)} = 100 \times A/(A + B + C) \tag{2}$$

where C = sum of carryover of PTH-*X* in cycles j + 1 to *n*. Carryover greatly affects accurate preview analysis of peptides that are rich in identical residues or that contain short repeating sequences. For such peptides, it may not be possible to obtain a precise preview for all residues in the sequence, since preview and carryover will not be distinguishable from each other *(11,15)*.

1.2.3. Internal Cleavages and Peptide Losses

All internal peptide bonds undergo partial acidolysis during the cleavage step of Edman degradation *(4,28–31)*, at an estimated *(30,31)* per step rate of 0.001–0.1%, depending on the sequence. A "background" PTH signal for every residue present in the peptide will therefore be generated, and will fluctuate in a sequence-dependent manner. The rise or fluctuation of the background signal decreases the accuracy and sensitivity of preview analysis *(7,8)*. In addition to internal cleavages, up to several percent of the C^{α}-terminus ester or amide bonds linking the peptide to the resin will be acidolyzed in each cleavage step, depending on the nature of the C-terminal residue and the type of linkage *(7–9,11,13,15)*. These internally or C-terminally cleaved peptides are likely to be washed out, causing a gradual weakening of the sequence signal. Extensive peptide losses owing to the C-terminal cleavages have been cited as the most serious deterrent to the widespread use of the Edman technique for sequencing resins from Fmoc/tBu SPPSs where acid-labile types of linkages are typically used *(32,33)*.

1.2.4. Losses of PTH Derivatives

Although most PTH derivatives are nearly quantitatively recovered during resin sequencing *(7,8,12)*, there are some that are either partially destroyed, form multiple PTH derivatives, or are only poorly extracted *(4,7,8,12,19,20–22,34)*. In most instances, however, the preview can still be determined for these derivatives since it is calculated on the basis of relative comparisons between cycles (*see* Eq. [2] and refs. *8,11,15*).

Clearly, many limitations of the Edman technique are cumulative in nature, and originate from inefficiencies in the chemistry and from the repeated exposure of the peptides to Edman reagents. These negative factors can be circumvented or significantly reduced by sequencing successive short segments of the entire sequence, e.g., in increments of 10–15 residues. This approach has proven useful especially during syntheses of longer (>30–40 residues) peptides *(8,24,25)*.

2. Materials

1. The model peptide, CG(21-40) (*see* Section 5.), was assembled using an Applied Biosystems, Inc. (ABI) (Foster City, CA) Model 430A Peptide Synthesizer on Boc-Leu-4-oxymethylphenylacetamidomethyl (PAM)-copoly(styrene/1% divinylbenzene) resin (substitution 0.75 mmol/g, ABI), using the manufacturer's Small Scale Rapid Cycles synthesis protocol (0.1 mmol synthesis scale, software version 1.40). All Boc-protected amino acids (ABI, Bachem Bioscience, Inc., Philadelphia, PA, or Bachem, Inc., Torrance, CA) were recoupled using the standard reaction vessel cycle, RECPL 22. Synthesis-grade reagents and solvents were purchased from ABI and from American Burdick and Jackson (Muskegon, MI).
2. Sequencing was done on an ABI Model 477A Protein Sequencer connected on-line with an ABI Model 120A PTH Analyzer, and equipped with a Model 900A Data Module (software version 1.61). All sequencing-grade reagents were purchased from ABI. The reaction vessel reagents were *(15)*: 5% (v/v) PITC in *n*-heptane (R1), 12.5% (v/v) aqueous trimethylamine (R2), and anhydrous TFA containing 0.001% dithiothreitole (DTT) (R3). The reaction vessel solvents, *n*-heptane (S1), ethyl acetate (S2), and *n*-butylchloride (S3), were used without the addition of DTT. (**Caution:** If DTT is added to S1, S2, and S3, its adduct with PITC will be present in every PTH chromatogram, coeluting with PTH-Pro and obscuring quantitation of proline.) The aqueous TFA (25%, v/v, Reagent R4), used for conversion, contained 0.001% DTT. The PTH amino acids were dissolved in 20% (v/v) aqueous acetonitrile containing 0.001% DTT (Reagent S4). The standard (coded) PTH amino acids were obtained from ABI and were stored desiccated at –20°C until they were reconstituted in sequencing-grade acetonitrile containing 0.001% DTT (Reagent R5) immediately before use. Standard PTH amino acids can also be obtained from Pierce Chemical Company (Rockford, IL) and from Sigma Chemical Co. (St. Louis, MO). The ABI PTH C18 Spheri-5 reversed-phase cartridge (2.1 × 220 mm, 5 mm particle size) was used in all PTH analyses.

3. Methods

3.1. Removal of the N^{α}-Terminal Boc Group

Samples of peptide resins may be removed for sequence analysis at any stage of SPPS. The peptide's N^{α}-terminus must, however, be in the amino or imino form in order to allow coupling with PITC and subsequent sequencing (Fig. 1). If the N^{α}-terminal Boc group is not completely removed, an initial carryover, proportional to the extent of residual Boc-protection, will be introduced in the first sequencing cycle.

Remove the N^{α}-terminal Boc group of the peptide resin by treating it with 50% (v/v) trifluoroacetic acid in dichloromethane for 20 min at room temperature. Wash the resin with dichloromethane, neutralize with 5% (v/v) diisopropylethylamine in dichloromethane, and wash with dichloromethane and methanol.

3.2. Staining and Loading of the Resin

Staining resin beads with bromophenol blue *(35)* prior to sequencing allows them to be easily seen and counted on the sequencer sample disk; this assures that the instrument will not be overloaded. Expect an average initial sequencing yield of 50–150 pmol of PTH amino acids/bead. Unstained beads are difficult to count on the white background of the disk. However, this can be done under a microscope *(14)*. The Boc-protected resins do not stain with bromophenol blue *(35)*.

Suspend 0.1–0.2 mg of the deprotected/neutralized resin in 200 µL of either 40% (v/v) methanol/dichloromethane or in 100% acetonitrile in an Eppendorf tube. Stain the beads navy blue by adding several microliters of 0.1% bromophenol blue in methanol; vortex. Remove any excess bromophenol blue by washing the resin several times with 1-mL aliquots of either 40% (v/v) methanol in dichloromethane or with 100% acetonitrile. Under a magnifying glass, load a suspension of resin beads onto the sample support disk. Remove the extra beads (*see* Table 1, "Practical Initial Load") from the disk using microforceps or a micropipet tip prewet with acetonitrile. (Preloading the disk with Polybrene/NaCl and precycling it prior to sequencing are not necessary for the resin-bound peptides *[11]*). Saturate the disk with 30 µL of acetonitrile; this will ensure that the beads adhere to the disk. Seal the reaction vessel cartridge in the sequencer, and dry the sample with argon. The following instructions assume that the operator is familiar with the principles of operation and with the chemistry of the ABI sequencer.

3.3. Selecting the Program and Conditions for Solid-Phase Sequencing

The reaction vessel and conversion flask cycles used for sequencing peptide resins on the ABI Model 477A Sequencer/120A PTH Analyzer sequencing system are listed in Table 3; conditions used to separate the PTH derivatives are in Table 4. In the first cycle, BEGIN-1, the standard PTH amino acids are analyzed. In the following cycles, REZ-1, the sample PTH amino acids are analyzed. The PTH amino acids are detected

Table 3
Sequencing Program *(11,15)*[a]

Cycle no.	Reaction cycle	Conversion cycle	Gradient	Analyte
1	BEGIN-1	BEGIN-1	REZ-1	PTH Standards
2-*n*	REZ-1	REZ-1	REZ-1	Sample PTH amino acids

[a]The reaction vessel temperature is set at 51 ± 2°C, depending on the application. The conversion flask temperature is set at 64°C.

Table 4
Conditions for HPLC Separation of PTH Amino Acids *(11,14,15,20,36)*[a]

Gradient REZ-1		
Time, min	% B	Flow rate, mL/min
0	13 ± 2	210
0.5	13 ± 2	210
18	38 ± 2	210
25	38 ± 2	210
38	65 ± 2	210
39	90	210
41	90	210

[a]Solvent A: ~60 m*M* aqueous sodium acetate pH 4.0 ± 0.05, 5% (v/v) in tetrahydrofuran. Aqueous trimethylamine (ABI Reagent R2) is added to solvent A to a final concentration of 0.2–2 m*M*. Trimethylamine helps stabilize the elution positions and increases the peak sharpness of the positively charged PTH amino acids (e. g., His, His[Me], Arg, and Har). The exact concentration of trimethylamine added depends on the ionic strength of solvent A, on the amount of the residual silanol groups of the column packing material, and on column age and usage. The exact composition of the gradient will also have to be found through experimentation.

Solvent B: HPLC-grade acetonitrile containing 0.5 μ*M N,N*-dimethyl-*N*′-phenylthiourea.

by monitoring the column eluent at 269 nm. If desired, the selective detection of serine and threonine can be achieved by monitoring the elution of their degradation products, PTH-dehydroalanine (S") and PTH-dehydro-2-aminoisobutyric acid (T"), at 313 nm *(8)* using an external UV detector and a strip chart recorder connected in series with the 120A PTH Analyzer. Gradient REZ-1 accomplishes separation of both the unprotected and the side-chain-protected PTH amino acids.

3.4. Establishing a Calibration File for Identification and Quantitation of PTH Amino Acids

PTH amino acids are identified and quantitated on-line in the ABI Data Module 900A of the sequencer *(22)*. The quantity of each PTH amino acid is calculated by dividing either its peak height or peak area by the peak height/pmol or peak area/pmol ratio for that PTH amino acid as found in the calibration file. This file consists of values predetermined by the investigator from sequencing known amounts of the pure PTH standards. Although 20 of these standards are available commercially, the PTH derivatives of side-chain-protected *(7,8,10–12,14,15,36,37)* and of noncoded amino acids are not available, and their retention times and peak height/pmol ratios must be determined by the investigator. This is done by Edman degradation of model peptide resins that contain both well-characterized amino acids and the amino acids for which the PTH derivatives are not identified *(8,10,11)*. The values obtained from these model resins are then used to create a calibration file that is used for analysis of the experimental resins. The procedure is as follows:

1. Couple the nonstandard or side-chain-protected Boc-amino acid to a previously assembled model peptide resin consisting of amino acids with stable PTH derivatives (e.g., Val-Leu-Phe-Ala-PAM-Resin). Remove the N^{α}-terminal Boc group, and sequence several resin beads using the above-described program. Determine the retention time of the new PTH derivative, and calculate its operational peak height/pmol ratio using the averaged picomole yields obtained for PTH-Val, PTH-Leu, and PTH-Phe in the same run. More than one PTH derivative of the new PTH amino acid may be present in the chromatogram, depending on its chemical nature (*see* Section 4.2.).
2. If only the retention time of the new PTH derivative is needed, it can be quickly determined by applying the Edman procedure to the free amino acid as follows. Remove the N^{α}-Boc group of the new Boc-protected amino acid by treating it with 50% (v/v) TFA in dichloromethane (1 mg/mL) in a sealed tube for 20 min at room temperature. Dilute the solution 10× with dichloromethane, and load 1–2 nmol of the deprotected amino acid onto a precycled sample support disk. (**Caution:** Unlike in resin sequencing, a Polybrene/NaCl pretreatment of the disk is needed to help retain the amino acid on the disk during sequencing. Preload the disk with 3 mg of Polybrene and 0.3 mg of NaCl, and precycle *[16,17,23]* using at least two manufacturer's FIL-1 reaction and conversion cycles prior to loading the

amino acid.) Conduct two REZ-1 cycles of Edman degradation, and determine retention time for the new PTH derivative. If the procedure needs to be repeated for other amino acids, pause the sequencer at the end of the second cycle, load the next amino acid, and continue sequencing.

3. Update the calibration file with the parameters obtained for the new PTH derivatives. Calculate the yields of PTH amino acids in each cycle of the experimental peptide resin, and identify the predominant amino acid sequence *(22)*. Using the background corrected data, calculate preview and carryover using Eq. (2). In practice, taking PTH yields from the two cycles preceding and two to three cycles following each residue is sufficient for most calculations *(7,8,11)*. The final (cumulative) preview at the end of the analysis estimates the total amount of deletion impurities in the peptide segment sequenced.

4. Notes

4.1. Sequencer Performance

The mechanical and chemical performance of the sequencer should be constantly monitored and kept at its maximum. Several excellent reviews have been written on this topic *(20,21,23,34,36)*. By far the most revealing indicator of overall performance is the magnitude of the initial and stepwise repetitive yield, and of the amount of carryover obtained from sequencing a standard protein or peptide. Thus, using 100 pmol of the β-lactalbumin standard, an initial yield higher than 70%, and an average stepwise repetitive yield *(22,29)* of 94–95% should be reproducibly obtained using the ABI 477A/120A sequencing system. Similar performance parameters have been reported for other makes and models of the current commercially available sequencers *(7–14,23,36)*.

4.2. Retention Times and Multiple PTH Derivatives

Retention times of PTH derivatives that have been identified in our laboratory are listed in Table 5. Despite the high resolution of the HPLC system, not all the derivatives listed in Table 5 can be completely separated from closely eluting PTH amino acids under standard gradient conditions. It is not, therefore, possible to identify these residues positively and/or calculate their preview if they follow each other in the sequence. Some of these residues, however, yield secondary PTH derivatives, which elute with distinct retention times and which can be used for positive identification instead (*see* Table 5). The yields of some PTH derivatives will vary depending on the performance of the sequencer and the

Table 5
Retention Times of PTH Amino Acids[a]

Retention time, min	PTH derivative or Edman chemistry by product[b]
4.7	Tyr(OP[OH]$_2$)
5.1	PTU, *N*-phenylthiourea (adduct of ammonia and PITC)[b,c]
5.2	Asp
5.6	Asn
6.4	Ser
6.6	Cys(Cm)
6.8	Gln
7.0	MPTU, *N*-methyl-*N'*-phenylthiourea[b,c]
7.1	Citrulline
7.2	Homoserine
7.2	Thr
7.7	Gly
8.1	Gln(Nγ-Me)
8.2	Cys(Cam)
8.5	Glu
8.8	DMPTU, *N,N*-dimethyl-*N'*-phenylthiourea[b]
9.5	Arg(NO$_2$)
9.9	Cys(Acm)
10.6	Lys(Ac)
11.1	Ala
11.6	3-(2-Pyridyl)-Ala
12.0	His
13.1	S', adduct of DTT and dehydroalanine
13.8	Lys(Biotinyl)
14.0	T', adduct of DTT and dehydro-2-aminoisobutyric acid
14.2	T', adduct of DTT and dehydro-2-aminoisobutyric acid
14.5	His(1-Me)
14.7	Tyr
14.8	His(3-Me)
15.3	S", dehydroalanine
15.6	S", dehydroalanine
16.1	T', adduct of DTT and dehydro-2-aminoisobutyric acid
16.3	Arg
16.3	T', adduct of DTT and dehydro-2-aminoisobutyric acid
16.3	N$^{\alpha}$-Me-Ala
16.8	T", dehydro-2-aminoisobutyric acid (dehydrothreonine)
17.5	Tyr(OP[OMe]$_2$)
17.8	Pro
17.8	DTT-PITC, adduct of DTT and PITC[b]
18.7	Met
19.1	Val

Table 5 *(continued)*

Retention time, min	PTH derivative or Edman chemistry by product[b]
19.1	Homoarginine
19.4	Cys(4-PyEt)
20.3	3,4-Dehydroproline
20.4	Norvaline
20.5	Asp', unidentified derivative of Asp
20.6	DPTU, *N,N'*-diphenylthiourea[b]
20.6	Arg(Tos)
21.5	Glu', Gln(NgPh)
21.5	His(2,4-Dnp)
22.1	DPU, *N,N'*-diphenylurea[b]
22.1	Trp
22.8	Phe
22.9	Trp(For)
23.0	Phe(4-NO_2)
23.3	Ile
23.5	His(3-Bzl)
23.8	Lys(Ptc)
24.2	Leu
24.6	Nle
25.5	N^{α}-Me-Phe
25.9	Ser(Bzl)
26.4	Arg(Mtr)
26.5	Arg(Mts)
26.8	Orn(Z)
28.1	His(3-Bom)
28.1	Thr(Bzl)
29.7	Homophenylalanine
29.9	Asp(OBzl)
30.2	Glu(OBzl)
30.2	Cys(4-MeOBzl)
31.5	Asp(OcHex)
31.6	3-(2-Naphthyl)-Ala
32.9	Phe(4-NHZ)
33.3	Thr(Bzl)', unidentified derivative of Thr(Bzl)
33.7	Lys(2-ClZ)
34.0	Lys(2,4-Dnp)
34.4	Glu(OcHex)
34.4	Hyp(Bzl)
34.7	Tyr(2,6-Cl_2Bzl)
35.2	Cys(4-MeBzl)
35.5	3-Cyclohexyl-Ala

(continued)

Table 5 *(continued)*
Retention Times of PTH Amino Acids[a]

Retention time, min	PTH derivative or Edman chemistry by product[b]
36.1	Trp', unidentified derivative of Trp
36.7	Trp', unidentified derivative of Trp
36.7	Hyp(Bzl)
37.1	Trp', unidentified derivative of Trp
37.3	Trp', unidentified derivative of Trp
39.1	Lys(Fmoc)
39.5	Trp', unidentified derivative of Trp
39.9	Tyr(2-BrZ)
40.0	Glu(OFm)
40.1	Trp, unidentified derivative of Trp
40.2	Trp, unidentified derivative of Trp
40.3	Tyr(3-[2,6-Cl_2Bzl])[d]

[a]*See* Table 4 for conditions. The column was operated at 56°C. Solvent A was 60 m*M* aqueous sodium acetate, 5% (v/v) in tetrahydrofuran and 0.4 µ*M* in trimethylamine, pH 3.95. Solvent B was acetonitrile containing 0.5 m*M* DMPTU. Retention times will vary between sequencing systems because of slight differences in chromatographic conditions. Note the closely eluting PTH derivatives.

[b]*See* refs. *4,20–22* for a general discussion of the origin of the Edman chemistry byproducts. The amounts of these compounds in the chromatograms vary, and depend on the sequencing program used and the performance of the instrument.

[c]PTU is generated when samples contain traces of ammonia. MPTU is generated when older lots of trimethylamine (Reagent R2) contain traces of methylamine.

[d]Tyr(3-[2,6-Cl_2Bzl]) is a side-reaction product that can be generated, e.g., during HF cleavage of the Tyr(2,6-Cl_2Bzl)-containing peptide resins (*see* Chapter 4, *PSP* and ref. *1*)

program used, and on the number of degradations preceding a particular residue in the sequence. The following observations may help during interpretation of the chromatograms (*see also* refs. *7–15,20,37,38)*.

Asp and Glu: Approximately 5–20% of Asp(OBzl) and Asp(OcHex), and 10–40% of Glu(OBzl), Glu(OcHex), and Glu(OFm) are recovered as PTH-Asp or PTH-Glu, respectively. PTH-Asp may coelute with *N*-phenylthiourea (PTU), an adduct of PITC and ammonia. The two compounds can be resolved by decreasing the pH of HPLC Solvent A by several tenths of a pH unit. This results in an increased retention of PTH-Asp owing to increased protonation of its β-carboxylate.

Asn and Gln: Approximately 5–10% of Asn and 15–20% of Gln are recovered as PTH-Asp or PTH-Glu, respectively.

Ser: PTH-Ser(Bzl) is only partially recovered. Ser(Bzl) partially β-eliminates, and is recovered as PTH-dehydroalanine (S") and its adduct with DTT, PTH-S'. In addition, 2–10% of PTH-Ser is typically recovered.

Cys: PTH-Cys(Acm) and PTH-Cys(4-PyEt) are stable during sequencing. PTH derivatives of Cys(4-MeOBzl), Cys(4-MeBzl), and Cys(Bzl) are not completely recovered since these residues undergo, to a lesser extent than Ser(Bzl), β-elimination, yielding PTH derivatives of S' and S" (*see* Ser). Between 0.2 and 1% of PTH-Ser, a product of rehydration of dehydroalanine, is also recovered. PTH-Cys(Cm) and PTH-Cys(Cam) are readily identified, the latter being partially deamidated (30–50%) to PTH-Cys(Cm) in the conversion flask.

Thr: PTH-Thr(Bzl) is not completely recovered. Thr(Bzl) partially β-eliminates to dehydro-2-aminoisobutyric acid (T") and to its adducts with DTT (four PTH-T' peaks). In addition to these derivatives, about 5–15% of PTH-Thr is typically recovered.

His: His(Dnp), His(Bzl), and His(Bom) form stable PTH derivatives. His(Z) is completely deprotected to His. PTH-His(Dnp) coelutes with the γ-anilide of PTH-Glu, a side reaction product of Edman degradation. PTH-His and PTH-His(Me) may not be completely extracted from the sequencer.

Tyr: Tyr(2-BrZ), Tyr(2,6-Cl_2Bzl), and Tyr(P) form relatively stable PTH derivatives (1–5% is recovered as PTH-Tyr). Tyr(OP[OMe]$_2$) is almost completely converted to Tyr(P) in the first few cycles. PTH-Tyr(P) is eluted in the elution front close to oxidized DTT, and may therefore escape positive identification if it is present in smaller (<100 pmol) quantities.

Arg and Homoarginine (Har): Arg(Tos) and Arg(NO_2) form stable PTH derivatives. Arg(Mts) is relatively stable, and typically 1–5% is recovered as PTH-Arg. Arg(Mtr) and Har(Pmc) are much less stable to anhydrous TFA, and are partially deprotected to Arg and Har. PTH-Arg and PTH-Har may not be completely extracted from the sequencer. PTH-Arg(Tos) coelutes with *N,N'*-diphenylthiourea (DPTU) from the sequencer.

Pro and Gly: PTH-Proline coelutes with the adduct of DTT and PITC if DTT is present in sequencer solvents S1, S2, and S3. At a low picomole level, this compound will obscure proline quantitation. Proline and glycine residues in the sequence may require extended cleavage times *(18)*. Incomplete cleavage at these residues can be the single most common source of carryover.

Met: Met(O) is completely reduced to Met by DTT, which is present in sequencer reagents R3, R4, and S4.

Trp: PTH-Trp is only partially recovered, yielding multiple unidentified PTH derivatives that elute late in the gradient. About 20–40% of Trp(For) is recovered as PTH-Trp. PTH-Trp(For) is difficult to separate from PTH-Phe.

Lys: Lys(2-ClZ) is a stable derivative (1–2% is recovered as PTH-Lys[Ptc]). Lys(Fmoc) is partially converted to Lys(Ptc) on exposure to vapors of trimethylamine (R2) during the coupling step.

Pyroglutamic acid (<Glu): The N-terminal pyroglutamyl residue prevents sequencing. The 2-pyrrolidone ring of <Glu can be quantitatively opened by aminolysis of the peptide in 20% aqueous methylamine for 20 h at 40°C *(39)*. Approximately equal amounts of PTH-Glu and PTH-Gln(N^{γ}Me) are recovered. The latter elutes between PTH-Gly and PTH-Glu (*see also* ref. *40*).

N^{α}-methyl amino acids: On sequencing peptides containing N^{α}-methyl amino acids, a premature cleavage of the N-terminus PTC-N^{α}-methyl-amino acid residue during the coupling step *(41)* creates a false preview (10–20% on the ABI sequencer) of the residue following the N^{α}-methyl amino acid. This preview will be carried into the following cycles.

PTC amino acids: Cyclization of a PTC amino acid intermediate to its PTH amino acid is the rate-limiting step in the conversion phase of Edman degradation. It is thus common that a fraction of the PTC amino acid remains unconverted and is present when the PTH derivatives are analyzed by HPLC. Although PTC derivatives are retained for a shorter time on reversed-phase columns in general (as compared to PTH derivatives), the PTC derivatives of less polar amino acids may be retained on the column long enough to coelute with some of the more polar PTH amino acids, thus obscuring their quantitation. For example, PTC-Val and PTH-Ser' coelute on our HPLC system. For this reason, complete conversion of PTC derivatives (by prolongation of the conversion time) is desirable, but will be accomplished at the expense of high recovery of the less stable amino acids (e.g., Ser, Thr).

5. Example: Analysis of a 20-Mer Peptide Resin

A 20-mer peptide resin, Boc-CG(21-40)-PAM-Resin, was deprotected and stained with bromophenol blue. Eight beads were sequenced using 21 REZ-1 cycles as described. The yields of the PTH derivatives are summarized in Table 6.

5.1. The Sequence and Sequential Purity

The PTH signal identifies unambiguously the CG(21-40) peptide as the predominant species in the product, indicating that no major acylation deficiencies occurred during synthesis. Note the commonly observed *(15)* very low yield of the C-terminal amino acid (leucine). The maximum cumulative preview of 6.5% (*see* Table 6) indicates that, besides the correct sequence, the product contains up to 6.5% deletion peptides. This cumulative preview translates into an average per-step acylation yield of 99.6% or higher for the synthesis.

5.2. Complications and Limitations

5.2.1. Preview Inaccuracies

Note that high coupling efficiencies during synthesis preclude accurate step-to-step measurements of acylation deficiencies—in this case, only a fraction of a percent for every coupling.

5.2.2. Carryover

The initial carryover of 2.4% for Gln-1 and the cumulative carryover of 17.0%, 16.3%, and 13.3% for residues 17, 18, and 19, respectively (i.e., <1% carryover/cycle), indicate the high efficiency of degradation in this run. This is particularly apparent from the low carryover observed at residue 3, proline, which is the most difficult residue to cleave.

5.2.3. Coeluting Derivatives

Because of the coelution of PTH-Glu(OBzl) and PTH-Cys(4-MeOBzl), the preview and carryover for Cys-9 and Glu-16 would have been obscured if these two residues had been closer together in the sequence. Note that the increased yields of PTH-Ser, PTH-Ser', and PTH-Ser" in cycle 9 reflect the presence of Cys(4-MeOBzl) in this position (*see* Section 4.2., Cys). In contrast, the increased yield of PTH-Ser' (but not of the other serine derivatives) in cycles 14 and 19 does not originate from serine or cysteine, but instead reflects the coeluting PTC-Val from Val-14 and Val-19. The presence of PTC derivatives indicates that the conversion time should have been increased before the sequencing was started. The incomplete conversion might also have caused, as previously observed *(18)*, lower yields of PTH glycine in cycles 10 and 11.

5.2.4. Repetitive Yields

The per-step repetitive yields, calculated for residues that are repeated in the sequence (Table 6), were, with the exception of PTH-Gly, higher than 94%. Note that very similar repetitive yields were obtained for both the primary and secondary PTH derivative. This demonstrates the validity of using secondary PTH derivatives in instances where coelution precludes use of the primary PTH derivative. Also note the artificially high repetitive yield for the three serine derivatives of Ser-2 and Ser-6, which is likely caused by variations in their inherent instability.

Table 6
Sequence Analysis of H_2N-CG(21-40)-PAM-Resin

Cycle no. sequence		PTH amino acid yield, pmol Ala	Arg[a]	Arg Mts	Asp[a]	Asp Bzl	Cys Mob[g]	Glu[a]	Glu Bzl	Gln	Gly	Ile
1	Gln	0.5	2.6	3.3	2.2	1.1	2.2	90	*h*	460	5.3	0.1
2	Ser	1.3	0.6	3.3	2.0	1.5	2.2	3.3	*h*	9.3	4.6	0.4
3	Pro	7.8	1.3	3.1	1.0	1.1	2.0	0.7	*h*	1.1	7.9	0.5
4	Ala	520	0.1	3.9	0.9	1.8	2.2	0.6	*h*	1.3	8.4	0.3
5	Gly	29	0.5	4.6	1.3	2.1	3.1	2.2	*h*	7.6	590	0.1
6	Gln	2.2	1.7	3.8	0.9	2.0	1.8	70	*h*	370	28	0.2
7	Ser	2.3	1.5	14	32	1.3	3.4	7.9	*h*	33	14	0.3
8	Arg	1.6	18	450	2.7	1.6	13	2.4	*h*	5.0	10	0.4
9	Cys	2.1	2.6	45	16	1.7	262	1.9	*h*	3.5	32	0.3
10	Gly	1.4	1.4	10	1.9	1.2	15	1.5	*h*	2.2	330	0.1
11	Gly	1.5	0.6	7.9	1.5	2.5	5.7	1.9	4.6	1.7	340	0.1
12	Phe	1.8	2.1	8.2	1.9	1.9	5.6	2.3	4.6	1.7	55	0.3
13	Leu	1.9	0.9	8.5	1.9	2.8	4.9	1.6	4.0	1.8	19	0.6
14	Val	1.5	1.8	26	1.6	3.5	*h*	2.5	4.3	1.3	13	0.4
15	Arg	1.3	13	253	1.4	3.8	*h*	6.0	13	0.8	6.5	0.7
16	Glu	1.5	3.4	53	4.8	11	*h*	120	221	0.9	7.9	0.2
17	Asp	1.4	2.0	15	65	270	*h*	21	39	0.9	7.7	0.7
18	Phe	1.2	2.3	11	14	51	*h*	6.5	12	1.2	7.8	0.1
19	Val	1.4	0.7	8.7	4.2	8.0	*h*	3.7	3.6	1.1	7.6	0.3
20	Leu	1.3	1.1	6.2	2.4	2.5	*h*	2.4	1.0	0.4	6.0	0.2
21	PAMR[d]	1.2	0.7	5.0	1.9	1.9	*h*	2.0	1.2	0.9	5.1	0.4

		Leu	Phe	Pro	Ser[a]	S'[a]	S''[a]	Ser Bzl	Val	P[b]	C/O[c]
1	Gln	0.4	2.4	1.6	4.3	3.4	3.1	6.6	0.4	n.d.	2.4
2	Ser	0.8	2.2	4.8	24	68	72	440	1.1	2.1	2.1
3	Pro	2.1	2.2	630	1.5	4.8	6.4	10	1.7	0.5	2.2
4	Ala	1.2	3.1	14	0.7	2.0	4.5	3.4	1.5	1.2	5.1
5	Gly	3.1	2.7	3.1	1.2	1.5	0.8	3.0	2.3	1.1	4.8
6	Gln	1.4	2.3	1.7	3.3	2.4	1.8	8.3	2.3	1.5	8.7

8	Arg	2.5	4.5	1.1	2.8	6.8	6.7	13	3.1	2.2	8.3
9	Cys	3.6	5.4	1.7	12	42	36	4.2	5.3	3.8	3.8
10	Gly	2.5	5.0	1.3	1.4	4.2	3.7	1.0	3.5	6.7	n.d.
11	Gly	3.3	18	1.1	1.3	1.6	2.4	0.2	4.4	n.d.	15.5
12	Phe	26	450	1.5	1.7	2.6	1.0	<0.1	5.6	4.3	10.0
13	Leu	440	45	1.5	1.5	4.1[e]	1.0	<0.1	22	4.9	11.5
14	Val	49	15	1.2	0.8	31[e]	0.5	<0.1	410	4.9	10.6
15	Arg	7.3	6.4	1.2	1.0	4.8[e]	0.3	<0.1	31	6.0	20.2
16	Glu	6.0	7.9	1.6	0.5	2.1[e]	1.5	<0.1	13	3.4	13.7
17	Asp	5.2	22	1.4	1.7	0.6[e]	0.7	<0.1	9.1	3.0	17.0
18	Phe	5.6	390	2.0	0.9	4.3[e]	2.5	<0.1	27	5.0	16.3
19	Val	7.0	67	1.5	1.1	34[e]	1.7	<0.1	390	6.5[f]	13.3
20	Leu	34	15	1.3	0.8	3.6[e]	1.5	<0.1	48	n.d.	n.d.
21	PAMR[d]	11	7.5	0.7	0.9	2.2[e]	3.0	<0.1	11	n.d.	n.d.

PTH amino acid repetitive yields

Gln (1,6)	95.7%	Gly (5,10,11)	90.0%
Glu (1,6)[a]	96.0%	Arg(Mts) (8,15)	94.7%
Ser(Bzl) (2,7)	102.8%	Arg (8,15)[a]	95.3%
Ser (2,7)[a]	108.5%	Phe (12,18)	97.6%
Ser' (2,7)[a]	105.1%	Val (14,19)	98.9%
Ser" (2,7)[a]	105.5%		

Average initial yield: ~70–80 pmol/bead

Average preview: <0.4%/cycle
Average carryover: <0.9%/cycle

[a]Secondary PTH derivative; [b]Preview; [c]Carryover; [d]PAM resin; [e]PTC-Val coelutes with PTH-Ser' (S'); [f]Cumulative preview; n.d., not determined. [g]Mob = 4-methoxybenzyl (4-MeOBzl). [h]Yields of PTH-Cys(4-MeOBzl) and PTH-Glu(OBzl) could not be determined independently because of their coelution (*see* Table 5); the yields of the expected PTH derivatives are underlined; Ile is not in the sequence, and the yields of PTH-Ile are reported for comparison; the side-chain-protecting groups used are indicated.

5.2.5. Background Levels of PTH Derivatives

Background levels of those PTH amino acids that are not present in the sequence should generally be below the 1-pmol level. Their presence reflects low-level contamination of sequencer reagents and hardware, and/or of the sample (*see*, for example, the PTH-Ile yields, Table 6). Similarly, background levels for the N-terminal amino acids (e.g., Gln, Ser, Pro, and Ala) dropped to the 1–2 pmol level once these residues were cleaved off the peptide during sequencing. In contrast, as expected, the yields for the C-terminal amino acid residues (e.g., Val, Phe) increased gradually, likely because of internal acidolytic cleavage. Note the persistent high level of PTH-Gly in the cycles following the second glycine residue, indicating possible sample contamination.

Acknowledgment

I thank Linda Jones for her thoughtful and constructive suggestions and careful editing of the manuscript.

References

1. Erickson, B. W. and Merrifield, R. B. (1976) Solid-phase synthesis, in *The Proteins,* vol. 2, 3rd ed. (Neurath, H. and Hill, R. L., eds.), Academic, New York, pp. 255–527.
2. Merrifield, R. B., Singer, J., and Chait, B. T. (1988) Mass spectrometric evaluation of synthetic peptides for deletions and insertions. *Anal. Biochem.* **174,** 399–414.
3. Smith, A. J., Young, J. D., Carr, S. A., Marshak, D. R., Williams, L. C., and Williams, K. R. (1992) State-of-the-art peptide synthesis: comparative characterization of a 16-mer synthesized in 31 different laboratories, in *Techniques in Protein Chemistry III* (Angeletti, R. H., ed.), Academic, San Diego, pp. 219–229.
4. Edman, P. and Henschen, A. (1975) Sequence determination, in *Protein Sequence Determination* (Needleman, S. B., ed.), Springer-Verlag, Berlin, pp. 232–279.
5. Niall, H. D., Tregear, G. W., and Jacobs, J. (1972) Automated Edman degradation monitoring of solid-phase peptide synthesis, in *Chemistry and Biology of Peptides,* Proc. Am. Pept. Symp. 3rd, pp. 695–699.
6. Tregear, G. W., van Rietschoten, J., Sauer, R., Niall, H. D., and Keutmann, H. T. (1977) Synthesis, purification, and chemical characterization of the amino-terminal 1-34 fragment of bovine parathyroid hormone synthesized by the solid-phase procedure. *Biochemistry* **16,** 2817–2823.
7. Margolies, M. N. and Matsueda, G. H. (1981) Solid-phase Edman degradation as an aid in evaluation of the homogeneity of peptidyl-resin intermediates obtained from Merrifield solid-phase synthesis, in *Chemical Synthesis and Sequencing of Peptides and Proteins* (Liu, T.-Y., Schechter, A. N., Heinrikson, R. L., and Condliffe, P. G., eds.), Elsevier North Holland, Inc., New York and Amsterdam. pp. 207–219.

8. Matsueda, G. R., Haber, E., and Margolies, M. N. (1981) Quantitative solid-phase Edman degradation for evaluation of extended solid-phase peptide synthesis. *Biochemistry* **20,** 2571–2580.
9. Kent, S. B. H., Riemen, M., LeDoux, M., and Merrifield, R. B. (1982) A study of the Edman degradation in the assessment of the purity of synthetic peptides, in *Methods in Protein Sequence Analysis* (Elzinga, M., ed.), Humana, Clifton, NJ, pp. 205–213.
10. Schlesinger, D. H. (1983) High-performance liquid chromatography of side-chain protected phenylthiohydantoins: application to solid-phase peptide synthesis. *Methods Enzymol.* **91,** 494–502.
11. Applied Biosystems User Bulletin #13 (1985) Sequence analysis of synthetic, side-chain protected, resin-bound peptides, 1–18.
12. Steinman, D. M., Ridge, R. J., and Matsueda, G. R. (1985) Synthesis of side-chain protected amino acid phenylthiohydantoins and their use in quantitative solid-phase Edman degradation. *Anal. Biochem.* **145,** 91–95.
13. Mc Cormick, D. J., Madden, B. J., and Ryan, R. J. (1986) Identification of side-chain protected L-phenylthiohydantoins on cyano HPLC columns: an application to gas-phase microsequencing of peptides synthesized on solid-phase supports, in *Modern Methods in Protein Chemistry* (L'Italien, J. J., ed.), Plenum, New York, pp. 403–413.
14. Sarin, V. K., Kim, Y., and Fox, J. L. (1986) Solid-phase sequencing on the gas-phase sequencer. *Anal. Biochem.* **154,** 542–551.
15. Kochersperger, M. L., Blacher, R., Kelly, P., Pierce, L., and Hawke, D. H. (1989) Sequencing of peptides on solid phase supports. *Am. Biotech. Lab.* **7,** 26–37.
16. Hewick, R. M., Hunkapiller, M. W., Hood, L. E., and Dreyer, W. J. (1981) A gas-liquid solid phase peptide and protein sequenator. *J. Biol. Chem.* **256,** 7990–7997.
17. Hunkapiller, M. W., Granlund-Moyer, K., and Whiteley, N. W. (1986) Gas-phase protein/peptide sequencer, in *Methods of Protein Microcharacterization* (Shively, J. E., ed.), Humana, Clifton, NJ, pp. 223–247.
18. Tarr, G. E. (1977) Improved manual sequencing methods. *Methods Enzymol.* **47,** 335–357.
19. Tarr, G. E. (1986) Manual Edman sequencing system, in *Methods of Protein Microcharacterization* (Shively, J. E., ed.), Humana, Clifton, NJ, pp. 155–194.
20. Hunkapiller, M. W. (1986) PTH amino acid analysis, in *Modern Methods in Protein Chemistry* (L'Italien, J. J., ed.), Plenum, New York, pp. 363–381.
21. Hunkapiller, M. W., Granlund-Moyer, K., and Whiteley, N. W. (1986) Analysis of phenylthiohydantoin amino acids by HPLC, in *Methods of Protein Microcharacterization* (Shively, J. E., ed.), Humana, Clifton, NJ, pp. 315–327.
22. Hunkapiller, M. W. (1987) Automated amino acid sequence assignment: development of a fully automated protein sequencer using Edman degradation, in *Methods in Protein Sequence Analysis* (Walsh, K. A., ed.), Humana, Clifton, NJ, pp. 367–384.
23. Geisow, M. J. and Aitken, A. (1989) Gas- or pulsed liquid-phase sequence analysis, in *Protein Sequencing. A Practical Approach* (Findlay, J. B. C. and Geisow, M. J., eds.), IRL, Oxford and New York, pp. 85–98.
24. Clark-Lewis, I., Aebersold, R., Ziltener, H., Schrader, J. W., Hood, L. E., and Kent, S. B. H. (1986) Automated chemical synthesis of a protein growth factor for hemopoietic cells, interleukin-3. *Science* **231,** 134–139.

25. Nutt, R. F., Brady, S. F., Darke, P. L., Ciccarone, T. M., Colton, C. D., Nutt, E. M., Rodkey, J. A., Bennett, C. D., Waxman, L. H., Sigal, I. S., Anderson, P. S., and Veber, D. F. (1988) Chemical synthesis and enzymatic activity of a 99-residue peptide with a sequence proposed for the human immunodeficiency virus protease. *Proc. Natl. Acad. Sci. USA* **85,** 7129–7133.
26. Sarin, V. K., Kent, S. B., Tam, J. P., and Merrifield, R. B. (1981) Quantitative monitoring of solid-phase peptide synthesis by the ninhydrin reaction. *Anal. Biochem.* **117,** 147–157.
27. Spande, T. F., Witkop, B., Degani, Y., and Patchornick, A. (1970) Selective cleavage and modification of peptides and proteins, in *Advances in Protein Chemistry,* vol. 24 (Anfinsen, C. B., Jr., Edsall, J. T., and Richards, F. M., eds.), Academic, New York, pp. 98–260.
28. Fontana, A. and Gross, E. (1986) Fragmentation of polypeptides by chemical methods, in *Practical Protein Chemistry. A Handbook* (Darbre, A., ed.), Wiley, Chichester and New York, pp. 67–120.
29. Smithies, O., Gibson, D., Fanning, E. M., Goodfliesh, R. M., Gilman, J. G., and Ballantyne, D. L. (1971) Quantitative procedures for use with the Edman-Begg sequenator. Partial sequences of two unusual immunoglobulin light chains, Rzf and Sac. *Biochemistry* **10,** 4912–4921.
30. Brandt, W. F., Henschen, A., and von Holt, C. (1982) The nature of non-specific peptide bond cleavage during the isothiocyanate degradation of proteins, in *Methods in Protein Sequence Analysis* (Elzinga, M., ed.), Humana, Clifton, NJ, pp. 101–110.
31. Inglis, A. S., Gillespie, J. M., Roxburgh, L. A., Whittaker, L. A., and Casagranda, F. (1986) Sequence of a glycine-rich protein from lizard claw: unusual dilute acid and heptafluorobutyric acid cleavages, in *Modern Methods in Protein Chemistry* (L'Italien, J. J., ed.), Plenum, New York, pp. 757–764.
32. Atherton, E. and Sheppard, R. C. (1989) Analytical and monitoring techniques, in *Solid Phase Peptide Synthesis. A Practical Approach.* IRL, Oxford and New York, pp. 107–130.
33. Fields, G. B. and Noble, R. L. (1990) Solid phase peptide synthesis utilizing 9-fluorenylmethoxycarbonyl amino acids. *Int. J. Peptide Protein Res.* **35,** 161–214.
34. Tempst, P. and Riviere, L. (1989) Examination of automated polypeptide sequencing using standard phenyl isothiocyanate reagent and subpicomole high-performance liquid chromatographic analysis. *Anal. Biochem.* **183,** 290–300.
35. Krchnák, V., Vágner, J., Safár, P., and Lebl, M. (1988) Non invasive continuous monitoring of solid-phase peptide synthesis by acid-base indicator. *Collect. Czech. Chem. Commun.* **53,** 2542–2548.
36. Crimmins, D. L., Grant, G. A., Mende-Mueller, L. M., Niece, R. L., Slaughter, C., Speicher, D. W., and Yuksel, K. U. (1992) Evaluation of protein sequencing core facilities: design, characterization, and results from a test sample, in *Techniques of Protein Chemistry III* (Angeletti, R. H., ed.), Academic, San Diego, pp. 35–43.
37. Sparrow, D. A. and Sparrow, J. T. (1990) Methods for the automated amino acid sequencing of peptide-resin conjugates, in *Peptides. Chemistry, Structure, Biology.* Proc. 11th Am. Peptide Symp. (Rivier, J. E. and Marshall, G. R., eds.), Escom, Leiden, pp. 446–448.

38. Pavlík, M. and Kostka, V. (1985) Modification of serine residues during sequential degradation of proteins and peptides by the phenylisothiocyanate method. *Anal. Biochem.* **151,** 520–525.
39. Marshall, R. C. and Inglis, A. S. (1986) Protein oligomer composition, preparation of monomers and constituent chains, in *Practical Protein Chemistry. A Handbook* (Darbre, A., ed.), Wiley, Chichester and New York, pp. 1–66.
40. Klotz, A. V., Thomas, B. A., Glazer, A. N., and Blacher, R. W. (1990) Detection of methylated asparagine and glutamine residues in polypeptides. *Anal. Biochem.* **186,** 95–100.
41. Chang, J. Y. (1978) A novel Edman-type degradation: direct formation of the thiohydantoin ring in alkaline solution by reaction of Edman-type reagents with *N*-monomethyl amino acids. *FEBS Lett.* **91,** 63–68.

CHAPTER 9

NMR Spectroscopy of Peptides and Proteins

Mark G. Hinds and Raymond S. Norton

1. Introduction

In recent years, a wide range of Nuclear Magnetic Resonance (NMR) techniques has become available for investigating the structure of peptides and proteins in solution, and their interactions with other molecules. These powerful methods have an important role to play in furthering our knowledge of the molecular basis of such processes as protein folding and molecular recognition. NMR spectroscopy is uniquely positioned to investigate these problems, being one of the few techniques available for determining high-resolution structures of biomolecules in solution.

The principal NMR experiments relevant to biological molecules have been discussed in several recent monographs and reviews *(1–6)*. NMR relies on the observation of nuclei with net spin; fortunately, the main constituents of biomolecules, namely, hydrogen, carbon, nitrogen, oxygen, and phosphorus, have at least one stable, NMR-active isotope. Most of the literature on NMR of biological molecules has been concerned with protons, but the increasing importance of isotopic labeling and new isotope-edited experiments has led to a burgeoning field in the NMR of isotopically enriched peptides and proteins *(3,5)*.

The basic one-dimensional (1D) pulsed NMR experiment *(7–10)* involves placing the sample in a strong magnetic field, pulsing it with radio-frequency radiation, and recording the response of the system. The

From: *Methods in Molecular Biology, Vol. 36: Peptide Analysis Protocols*
Edited by: B. M. Dunn and M. W. Pennington

resonant frequency of a given nucleus is determined by the strength of the magnetic field and the magnetogyric ratio of the nucleus, as shown by Eq. (1).

$$\nu = (\gamma/2\pi)B_0 \quad (1)$$

where ν is the frequency, γ the magnetogyric ratio, and B_0 the static magnetic field strength. Increasing the field strength leads to greater spectral dispersion and increased sensitivity. Both of these factors are important in the study of biomolecules, since increased spectral dispersion eases the problem of overlapping signals, making resonance assignment easier, whereas increased sensitivity allows less sample to be used.

The positions of resonances in an NMR spectrum are sensitive to the chemical environment of the relevant nuclei, the parameter that reflects this being the chemical shift. Differences in chemical shift of nuclei arise from differences in their shielding from the external magnetic field by the electronic environment, thus allowing the nuclei to resonate at slightly different frequencies (Eq. [2]).

$$\nu = (\gamma/2\pi)B_0 (1 - \sigma) \quad (2)$$

where σ is the shielding constant. After a radio-frequency pulse, the NMR signal (the free induction decay or FID) is converted from an analog to a digital signal by an analog-to-digital converter (ADC) and stored in a computer. The magnetization of the sample is then allowed to relax before the next scan commences. This method of data accumulation allows the signal-to-noise ratio of the spectrum to grow as the square root of the number of scans acquired. This is an important consideration when dealing with small sample quantities, since reducing the concentration increases considerably the time required to obtain a spectrum with the same signal-to-noise ratio. Fourier transformation converts the data acquired from the time domain (FID) to frequency domain (spectrum). Various forms of data manipulation are used to improve the signal-to-noise ratio and resolution of the spectra, as discussed in Section 3.3.3.

There have been a large number of NMR studies of the solution properties of peptides and proteins *(1,5,9)*. Proteins generally have a better-defined overall structure than short linear peptides, which frequently adopt random-coil conformations. Nevertheless, NMR has provided evidence for the presence of stable secondary structure, including β-turns and regions of α-helix in peptides in aqueous solution *(11)*, even though these flexible peptides are probably undergoing rapid interconversion between many low-energy conformations. At the other extreme, the molecular mass limit of a molecule for which NMR is capable of determining a complete 3D structure is currently of the order of 25 kDa, provided isotopic labeling and heteronuclear techniques are used *(12)*. Homonuclear techniques alone can be used to determine the structure of molecules up to approximately half this molecular mass.

The purpose of this chapter is to give the experimenter who is less familiar with NMR spectroscopy a brief introduction to the area, with emphasis on methods of acquisition, analysis, and interpretation of NMR spectra of peptides and proteins.

2. Materials

Two critical parameters concerning any sample for NMR analysis are its purity and the choice of solution conditions.

2.1. Sample Purity

The sample to be analyzed should be chemically homogeneous and free from low-molecular-mass protiated impurities. The most common contaminants of this type are biological buffers, such as TRIS, MOPS, TEA, and so on, that are routinely used in biochemical preparations. These compounds have very intense ^{1}H-NMR signals and their molar concentrations are usually orders of magnitude higher than the solute under investigation, leading to problems with overlap of signals and the dynamic range of the ADC. These small molecule contaminants are best removed by chromatographic methods, such as HPLC. Dialysis of the sample against the NMR solvent is usually insufficient by itself as a method for removal of these low-molecular-mass molecules, since they often bind to the protein. If it is necessary to include a buffer, it is generally desirable to keep the salt concentration of the sample as low as possible to prevent sensitivity loss and excessive sample heating during the experiment owing to dielectric loss of the radio-frequency pulses employed. The buffer should be chosen to avoid signals that interfere with resonances of the sample one wishes to observe. Typically, a phosphate buffer is used for pH values near neutral, with sodium or potassium salts to raise the ionic strength. Deuteroacetate is suitable as a buffer for lower pH values.

Table 1
Properties of Common NMR Solvents

Solvent	bp[a]	mp[b]	ρ[c]	ε[d]	μ[e]	δ[1]H[f,g]	δ[1]H[f,h]
H_2O	100	0	0.997	78.5	1.8	4.8	4.8
^2H_2O	101.4	3.8	1.56	78.5	1.8		
CH_3OH	64.7	–97.7	0.787	32.7	1.6	3.35	3.35
CH_3CN	81.6	–44	0.777	37.5	3.5	2.0	2.05
$(CH_3)_2SO$	189	18.5	1.096	46.7	3.9	2.6	2.70
CF_3CH_2OH	77	–43.5	1.382	26.1	2.5	3.6	

[a]bp, Boiling point (°C).
[b]mp, Melting point (°C).
[c]ρ, Density (gcm^{-3}).
[d]ε, Dielectric constant.
[e]μ, Dipole moment (D).
[f]δ, Chemical shift (ppm) of residual protons relative to TMS.
[g]Neat.
[h]In aqueous solution. The exact resonance position may vary with pH and temperature.

2.2. Solvents

Solvents commonly used in NMR studies of peptide and protein samples are H_2O, DMSO, MeOH, and MeCN. Some of their relevant properties are summarized in Table 1.

The choice of solvent can be critical in determining whether any structure is formed in peptides, where trifluoroethanol (TFE) and DMSO are commonly used to induce structure formation (*see* Section 3.3.6.). The prime concern for the solvent is that it must be of high chemical purity, particularly when dealing with low concentrations of solute, where impurities may dominate the spectrum. There must also be a certain percentage of deuterium present in the solvent, since this is necessary for the deuterium field-frequency lock on the spectrometer, which is required for the field stability of the experiment *(7)*. Under protic (H_2O) conditions, it is sufficient for 5–10% of the solvent to be deuterated.

Deuterated solvents may be used to remove the large resonances that arise from solvent protons in ^{1}H-NMR spectra. The required level of isotopic enrichment of the deuterated solvents depends on the experiment. Typical enrichments available commercially are between 99.8 and 99.996%, the higher grades being necessary for samples that require examination under more rigorously aprotic conditions. Because deuterated solvents are hygroscopic, it is preferable to store them under an inert atmosphere in a desiccator. Deuterated solvents can also be purchased in ampules containing sufficient volume for a single NMR sample, which is a preferable method of obtaining the higher grades (99.96% or greater isotopic purity).

2.3. NMR Tubes

The choice of NMR tube diameter will depend on the instrumentation available, the most common probe diameters for high-resolution instruments being 5 and 10 mm. The higher the field strength of the spectrometer, the more critical the choice of tube becomes, and at the highest field strengths, the highest-quality thin-walled NMR tubes are recommended in order to maximize the amount of sample within the probe and minimize contributions from the tube. Flat-bottomed tubes should be avoided for high-resolution NMR work, since the abrupt cutoff at the base of the tube makes shimming extremely difficult. The tube should be free of scratches, marks, and chips, because these may adversely affect both the shimming and spinning characteristics of the sample. For samples where the quantity is very low, a microprobe designed especially for small samples can be used to increase the solute-to-solvent ratio. Alternatively, constant susceptibility plugs may be placed in the NMR tube to displace the sample in the base of the tube so that it lies within the coil of the probe *(9)*. Prior to use, it is advisable to clean the tube with a detergent suitable for biological samples, followed by thorough rinsing with high-purity water and drying. It is usually inadvisable to use organic solvents to rinse the tubes because these can lead to contamination of the sample, but high-purity ethanol may be used if necessary.

3. Methods

3.1. Sample Preparation

Only some general points of sample preparation will be mentioned here because this topic has been discussed in detail elsewhere *(13)*. The sample under study must be treated with care in order not to introduce any contaminants. Paramagnetic impurities must be rigorously avoided since they will degrade the spectra by increasing the observed linewidths by providing additional relaxation pathways *(7)*. In this regard, it is desirable to avoid or minimize contact of the sample with metal surfaces and implements (e.g., spatulas, HPLC tubing) during preparation. Similarly, only analytical-grade reagents should be used in the preparation of buffers or salts needed for the sample.

When preparing protein samples, it is advisable to avoid excessive agitation, which may cause degradation of the sample because of denaturation. The removal of any particulate matter is of importance since these particles will make shimming more difficult. The most efficient way to remove particulate matter from NMR samples is by centrifugation in a minicentrifuge tube capable of holding approx 500 µL of sample. The likely sample heating during centrifugation must be considered, and a thermostated centrifuge used where necessary. An alternative to pelleting is to filter the solution through a 0.2-µm low-protein binding membrane in a centrifuge filter unit, which has the added advantage of sterilizing the sample.

Integral membrane peptides or proteins often require the presence of detergents or lipids that solubilize these molecules in an environment that mimics the membrane. However, the resulting micelles tumble slowly in solution because of their large size, causing the peptides or proteins to behave as larger molecules, having broad resonance lines and the problems associated with this. The detergent used to prepare the micelles must be perdeuterated to ensure that its resonances do not dominate the spectra, and to minimize the loss of magnetization from the peptide or protein to the micelle by spin-diffusion processes *(14)*.

3.1.1. pH

In peptides and proteins, the backbone and side-chain amide protons may exchange with water, leading to exchange broadening of their resonances *(1)*. The exchange process is at a minimum at a pH of approx 3 for backbone amide protons *(1)*, but the actual pH used for NMR studies will also depend on the isoelectric point, because peptides and proteins are least soluble around this point. The stability of the protein or peptide at low pH must also be considered. High pH values are best avoided in order to minimize disulfide bond exchange and degradation. The pH is most simply adjusted by microliter additions of dilute solutions (ca. 10–100 m*M*) of either ^{2}HCl or NaO^{2}H, keeping the desirability of low salt concentration in mind. Narrow-bore pH electrodes suitable for 5-mm NMR tubes are available.

3.1.2. Volume and Concentration

The sample concentration should be of the order of 1–5 m*M* for two-dimensional (2D) NMR studies. Because the signal-to-noise ratio increases as the square root of the number of scans, the higher the sample concentration, the more rapidly the data can be collected. This bears consideration when dealing with dilute solutions, where the experiment time increases significantly. This can bring other factors into play, such as excessive t_1 noise arising from instrumental stabilities *(7,15)*, which degrades the quality of the spectra by giving bands of noise parallel to the w_1 axis in 2D-NMR spectra. Clearly the sample concentration used will depend not only on the availability of material, but also on the solution properties of the sample, its propensity for aggregation, and so forth.

Where it is necessary to study a sample in a fully deuterated solvent, the most convenient method to remove exchangeable protons and concentrate the sample is to lyophilize it several times from the deuterated solvent. If repeated cycles of freezing are detrimental to the sample (especially for proteins where denaturation may occur), the sample may be concentrated by centrifuging in a small ultrafiltration cell. Prior to use, the concentrators must be thoroughly cleaned to remove preservatives from the membrane, such as glycerol, which will interfere with the sample. The exact sample volume will be dependent on the specifications of the probe to be used. It is important that the sample height in the NMR tube be greater than the receiver coil height in order to facilitate shimming.

3.1.3. Temperature

The choice of temperature will be influenced by the thermal stability of the sample. Since line-widths are narrower at higher temperatures because of a decrease in the overall rotational correlation time of the molecule *(1,10)*, it is generally desirable to record spectra at the highest temperatures compatible with sample stability and the maintenance of native structure. On the other hand, chemical exchange is more rapid at higher temperatures, and may lead to broadening or even loss of signals, particularly for backbone amide protons of peptides and proteins, which are vitally important for the assignment of spectra *(1,5)*.

3.1.4. Potential Problems

Two problems that may arise for peptides and proteins are low solubility and a propensity to aggregate. Both may be alleviated by adjusting parameters, such as the pH, ionic strength, solute concentration, and so on. Addition of a detergent, such as CHAPS, has been shown to reduce aggregate formation *(16)*. Aggregation can also be reduced by acquiring the data at the lowest practical sample concentration, in which case larger

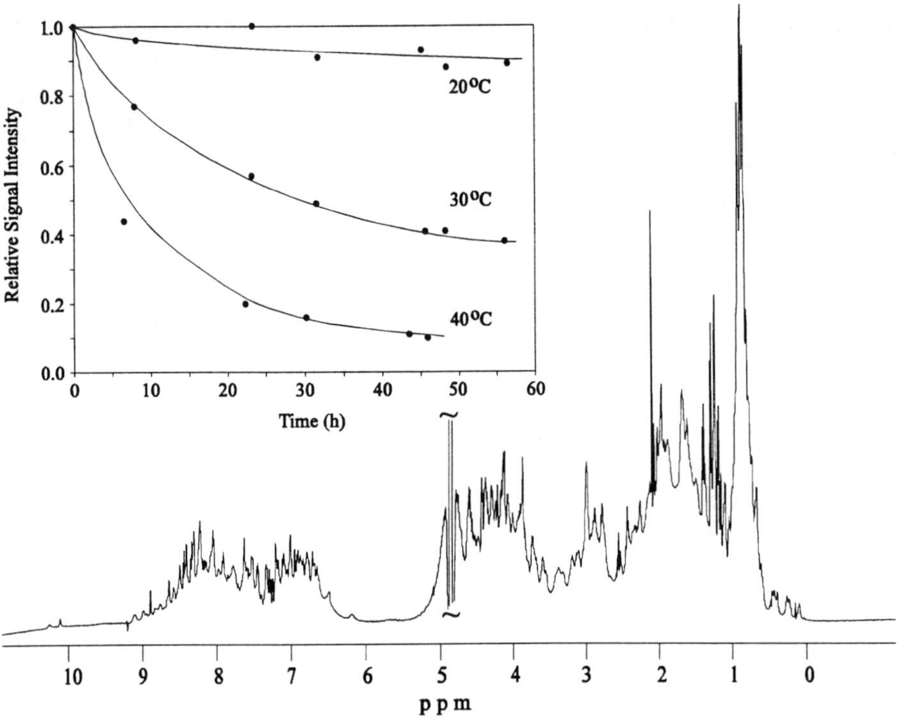

Fig. 1. 600 MHz ^1H-NMR spectrum of 1.4 m*M* interleukin-6 in 90% H_2O/ 10% ^2H_2O at pH 3.5 and 20°C. Inset shows the plot of relative signal intensity vs time for three different temperatures.

sample volumes and wider bore tubes may be preferable. If aggregation is caused by hydrophobic interactions, then the temperature at which the spectra are acquired should be as low as possible. Figure 1 shows the ^1H-NMR spectrum of interleukin-6 (IL-6), a cytokine of molecular mass 21 kDa, which undergoes a temperature-dependent aggregation to a state with line-widths too broad to observe at 600 MHz. The aggregation was found to be more rapid at higher temperatures, with a half-life of approx 10 h at 40°C compared to many days at 20°C.

3.2. Instrumentation

Because of their increased sensitivity and spectral dispersion, the highest fields available are usually preferable for the investigation of biomolecules. It is difficult to give any definitive rules concerning the relationship between the size of the molecule and the operating frequency required for complete analysis by NMR methods, but spectrometers operating below 500 MHz (11.74 T) are generally suitable only for investigation of smaller molecules, such as peptides up to about ten amino acid residues in length.

The other advantage of increased field strength is that for certain rotational correlation times, negative nuclear Overhauser effects (NOEs) will be observed where near-zero NOEs would be found at lower field (*see* Section 3.4.2.).

Except for the acquisition of 1D-NMR spectra, sample spinning is usually avoided when using higher field spectrometers because of the artifacts that are introduced into the spectra. Rotating the sample introduces radio-frequency phase modulations into the NMR signals, which are manifested as increased t_1 noise in the spectra *(17)*. The choice of using either simultaneous or sequential acquisition is determined by the type of instrument available. These terms refer to the timing of the acquisition of complex data points by the receiver, and the advantage of using simultaneous acquisition mode is that flatter baselines are produced *(18)*. If sequential acquisition is to be used, both the receiver phase and the delay before acquisition of the first data point should be adjusted to optimize the flatness of the baseline *(18,19)*.

3.3. One-Dimensional Techniques

Prior to acquisition of a 1D-NMR spectrum, some time must be allowed for equilibration of the sample and probe temperatures. It is usually better to control the temperature of the probe actively rather than use ambient temperature since temperature fluctuations may lead to spectra with poor line shape, especially when ^2H_2O is used as a lock signal. Once the sample has equilibrated thermally and the field is locked onto the deuterium signal, shimming the probe to improve the magnetic field homogeneity can commence. Suggested procedures for improving the shimming can be found in references *(7)* and *(20)*. The probe should also be tuned and matched to obtain the maximum sensitivity. Two important considerations for acquiring an NMR spectrum are the choice of sweep width and receiver gain. Initially a wide sweep width should be chosen, which can then be reduced to obtain the highest digital resolution possible without causing folding of the signals *(7)*. The receiver gain must be adjusted to avoid receiver overload, which is manifested as a severe baseline distortion in the frequency domain spectrum. The highest receiver gains attainable without receiver overload are desirable for optimal sensitivity.

3.3.1. Pulse Calibration

Pulse widths are best calibrated on the sample under study. This is achieved by first using a short pulse width to acquire a spectrum, which is phase corrected in the usual manner. The pulse width is then increased, the spectrum reacquired, and the phase corrected with the same parameters. As the pulse width is increased, the signal will go through a maximum corresponding to a 90° pulse and then a minimum, the first minimum encountered (provided that the starting pulse width was short enough) corresponding to a 180° pulse. This value is halved to give the corresponding value of the 90° pulse. Ideally, the 180° pulse width will be half that of a 360° pulse, but because of a finite rise time of the pulses, this is usually not the case. The 360° pulse has some advantages in terms of measuring the pulse width because the recycle time can be shorter since the magnetization has not been moved far from its equilibrium position and, hence, does not require as much time to relax.

3.3.2. Solvent Suppression

Samples that are dissolved in protic solvents will require suppression of the large solvent resonance to avoid exceeding the dynamic range of the ADC. The simplest and most direct way to reduce the intensity of the solvent signal is by irradiation with a low-power radio-frequency field prior to excitation and acquisition. The radio-frequency field used in the presaturation of the solvent should be the lowest power possible in order to minimize the partial saturation of signals close to the solvent resonance (principally $C^{\alpha}H$ resonances). These signals may be observed by altering the temperature of the sample in order to move the solvent resonance with respect to the solute resonances, so that the latter are no longer saturated. Those signals that are in fast or intermediate chemical exchange with the solvent will also be affected by presaturation. This is of particular importance for amide protons and is one reason for acquiring the spectra at low pH values if possible, where this exchange is minimized (*see* Section 3.1.1.).

To obtain the best suppression, it is necessary to have a well-shimmed sample, and to have optimized both the power and frequency of the irradiation. Another important factor in presaturation is that the irradiation should have the same phase as the high-power pulses in order to reduce spectral distortion *(21)*. If fast power switching between the low and high power levels is not possible on the spectrometer, as is the case with many older instruments, then it may be advantageous to use the DANTE pulse train *(21)* for solvent suppression.

There are several other methods of solvent suppression available that have been reviewed recently *(22)*. These make use of selective excitation methods and rely on explicitly not exciting the solvent resonance, thus avoiding magnetization transfer from the solvent resonance to the solute. The two methods that are most commonly used are the jump-return (1-1) method and the 1-3-3-1 pulse sequence, since these are comparatively easy to implement and give relatively flat baselines. Both procedures have the disadvantages of suppressing resonances close to solvent resonance and giving spectra with a phase inversion at the solvent frequency.

3.3.3. Data Processing

After acquisition, the data must be processed *(23)* in order to obtain the frequency information from the time domain. Fourier transformation is usually used to achieve this. Prior to Fourier transformation, the first point of the FID should be corrected for incorrect weighting, either by reducing the value of the first data point *(24)* or by linear prediction *(25)*. The FID is then multiplied by a window function in order to improve the signal-to-noise ratio or increase the resolution of the spectrum before Fourier transformation. The choice of the window function will be dependent on the data, in particular the intrinsic resonance line widths. An exponential line broadening factor is usually employed to increase the signal-to-noise ratio, and a Gaussian or phase-shifted sine-bell function to improve the resolution. It must be borne in mind that improved resolution is at the cost of a reduced signal-to-noise ratio.

Spectra have traditionally been referenced internally with compounds added to the sample. Typical reference compounds used with biological samples include tetramethylammonium salts, 1,4-dioxan, *t*-butanol, sodium 2,2-dimethyl-2-silapentane-5 sulfonate (DSS), and 3-(trimethylsilyl)-propionic acid-$^2H_4$ sodium salt (TSP). If the sample is being studied under nonaqueous conditions, then tetramethylsilane (TMS) may be used as an internal reference. The residual solvent signal can also be used as a reference, although its chemical shift may change with temperature or pH. It has the advantage, however, that no foreign compounds need be added to the sample. The main disadvantage of adding substances to the sample is that they may require removal at a later stage. The common referencing agents for aqueous solutions, such as TSP and DSS, are not volatile and may bind to the protein or peptide, making them unsuitable as a reference and difficult to remove by dialysis or lyophilization. The use of an external standard is less desirable because the reference must necessarily be examined under different conditions from the sample, but this is less of a problem with NMR of heteronuclei because of the larger chemical shift range and thus smaller error relative to the total chemical shift range.

Under ideal conditions, the areas of the signals in an NMR spectrum are proportional to the number of nuclei giving rise to those signals. To obtain accurate integrals from spectra, however, it is necessary to ensure that the spectra are acquired with sufficient time between successive scans for complete relaxation, usually three to five times the T_1 value. This is especially important when analyzing data for structural calculations. There are two common methods for determining T_1 values, the inversion recovery Fourier transform (IRFT) and progressive saturation Fourier transform (PSFT) methods *(26)*. PSFT is a relatively rapid and straightforward technique to apply, but is of limited accuracy and has a lower sensitivity compared to the IRFT method. Either method is suitable for quickly determining the T_1 values in a molecule before data acquisition.

3.3.4. Chemical Characterization

NMR is a useful method for chemical characterization of a sample, even though it requires larger quantities of material than such techniques as amino acid analysis and mass spectrometry. The presence of peptide or protein impurities can be distinguished by the appearance of more than the expected number of signals for the sample. As an example, Fig. 2 shows the 500-MHz ^{1}H-NMR spectra of synthetic peptides that correspond to the C-terminal 19 residues of the cytokine interleukin-6. The upper spectrum in Fig. 2A shows the amide region of the 1D ^{1}H-NMR spectrum of the peptide with an oxidized methionine residue, the Met-19 and Glu-18 amide resonances being split because of the formation of the diastereomeric center at the sulfur atom. Reduction of the peptide with dithiothreitol restores these signals to their expected positions, as shown in the lower spectrum of Fig. 2A. The Met-19 methyl resonance is also affected, as shown in Fig. 2B.

3.3.5. pH Dependence

Useful information on the solution properties and the local environment of specific ionizable groups can be obtained from a pH titration. The ionizable groups of peptides and proteins have characteristic pK_a

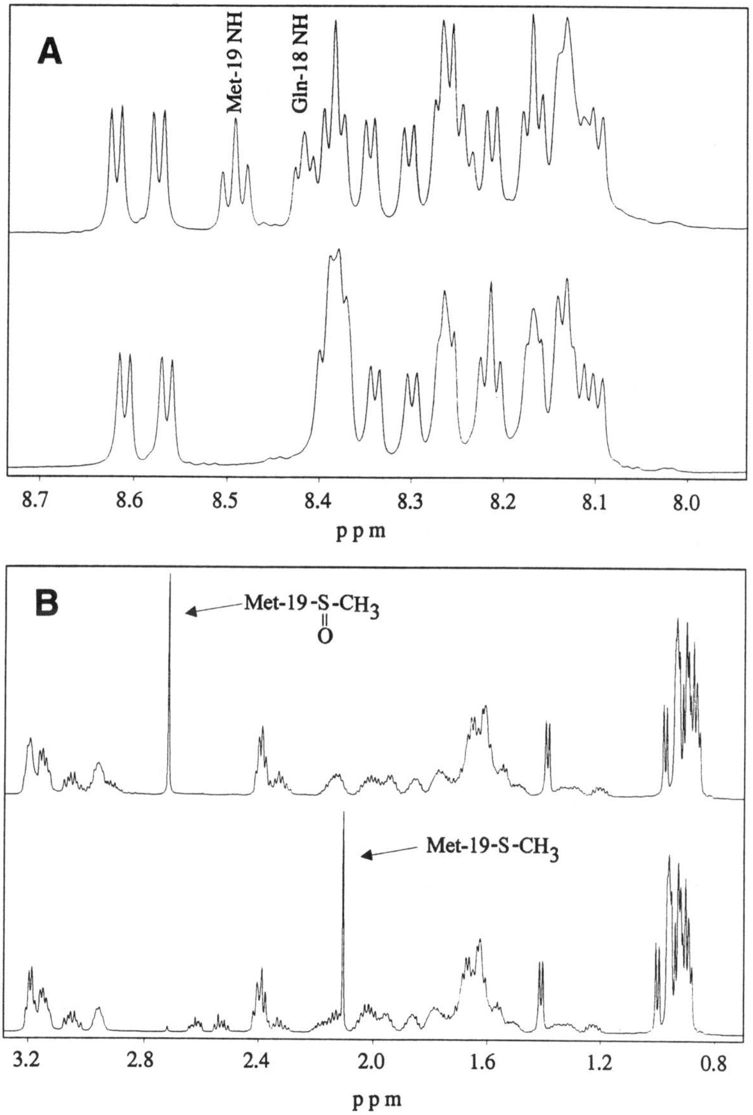

Fig. 2. 500 MHz ^{1}H-NMR spectra showing the effects of reducing the oxidized C-terminal methionine residue in the synthetic peptide ILRSFKE FLQSSLRALRQM. The backbone amide resonances of both Met-19 and Gln-18 are split owing to the formation of a chiral center at the oxidized side-chain sulfur, as is shown in the upper spectrum of **(A)**. The upfield methionine methyl group also has an altered chemical shift in the oxidized form, as shown in **(B)**. These effects disappear on reduction of the sulfoxide as shown in the lower spectra in **(A)** and **(B)** *(57)*.

values *(27)*, any deviation from the "random-coil" values being indicative of an unusual chemical environment. Historically, histidine residues have been investigated because their side-chain protons C(2)H and C(4)H have well-resolved signals, making the protonation state of the side chain easy to follow even using 1D-NMR *(9)*.

The pH should be measured both before and after the NMR measurement, and these readings should agree to within 0.05 of a pH unit or the measurement should be repeated. pK_a values may be obtained by nonlinear least-squares fitting of the observed chemical shift changes to the Henderson-Hasselbalch equation for a single ionization *(9)*, assuming fast exchange between the conjugate acids and bases. Where the conjugate acid and base are in slow exchange, separate signals from the protonated and nonprotonated forms will be present, and their relative intensities will be pH-dependent.

3.3.6. Trifluoroethanol Titration

Trifluoroethanol (TFE) has often been used to induce structure (usually helix) in peptides that otherwise do not have any stable structure in aqueous solution *(11)*. It has been shown, however, that TFE is not simply a helix-inducing solvent, since not all peptides become helical on addition of TFE *(28,29)*. Rather, it can be considered more a helix-enhancing cosolvent, stabilizing helices in peptides that have a helix-forming propensity. 2,2,2-Trifluoroethanol-$^2H_3$ should be used in the titration to avoid problems with the large signals from nonexchangable protons of protiated TFE.

3.3.7. Hydrogen Exchange

Exchange-rate measurements give information on the solvent accessibility of exchangeable protons, which can be important in studying such processes as protein folding and dynamics. The formation of hydrogen bonds in ordered structures, such as α-helices and β-sheets, dramatically reduces the hydrogen-deuterium exchange rates of the amide protons. Typically, the sample to be studied will be lyophilized from protic solution (H_2O), rapidly dissolved in deuterated solution (^2H_2O), and then placed in the spectrometer, where a time-course study measures the disappearance of those protons that are readily exchangeable with the solvent. Usually, a 2D-NMR experiment is necessary to confirm the assignment of the slowly exchanging protons *(1)*. Alternately, a fully exchanged sample can be lyophilized from ^2H_2O before being redissolved in H_2O, and the reappearance of the rapidly exchanging resonances in the ^{1}H-NMR spectrum can be studied.

3.3.8. Amide Temperature Dependence

The temperature dependence of the amide proton chemical shifts in peptides has been used as an indicator for those residues that are hydrogen-bonded *(9,30)*. An amide proton involved in a stable hydrogen bond or one inaccessible to solvent typically shows a reduced temperature coefficient of greater than about -6.0×10^{-3} ppm°C^{-1}. These data must be interpreted cautiously, however, and should not be used in isolation as evidence for hydrogen bond formation.

3.3.9. Other Nuclei

^{13}C and ^{15}N are commonly used isotopes in the investigation of protein structure. If the proteins have been overexpressed in bacteria, it is possible to incorporate these isotopes into the proteins by using minimal media where the sole nitrogen or carbon source is isotopically enriched *(31)*. The presence of ^{13}C and ^{15}N helps alleviate peak overlap, owing to the large chemical shift dispersion of these nuclei *(12,32,33)*.

Other useful NMR-active nuclei are ^{2}H, ^{3}H, ^{19}F, and ^{17}O. ^{19}F-labeled amino acids are commercially available and may be fed to suitably auxotrophic bacteria for incorporation studies. Among the naturally occurring isotopes, ^{19}F is second only to the proton in sensitivity. The absence of fluorine from naturally occurring proteins together with the presence of relatively few signals and a wide chemical shift dispersion often results in well-resolved signals. The ^{19}F chemical shifts are sensitive to the nuclear environment and can be interpreted *(34)*. ^{2}H and ^{3}H are also useful substitutions for protons, since they do not introduce any polar or steric effects.

3.4. Two-Dimensional NMR

2D NMR is an extension of the 1D method into a second frequency axis, thus spreading the resonances in two dimensions and facilitating spectral analysis. The method involves a preparation period, followed by evolution and mixing periods in which frequency labeling occurs, and finally data acquisition *(7,10)*.

As in the case of 1D-NMR, postacquisition data processing may be used to improve 2D-NMR spectra. Baseline correction *(35)*, removal of the solvent resonance *(36)*, and t_1 noise reduction *(37)* can be used to improve the visual appearance of 2D spectra, and make analysis using either manual or automated techniques more straightforward. Linear prediction methods can be used to improve the resolution of 2D and more importantly 3D and 4D spectra, where the data are truncated to reduce acquisition time *(38)*. All spectra should be acquired in the phase-sensitive mode owing to the improved resolution that this affords *(39)*. The alternative of magnitude mode processing gives crosspeaks with broad wings, which may require severe resolution enhancement in order to improve the resolution, thus decreasing the signal-to-noise ratio.

The principal 2D-NMR experiments used for peptides and proteins are DQF-COSY (double-quantum filtered correlation spectroscopy), TOCSY (total correlation spectroscopy, also referred to as HOHAHA, for homonuclear Hartmann Hahn spectroscopy), NOESY (nuclear Overhauser effect spectroscopy), and ROESY (rotating frame Overhauser effect spectroscopy). The first two methods rely on through-bond correlations via J-coupling, and the latter two on through-space correlations.

3.4.1. DQF-COSY and TOCSY

DQF-COSY spectra correlate J-coupled spin systems *(40)*, giving crosspeaks with antiphase line shape compared to COSY (correlated spectroscopy) spectra, which give both antiphase and dispersive line shapes. The dispersive nature of the lines in the COSY experiment makes analysis of these spectra more difficult since the large dispersive tails of the diagonal peaks obscure many crosspeaks. Although a DQF-COSY experiment is less sensitive than a COSY, the antiphase nature of both the crosspeak and diagonal signals gives a spectrum with higher resolution.

In contrast to the DQF-COSY experiment, the TOCSY *(41)* experiment gives both short- and long-range J-coupled correlations and crosspeaks with in-phase line shape. The main advantage of having crosspeaks with in-phase line shape over those that are antiphase is that it avoids the mutual cancellation of crosspeaks that occurs for large molecules with broad line widths. Thus, the TOCSY spectrum is useful over a wider molecular mass range than the DQF-COSY. The length of the TOCSY spin-lock mixing period determines how far along the spin system magnetization is transferred; in favorable cases, it is possible to find all correlations in a spin system *(42)*. Figure 3 shows the fingerprint region of a TOCSY spectrum of the 27-residue polypeptide ω-conotoxin, a calcium channel blocker isolated from the cone shell *Conus geographus (43)*.

3.4.2. ROESY and NOESY

Both the ROESY *(44)* and NOESY *(45,46)* experiments give information on the distances between nuclei. The NOESY experiment uses the exchange of magnetization in the laboratory frame between two dipolar coupled nuclei to establish spatial connectivities. The rate at which the exchange occurs is known as the crossrelaxation rate, and is dependent on both the internuclear distance and the dynamics of the internuclear vector relative to the external magnetic field. If long mixing times are used in the NOESY experiment, one must take into account spin diffusion *(47)*, a process that involves the relay of magnetization through an intermediate nucleus, thus giving a spectrum where crosspeak intensities are not directly related to single internuclear distances.

The sign of the NOE changes according to the rotational correlation time of the molecule *(1,46)*, small molecules having positive NOEs and large, slowly tumbling molecules having negative NOEs. When the correlation time of the molecule is of the order $\sqrt{5}/4\pi\nu_0$, where ν_0 is the observation frequency, the NOE is near zero. On current high-field spectrometers, this occurs for peptides of about 5–10 residues in length. Hence, little or no NOE information can be obtained for such molecules using the NOESY experiment. In contrast to the NOESY, the crossrelaxation rate in the ROESY is always positive, as a result of which signals arising from crossrelaxation are always positive and distance information can be obtained. A disadvantage of the ROESY technique, however, is that the artifacts from TOCSY transfer may reduce crosspeak intensities and make quantitative interpretation difficult. The point at which the NOE becomes zero is dependent on the field strength of the spectrometer, so using a higher field strength may make the NOE observable. Alternatively, the temperature may be lowered to increase the effective correlation time and thus move into the negative NOE regime.

4. Structural Characterization

4.1. Assignment and Structure Determination

On completion of data acquisition, sequence-specific resonance assignments *(1)* have to be undertaken before structure calculations can begin. First the complete spin systems of the amino acid residues are identified using their proton-proton J-couplings, and then these spin systems are located within the peptide sequence by the use of through-space, sequential NOE connectivities between adjacent residues *(1,42)*.

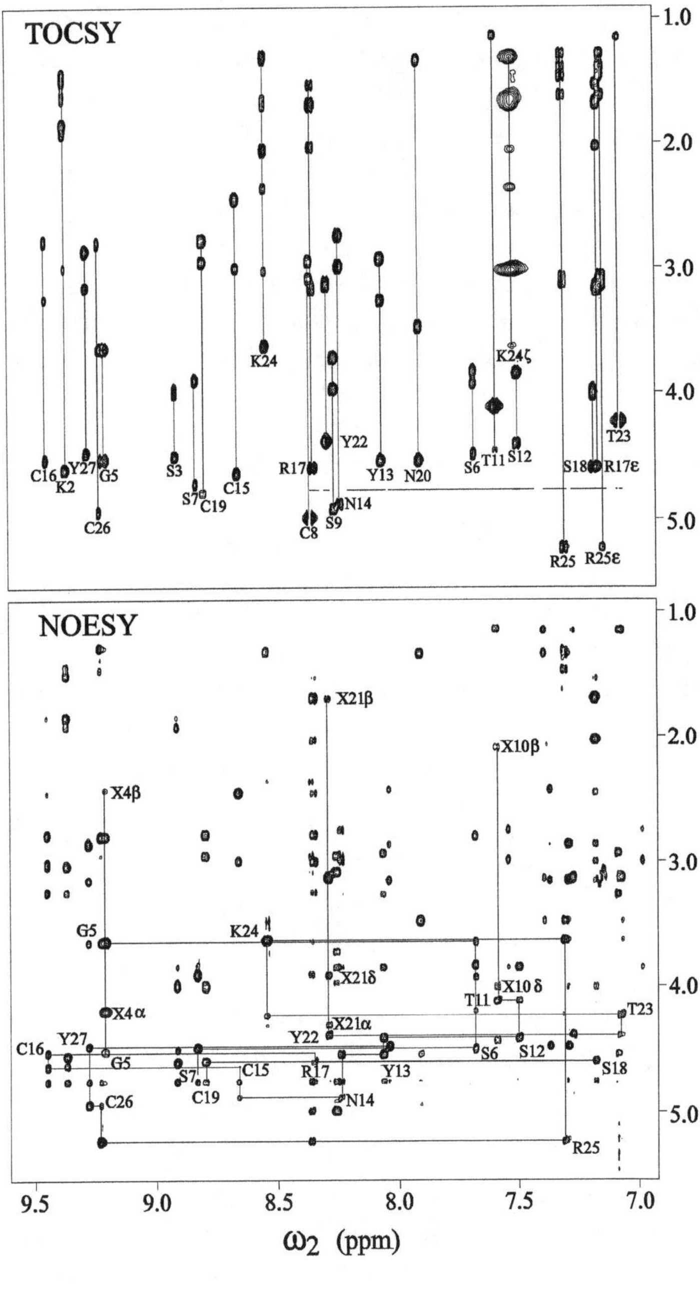

Fig. 3. Fingerprint regions of TOCSY and NOESY spectra of 3 m*M* ω-conotoxin in 90% H_2O/10% ^2H_2O at pH 3.4 and 26°C, showing connectivities from backbone and side-chain NH resonances *(43)*. Spectra were recorded at

The key resonances in this process are those from the amide protons, so spectra must be acquired under conditions where they are present. Intraresidue connectivities of the spin systems are determined from DQF-COSY and TOCSY spectra, and these are connected to their adjacent residues by the interresidue sequential NOEs $d_{NN}(i,i + 1)$ $d_{\alpha N}(i,i + 1)$, and $d_{\beta N}(i,i + 1)$ (where $d_{\alpha N}[i,i + 1]$ indicates an NOE between the $C^{\alpha}H$ proton on residue i and the NH of residue i + 1). Sequential connectivities for ω-conotoxin are shown in the NOESY spectrum of Fig. 3.

Following sequence-specific resonance assignment, crosspeak volumes in NOESY spectra can be converted into distances using the r^{-6} dependence of the volume on the internuclear distance, r *(1,46)*. Distance geometry programs are then used to generate a family of structures that are consistent with the covalent structure and the distances inferred from the NOESY spectra *(48)*. Further constraints on the structure can be obtained from coupling constants in conjunction with the Karplus relationship to define dihedral angles *(1,5)*. Once a set of structures has been generated that conform to the distance and angle constraints, as monitored by some form of penalty function, the structures can be further refined using techniques, such as restrained energy minimization or restrained molecular dynamics, usually including a simulated annealing step *(5)*. The structures can be further refined using an iterative back-calculation procedure, where the NOESY spectrum is calculated from the internuclear distances generated from the initial structure determination. The distances are then adjusted and the structure calculations repeated until there is good agreement between the experimental NOESY spectrum and the calculated spectrum *(49,50)*.

4.2. Methods Applicable to Flexible Peptides

Short linear peptides are usually present in solution as an ensemble of rapidly interconverting conformations *(11)*, but such peptides are capable of adopting stable secondary structure in solution *(51)*. Caution is required when interpreting NOESY or ROESY data for short peptides,

---

600 MHz with a spin-lock time of 90 ms in the TOCSY and a mixing time of 300 ms in the NOESY spectrum. In the NOESY spectrum, $d_{\alpha N}$ (i, i + 1) connectivities are shown for residues 4–7, 10–19, and 21–27; $d_{\beta N}$ and $d_{\delta N}$ (i, i + 1) connectivities are also shown for hydroxyproline (X) residues. Intraresidue NH-$C^{\alpha}H$ crosspeaks are marked with the residue number.

since the information obtained is the population-weighted average over all structures in the conformational ensemble. Intraresidue and sequential NOEs provide information on the local backbone dihedral angle populations, but in isolation do not provide enough evidence for the presence of folded conformations. Since there is little possibility of long-range NOEs in linear peptides, the NMR evidence that best describes structure in linear peptides is medium-range NOEs. For example, an α-helix is characterized by strong $d_{NN}(i,i + 1)$ and weak $d_{\alpha N}(i,i + 1)$ sequential NOEs, in conjunction with small coupling constants $^3J_{HN\alpha}$ and the medium-range NOEs $d_{NN}(i,i + 2)$, $d_{\alpha N}(i,i + 3)$, $d_{\alpha\beta}(i,i + 3)$, and $d_{\alpha N}(i,i + 4)$ *(1)*. Because of the conformational flexibility of peptides, structural calculations may be of limited value since they are meaningful only for molecules that adopt a stable structure in solution.

4.3. Methods Applicable to Polypeptides and Proteins

Proteins usually adopt more stable structures than those of small, linear peptides, and it is usually possible to observe a greater number of NOEs, in particular long-range NOEs, for these molecules. Although the approaches employed for polypeptides and proteins are fundamentally similar to those described for peptides *(52)*, the larger molecular mass brings a new set of problems. The increased line-widths and large number of resonances associated with larger molecules require the application of new techniques to help assign the resonances in these molecules. It is usually necessary to use proteins that have been overexpressed in bacteria and isotopically labeled, in conjunction with an array of heteronuclear 2D, 3D, and 4D methods directed toward simplifying sequence-specific assignment *(12)*. Spreading the chemical shifts in the heteronuclear dimension removes much of the overlap of signals that is present in homonuclear ^{1}H-NMR spectra. Furthermore, heteronuclear coupling constants obtained from these methods enable further structural constraints to be used in distance geometry calculations.

One of the advantages of NMR over crystallographic methods of structure determination is that it is possible to obtain information on the dynamic behavior of molecules. NMR relaxation studies have provided evidence of internal mobility within proteins spanning a wide range of frequencies and amplitudes. The parameters characterizing these molecular motions are obtained principally from heteronuclear relaxation measurement experiments *(53,54)*. From a plot of the relaxation parameters vs the sequence position, it is possible to determine which parts of the protein are more mobile than others.

Chemical shifts are also useful indicators of structure in certain cases. The chemical shifts of protein $C^{\alpha}H$ resonances have been correlated to the common secondary structural motifs *(55)*, with $C^{\alpha}H$ resonances in α-helices generally lying upfield of their random-coil chemical shifts and those in β-sheet being downfield. Similar findings have been reported for the ^{13}C nucleus, where the chemical shifts of C^{α} and C^{β} carbons have been correlated to backbone conformation *(56)*. Although this method of structural analysis is potentially rapid, requiring only the chemical shift information, it is not intended to replace rigorous structure calculations based on NOEs and other NMR-derived constraints.

5. Concluding Remarks

Improvements in the methodology of molecular biology that have led to the ability to overexpress almost any protein in bacteria and advances in methods for the synthesis and purification of peptides have resulted in an ever-increasing supply of biological molecules for study. Concurrent advances in NMR methodolgy and instrumentation have led to the application of this method to molecules of a wider molecular-mass range. The main requirements for NMR analysis are that the molecule is soluble, homogenous, and does not aggregate. Within these limitations, NMR can be expected to provide valuable information about most peptides and proteins.

Acknowledgments

We are grateful to C. J. Morton for providing data for Figs. 1 and 2, and P. K. Pallaghy and B. M. Duggan for Fig. 3.

References

1. Wüthrich, K. (1986) *NMR of Proteins and Nucleic Acids*. Wiley-Interscience, New York.
2. Croasmun W. R. and Carlson R. M. K. (eds.) (1987) *Two-Dimensional NMR Spectroscopy: Applications for Chemists and Biochemists.* VCH, New York.
3. Oppenheimer, N. J. and James, T. L. (eds.) (1989) Nuclear magnetic resonance. Part A. Spectral techniques and dynamics, in *Methods in Enzymology,* vol. 176, Academic, San Diego.
4. Oppenheimer, N. J. and James, T. L. (eds.) (1989) Nuclear magnetic resonance. Part B. Structure and mechanism, in *Methods in Enzymology,* vol. 177, Academic, San Diego.

5. Clore G. M. and Gronenborn A. M. (1989) Determination of three-dimensional structures of proteins and nucleic acids in solution by nuclear magnetic resonance spectroscopy. *CRC Crit. Rev. Biochem. Mol. Biol.* **24,** 479–564.
6. Fesik S. W. (1991) NMR studies of molecular complexes as a tool in drug design. *J. Med Chem.* **34,** 2937–2945.
7. Derome, A. E. (1987) *Modern NMR Techniques for Chemistry Research.* Pergamon, Oxford.
8. Homans S. W. (1992) *A Dictionary of Concepts in NMR.* Clarendon, Oxford.
9. Wüthrich, K. (1976) *NMR in Biological Research: Peptides and Proteins.* North-Holland Publishing, Amsterdam.
10. Ernst, R. R., Bodenhausen, G., and Wokaun, A. (1987) *Principles of Nuclear Magnetic Resonance in One and Two Dimensions.* Clarendon, Oxford.
11. Dyson, H. J. and Wright, P. E. (1991) Defining solution conformations of small linear peptides. *Annu. Rev. Biophys. Chem.* **20,** 519–538.
12. Clore, G. M. and Gronenborn, A. M. (1991) Applications of three- and four-dimensional heteronuclear NMR spectroscopy to protein structure determination. *Prog. NMR Spectroscopy* **23,** 43–92.
13. Oppenheimer, N. J. (1989) Sample preparation. *Meth. Enzymol.* **176**, 78–89.
14. Brown, L. R. and Wüthrich, K. (1981) Melittin bound to dodecylphosphocholine micelles ^{1}H-NMR assignments and global conformational features. *Biochim. Biophys. Acta* **647,** 95–111.
15. Mehlkopf, A. F., Korbee, D., Tiggelman, T. A., and Freeman, R. (1984) Sources of t_1 noise in two-dimensional NMR. *J. Magn. Reson.* **58,** 315–323.
16. Anglister, J., Grzesiek, S., Ren, H., Klee, C. B., and Bax, A. (1993) Isotope-edited multidimensional NMR of calcineurin B in the presence of the non-deuterated detergent CHAPS. *J. Biomol. NMR* **3,** 121–126.
17. Morris, G. A. (1992) Systematic sources of signal irreproducibility and t_1 noise in high-field NMR spectrometers. *J. Magn. Reson.* **100,** 316–328.
18. Marion, D. and Bax, A. (1988) Baseline distortion in real Fourier transform NMR spectra. *J. Magn. Reson.* **79,** 352–356.
19. Hoult, D. I., Chen, C.-N., Eden, H., and Eden, M. (1983) Elimination of baseline artifacts in spectra and their integrals. *J. Magn. Reson.* **51,** 110–117.
20. Conover, W. W. (1984) Practical guide to shimming superconducting NMR magnets, in *Topics in Carbon-13 NMR Spectroscopy,* vol. 4 (Levy, G., ed.), Wiley, New York, pp. 37–57.
21. Zuiderweg, E. R. P., Hallenga, K., and Olejniczak, E. T. (1986) Improvement of 2D NOE spectra of biomacromolecules in H_2O solution by coherent suppression of the solvent resonance. *J. Magn. Reson.* **70,** 336–343.
22. Hore, P. J. (1989) Solvent suppression. *Meth. Enzymol.* **176,** 64–77.
23. Lindon, J. C. and Ferrige, A. C. (1980) Digitisation and data processing in Fourier transform NMR. *Prog. NMR Spectrosc.* **14,** 27–66.
24. Otting, G., Widmer, H., Wagner, G., and Wüthrich, K. (1986) Origin of t_1 and t_2 ridges in 2D NMR spectra and procedures for suppression. *J. Magn. Reson.* **66,** 187–193.
25. Marion, D. and Bax, A. (1989) Baseline correction of 2D FT NMR spectra using a

simple linear prediction extrapolation of the time-domain data. *J. Magn. Reson.* **83,** 205–211.
26. Martin, M. L., Delpuech, J.-J., and Martin G. J. (1980) *Practical NMR Spectroscopy.* Heyden, London, pp. 244–290.
27. Bundi, A. and Wüthrich, K. (1979) ^{1}H-NMR parameters of the common amino acid residues measured in aqueous solutions of the linear tetrapeptides H-Gly-Gly-X-L-Ala-OH. *Biopolymers* **18,** 285–297.
28. Sönnichsen, F. D, Van Eyk, J. E., Hodges, R. E., and Sykes, B. D. (1992) Effect of trifluoroethanol on protein secondary structure: an NMR and CD study using a synthetic actin peptide. *Biochemistry* **31,** 8790–8798.
29. Dyson, H. J., Merutka, G., Waltho, J. P., Lerner, R. A., and Wright, P. E. (1992) Folding of peptide fragments comprising the complete sequence of proteins. Models for initiation of protein folding. I. Myohemerythrin. *J. Mol. Biol.* **226,** 795–817.
30. Jardetzky, O. and Roberts, G. C. K. (1981) *NMR in Molecular Biology.* Academic, New York, p. 166.
31. McIntosh, L. P. and Dahlquist, F. W. (1990) Biosynthetic incorporation of ^{15}N and ^{13}C for assignment and interpretation of nuclear magnetic resonance spectra of proteins. *Q. Rev. Biophys.* **23,** 1–38.
32. Kay, L. E., Clore, G. M., Bax, A., and Gronenborn, A. M. (1990) Four-dimensional heteronuclear triple-resonance NMR spectroscopy of interleukin-1β in solution. *Science* **249,** 411–414.
33. Bax, A., Sparks, S. W., and Torchia, D. A. (1989) Detection of insensitive nuclei. *Meth. Enzymol.* **176,** 134–150.
34. Hull, W. E. and Sykes, B. D. (1976) Fluorine-19 nuclear magnetic resonance study of fluorotyrosine alkaline phosphatase: the influence of zinc on protein structure and a conformational change induced by phosphate binding. *Biochemistry* **15,** 1535–1543.
35. Güntert, P. and Wüthrich, K. (1992) FLATT—A new procedure for high-quality baseline correction of multidimensional NMR spectra. *J. Magn. Reson.* **96,** 403–407.
36. Marion, D., Ikura, K., and Bax, A. (1989) Improved solvent suppression in one and two dimensional NMR spectra by convolution of time domain data. *J. Magn. Reson.* **84,** 425–430.
37. Manoleras, N. and Norton, R. S. (1992) Spectral processing methods for the removal of t_1 noise and solvent artifacts from NMR spectra. *J. Biomol. NMR* **2,** 485–494.
38. Zhu, G. and Bax, A. (1992) Two-dimensional linear prediction for signals truncated in both dimensions. *J. Magn. Reson.* **98,** 192–199.
39. Keeler, J. and Neuhaus, D. (1985) Comparison and evaluation of methods for two-dimensional NMR spectra with absorption mode lineshapes. *J. Magn. Reson.* **63,** 454–472.
40. Müller, N., Ernst, R. R., and Wüthrich, K. (1986) Multiple-quantum-filtered two-dimensional correlated NMR spectroscopy of proteins. *J. Am. Chem. Soc.* **108,** 6482–6492.
41. Braunschweiler, L. and Ernst, R. R. (1983) Coherence transfer by isotropic mixing: application to proton correlation spectroscopy. *J. Magn. Reson.* **53,** 521–528.
42. Chazin, W. and Wright, P. E. (1987) A modified strategy for identification of ^{1}H spin systems in proteins. *Biopolymers* **26,** 973–977.

43. Pallaghy, P. K., Duggan, B. M., Pennington, M. W., and Norton, R. S. (1993) Three-dimensional structure in solution of the calcium channel blocker ω-conotoxin. *J. Mol. Biol.* **234,** 405–420.
44. Bax, A. and Davis, D. G. (1985) Practical aspects of two-dimensional transverse NOE spectroscopy. *J. Magn. Reson.* **63,** 207–213.
45. Macura, S., Huang, Y., Suter, D., and Ernst, R. R. (1981) Two-dimensional chemical exchange and cross relaxation spectroscopy of coupled nuclear spins. *J. Magn. Reson.* **43,** 259–281.
46. Neuhaus, D. and Williamson, M. (1989) *The Nuclear Overhauser Effect in Structural and Conformational Analysis.* VCH, New York.
47. Kalk, A. and Berendsen, H. J. (1975) Proton magnetic relaxation and spin diffusion in proteins. *J. Magn. Reson.* **24,** 343–366.
48. Havel, T. F. (1991) An evaluation of computational strategies for use in the determination of protein structure from distance constraints obtained by nuclear magnetic resonance. *Prog. Biophys. Molec. Biol.* **56,** 43–78.
49. Nilges, M., Habazettl, J., Brünger, A. T., and Holak, T. A. (1991) Relaxation matrix refinement of the solution structure of squash trypsin inhibitor. *J. Mol. Biol.* **219,** 499–510.
50. Wilcox, G. R., Fogh, R. H., and Norton, R. S. (1993) Refinement of the solution structure of the sea anemone neurotoxin ShI. *J. Biol. Chem.* **268,** 24,707–24,719.
51. Dyson, H. J., Rance, M., Houghton, R. H., Lerner, R. A., and Wright, P. E. (1988) Folding of immunogenic peptide fragments of proteins in water solution. I. Sequence requirements for the formation of a reverse turn. *J. Mol. Biol.* **201,** 161–200.
52. Ciba Foundation Symposia 161 (1991) Proceedings of the symposium on protein conformation, held Jan. 22–24, 1991, at the Ciba Foundation, London, England. John Wiley and Sons, Chichester, England.
53. Barbato, G., Ikura M., Kay, L. E., Pastor, R. W., and Bax, A. (1992) Backbone dynamics of calmodulin studied by ^{15}N relaxation using inverse detected two-dimensional NMR spectroscopy: the central helix is flexible. *Biochemistry* **31,** 5269–5278.
54. Peng, J. W. and Wagner, G. (1992) Mapping of the spectral densities of the N—H bond motions in eglin *c* using heteronuclear relaxation experiments. *Biochemistry* **31,** 8571–8586.
55. Wishart, D. S., Sykes, B. D., and Richards, F. M. (1992) The chemical shift index: a fast and simple method for the assignment of protein secondary structure through NMR spectroscopy. *Biochemistry* **31,** 1647–1651.
56. Spera, S. and Bax, A. (1991) Empirical correlation between protein backbone conformation and Cα and Cβ ^{13}C nuclear magnetic resonance chemical shifts. *J. Am. Chem. Soc.* **113,** 5490–5492.
57. Morton, C. J., Simpson, R. J., and Norton, R. S. (1994) Solution structure of synthetic peptides corresponding to the c-terminal helix of interleukin-6. *Eur. J. Biochem.* **219,** 97–107.

CHAPTER 10

Techniques for Conjugation of Synthetic Peptides to Carrier Molecules

J. Mark Carter

1. Introduction

1.1. Rationale

1.1.1. Basic Immunology/Immunochemistry

There are two common purposes for conjugation of peptides. The most common is induction of humoral immunity. This is the production of antibodies capable of binding to the peptide immunogen. The antibodies are elaborated by plasma cells, which are terminally differentiated B-lymphocytes. However, in order for immunity to be successfully induced in a secondary anamnestic response, the immunogen must also react with T-lymphocytes. Many peptides contain B-cell epitopes, but not T-cell epitopes. In immunological terms, these peptides and other such molecules are called haptens. Coupling these molecules to a large carrier protein containing T-cell epitopes allows the induction of a B-cell response to the entire immunogen, including the peptide *(1)*. New synthetic peptides offer promise as vaccines.

The next most important reason for conjugation of peptides is to create an effective "capture antigen" (e.g., ref. 2). Capture antigens are compounds used in antigen-binding assays, such as EIA. In addition to binding antibodies, these compounds must also be capable of binding to a plastic microtiter plate in a nonspecific fashion. Whereas larger peptides (>20 residues) frequently bind well to the ELISA plates, small- to

From: *Methods in Molecular Biology, Vol. 36: Peptide Analysis Protocols*
Edited by: B. M. Dunn and M. W. Pennington

medium-sized peptides often do so poorly. The carrier proteins used for capture antigen production are large enough to convey good nonspecific binding, and the peptides carried on the surfaces of the conjugates remain accessible for specific binding to antibody molecules.

This chapter does not purport to be a comprehensive index of conjugation techniques. Rather, it is a list of a few extremely versatile and well-demonstrated reactions capable of performing most conjugations for immunogen preparation. For a more exhaustive list of peptide conjugation recipes, *see* ref. *3*.

1.1.2. Antigen Orientation/Conformation for Native-Like Presentation

For either of the above functions of peptide conjugation, people generally believe it is necessary that the peptide attain a native-like orientation and conformation. Indeed, in most cases in which peptides are used as immunogens, they must mimic certain regions of larger intact protein molecules. The goal of experiments with these molecules is usually production and/or detection of antibodies reactive to the native protein. Although a free peptide in solution has many more conformational possibilities than the corresponding region of the native protein, it is suggested that the conjugate is able to assume the immunologically important structures normally exhibited by the native protein. However, this is not always true, and that same sequence, isolated from the rest of the parent protein molecule, may take up a different conformation when free in solution. For this reason, it is sometimes very important to conjugate the peptide in a manner that will encourage appropriate native-like conformations for antibody induction and binding.

1.1.3. Coupling at Peptide Termini

There are two good reasons for conjugating a peptide through its terminal amino acids. The first reason is to avoid steric problems in epitope presentation and antibody binding of the folded peptide structure. Indeed, studies indicate that the free (nonconjugated) end of a peptide molecule thus coupled through one of its termini is much more antigenic than the bound end (site of conjugation) *(4)*.

The second reason is that some purification of the peptide product from the impurities in a crude synthesis mixture can be achieved by selection of particular N-terminal amino acids and conjugation protocols. If the N-terminal amino acid contains a moiety for conjugation not expressed in the remainder of the molecule, then only those molecules bearing this amino acid will be conjugated. To effect purification, this conjugation strategy is employed after a capping (acetylation) synthesis protocol. Incomplete couplings during peptide synthesis are terminated via acetylation, so that only full-length molecules bear the N-terminal amino acid. If the N-terminal amino acid is the one used for conjugation, then only full-length peptides will be coupled to the carrier protein *(5)*. After dialysis, these conjugates of purified full-length peptides are used for immunogenesis.

1.1.4. Circular Peptides

One successful approach to stabilization of native-like antigenic conformation in synthetic peptide bend and loop structures involves the use of circular peptides (*see* Chapters 6,7, and 11, *PSP*)]. Circularization is generally most easily achieved via formation of a disulfide bond between the N- and C-termini of the peptide. This is easily accomplished by synthesis of the peptide with an extra (nonphysiological) cysteine at each end of the sequence. Following cleavage, gentle air oxidation of a dilute solution of the peptide will generate the desired intramolecular bonds. On the other hand, if exactly one cysteine residue is naturally contained in the peptide sequence, then it is probably involved in a disulfide bond in the native protein structure. In the peptide molecule, this cysteine should therefore be utilized in conjugation to a carrier protein in order to mimic involvement in an internal disulfide bond.

Other possibilities for circularization and stabilization of structure in a molecule include formation of a peptide bond between the N- and C-termini of the molecule, as well as creation of a thioether linkage between the N- and C-termini (qv). Alternatively, the hydrogen bonds normally stabilizing a loop or bend configuration in the native protein may be replaced by covalent bonds, such as hydrazone-ethane *(6)*.

1.1.5. Noncovalent Stabilization

Finally, noncovalent forces must not be overlooked. One of the earliest generally successful immunization protocols for peptides (Freund's adjuvant; *see* Section 8.) incorporated a lipid emulsion. Strangely enough, one of the latest great breakthroughs hailed in vaccine development (the liposome, again detailed in Section 8.) also incorporates lipid emulsion. These methods are probably successful at encouraging the peptide to attain a native-like conformation via their water–lipid interface.

Because the native protein comprises an amphipathic environment, with a hydrophobic interior and a hydrophilic exterior, a peptide in a lipid emulsion probably experiences many of the same forces as in its native protein. In emulsion, the peptides probably align and insert themselves along the water–lipid interface much the same way as they do in the native protein structure.

1.2. Optimization

1.2.1. Substitution Determination

Several studies suggest that maximal antibody response to a peptide–carrier complex will occur if there are about 20 peptide molecules per carrier protein molecule. However, choice of coupling chemistry may not facilitate such a level of substitution. Consider, for example, a peptide to be coupled through free thiol groups on a protein, where the protein contains only four disulfide bonds per molecule. Clearly no coupling can occur until the disulfide bonds in the carrier protein are first reduced, but even afterwards, only eight free sulfhydryl groups will be present, and some of them may be sterically inconvenient for the conjugation reaction. In such cases, the carrier may be derivatized (qv) to create an increased number of reactive moieties if optimum immunogenicity is desired. Note that for most peptide–carrier protein combinations, a 20:1 ratio is achieved with nearly equal masses of peptide and carrier.

After the conjugation reaction, most investigators choose to immunize without determining the precise level of substitution of the immunoconjugate. This simplistic approach is appropriate for experiments requiring only the generation of antibodies to the peptide, without comparisons of different immunogen preparations. However, especially when peptide vaccines are being studied, the investigator should bear in mind the importance of reproducibility of the immunogen preparation for concurrent reproducibility of experimental results. In these cases, substitution may be important and must be documented.

There are a number of means of determining the number of peptide molecules coupled to a carrier protein. Regardless of which method is employed, it is critical that the conjugate preparation be free of unconjugated peptide. This may be achieved easily, either by extensive dialysis or by gel filtration. One of the most common techniques for substitution determination is quantitative amino acid analysis (AAA) of the conjugate. Comparing the amino acid profiles of the unconjugated carrier with that of the peptide conjugated carrier allows quantitative differences to be established.

1. The amino acid composition of the carrier protein, peptide, and conjugate must be known. This is generally determined via quantitive AAA after complete acid hydrolysis. Subsequent calculations are greatly facilitated if some amino acids present in the carrier are not also present in the peptide.
2. Divide the quantity of amino acid in the conjugate by the quantity of the same amino acid in the carrier (Eq. [1]). This is performed for several amino acids in order to obtain an average. This mean indicates the ratio of the quantity of conjugate in the conjugate AAA hydrolyzate to the quantity of carrier in the carrier AAA hydrolyzate sample:

$$\text{(nmol AA1 in conjugate/nmol AA1 in carrier)} = \text{mean ratio of (conjugate/carrier)} \quad (1)$$

3. Subtract the quantity of amino acid present in the carrier from that of the conjugate to give the quantity of peptide present in the conjugate (Eq. [2]). Again, this is performed for several amino acids in order to obtain an average. In making this calculation, consider the number of times the particular amino acid is present in the peptide, as well as the relative amount of each molecule in its respective hydrolyzate sample, as shown:

$$\text{(nmol AA1 in conjugate/number of AA1/peptide)} - \text{nmol AA1 in carrier * mean ratio from step 1} \quad (2)$$

4. Then determine the amount of protein present in the conjugate hydrolyzate sample by multiplying by the average formula weight of an amino acid (110), and then dividing by the total amount of all amino acids in the conjugate, and by the molecular weight of the carrier (Eq. [3]). Again consider the relative amounts of each molecule in its respective hydrolyzate sample, as shown:

$$\text{(total nmol all carrier protein amino acids/mol wt of carrier protein)} \text{ * mean ratio from step 1/110} \quad (3)$$

5. Finally, simply divide the amount of peptide present in the conjugate, as determined in step 3, by the amount of carrier protein present in the same preparation, as determined in step 4 (Eq. [4]):

$$\text{Substitution ratio} = \text{(nmol peptide in conjugate/nmol carrier protein present in conjugate)} \quad (4)$$

6. A possibility for simplification of quantitation of substitution via AAA is the introduction of nonnatural amino acids into the peptide as an internal standard. β-alanine and norleucine have commonly been utilized for this approach. Obviously, the nonnatural amino acids are not present in the car-

rier. They may be readily quantitated in the conjugate hydrolyzate without interference from the carrier. Thus, step 3 above becomes simply (Eq. [5]):

$$\text{(nmol AA1 in conjugate/number of AA1/peptide)} \qquad (5)$$

The remainder of the calculation is performed as usual.

Some conjugation methods introduce a novel amino acid via unique reaction chemistry (*see* Section 1.3.), but AAA may not discriminate between these reaction products and the products of intramolecular reaction within the activated carrier molecule, which do not actually incorporate the peptide.

Another possibility for quantitation utilizes the release of a chromophore by the coupling reaction. This method is usually used for those chemistries driven by the release of 2-pyridylthione. One likely drawback of this type of quantitation is its sensitivity to interference from other chromophores in the assay buffer. Also, the chromophore may be released by intramolecular reaction.

A similar approach involves titration of reactive groups on the carrier, both before and after the reaction. This method is particularly useful in titrating remaining free thiols via Ellman's reagent. Unfortunately, it also suffers from susceptibility to interference from the consumption of thiols by intramolecular reaction.

By far the fastest and simplest method of quantitation involves the incorporation of radiolabeled peptide into the conjugates. The carrier is quantitated via AAA, and the peptide in the conjugate is quantitated via trace label calculation from the specific activity of the labeled peptide preparation. Unfortunately, the simplest radiolabeling methods use ^{125}I iodination of tyrosine moieties, and this modification produces a neo-antigen. The neo-antigen is not the same as the native tyrosine-containing antigen, and it may often prove immunodominant, obscuring the immunogenicity of the unmodified peptide in the preparation.

Another problem stems from the persistence of the radionuclide. "Hot" conjugates produce "hot" experimental animals. The "hot" animals, in turn, produce "hot" sera. The radioactive sera and animal wastes require special handling and disposal.

1.2.2. Storage

After the conjugate is dialyzed and the substitution level is determined (if desired), a method of storage must be considered. This will depend on the character of the conjugate preparation and the storage length requirement.

Many conjugates will precipitate during the coupling reaction. Although some will redissolve on dialysis, at least some compound often remains insoluble. Experience shows that when suspended along with the soluble conjugate, this precipitate causes no problems with animal health or peptide immunogenicity. However, it is conceivable that the aggregated form of the compounds may exhibit steric interference in regard to some epitopes. Also, it is unlikely that reproducible results would be obtained in immunization trials with such an inherently nonhomogenous mixture, and this may prevent such vaccines from being approved for use in humans. Finally, this irreproducibility would extend to use of the conjugate as a capture antigen for binding assays.

If the conjugate is mostly soluble, it may be sterilized and clarified (before substitution determination) via filtration on a 0.4-μ syringe filter. This will enhance its stability with respect to microbial degradation. For brief storage of filter-sterilized preparations, conjugates may be stored at 4°C. In a typical buffer, under aseptic conditions, they will last for at least 2 wk.

For longer storage of either soluble or insoluble preparations, freezing is appropriate: –20°C is sufficient for periods of a few months, whereas –70°C or lower is probably better for longer storage. Often, frozen conjugates exhibit increased precipitate on thawing, especially with repeated freeze–thaw cycles.

For extremely long storage (over 1 yr), lyophilization is recommended. Unfortunately, on redissolving these compounds, formation of some amount of precipitate is virtually inevitable. This is probably a result of irreversible denaturation of the carrier protein via removal of structural water.

1.2.3. Comparative Stability of Different Coupling Types

At least one study has been performed to examine the stability of different linking groups in conjugates, under different conditions of preparation and storage *(7)*. Although intuitive, the results suggest that increasing the number of bonds between the peptide and the carrier increases the temporal stability of the conjugate. Also, certain types of bonds are inherently more stable than others. For example, peptide bonds are more stable than disulfide bonds. For preparation of immunogens that must be stored for long periods of time, these considerations may be important. However, for most experiments, involving a limited number of animals studied over a period of a few months, the long-term chemical stability of the conjugate is probably not an important issue.

1.3. Chemical Requirements for Conjugation

1.3.1. Carrier Choice

For practical and immunological reasons, the choice for carrier proteins usually ends at bovine serum albumin (BSA), ovalbumin (OVA), or keyhole limpet hemocyanin (KLH). BSA or OVA are usually chosen when low price is important. As albumins, they have good solubility characteristics and are well-described molecules. BSA crossreacts immunologically with other serum albumin proteins, and so it is not generally used in immunogens, but it has high nonspecific binding, which makes it ideally suitable for use in immunoassay capture antigens (qv). OVA does not crossreact as strongly with serum albumins as BSA, and it has been successfully used with immunogens as well as capture antigen.

KLH is the preferred choice for good immunogenicity. This is probably because of its relatively distant phylogenetic removal from the vertebrates generally used for antibody production. Other invertebrate proteins (e.g., *Limulus polyphemus* hemocyanin) are similarly immunogenic in experimental animals. KLH functions for the mollusc as an oxygen-transport metalloprotein, existing as a family of soluble oligomers in the animal's hemolymph. As commercially obtained, KLH is infamous for its poor lot-to-lot reproducibility and difficult aqueous solubility.

KLH is usually purchased as a green-gray lyophilized powder displaying variable solubility. KLH is often virtually insoluble in water and does much better in saline. Solubility is maximized by dissolving 5 mg/mL in 6*M* guanidine followed by overnight dialysis. As with any protein solution, do not mix vigorously (e.g., on a high-speed vortex mixer) in an attempt to increase dissolution. This will cause denaturation of the protein at the air interface, observed as foaming, which may lead to a substantial loss of soluble material. Quantitation of the insoluble material is easiest by weighing it after it is filtered off and dried on filter paper. This allows an indirect determination of the amount of KLH actually dissolved and available for conjugation.

Thiols in these proteins are generally locked up as disulfides. In order to make them available for conjugation reactions, these proteins should be reduced to the free thiol form by incubation with an appropriate reducing reagent. Cysteine-containing peptides are treated the same way.

Effective reducing agents include: 2-mercaptoethanol, dithiothreitol, glutathione, and tributylphosphine *(8)*. Any of these compounds is effective at 0.1%, in aqueous or organic solution, for 60 min at room temperature. If the protein or peptide must be kept cold, allow the disulfide reduction to proceed overnight. After reduction is complete, these reagents must be thoroughly removed, for example, by gel filtration (*see* Chapter 1) or extensive dialysis, before proceeding with the conjugation. The exception to this is tributylphosphine, which effectively reduces peptides and proteins, but does not subsequently interfere in thiol-based couplings. Extent of reduction to the free thiol form can be monitored by means of Ellman's reaction.

1.3.2. Addition of Extra Amino Acids

Many synthetic peptides do not contain a cysteine as a thiol donor. When such molecules are to be conjugated via thiol-type chemistry, a cysteine may be incorporated at either the N- or C-terminal. This will generally not adversely affect the immunogenicity of the sequence. Indeed, addition of the cysteine at the N-terminal affords a simple scheme for purification via conjugation (qv).

If the synthesis is performed with acetylation after each amino acid coupling, only full-length molecules will bear the N-terminal cysteine thiol. The unconjugated truncated synthesis side products may be easily removed from the conjugate by gel filtration or dialysis. Occasionally, a proposed synthetic peptide sequence may contain three or more odd numbers of cysteines, with all but one involved in intramolecular disulfide bonds. Of course, such compounds are difficult enough to prepare and purify, if only by virtue of their numerous reactive thiols. Many such peptides will fold into a native-like conformation and spontaneously form primarily native-like disulfide bonds. However, when an odd number of thiols is present, it is often necessary to eliminate the cysteine not involved in intramolecular disulfide bonding. In such cases, a substitution of serine for the deleted cysteine preserves a similar size, charge, and hydrogen-bonding preference, while omitting the reactive thiol group.

Quite often it is necessary to introduce a thiol group into a peptide after synthesis in order to effect conjugation. For example, a peptide may bear several lysine amino groups throughout its sequence, although a single conjugation bond through one terminus is intended. Three different methods are useful. By far, the easiest is introduction of a cysteine

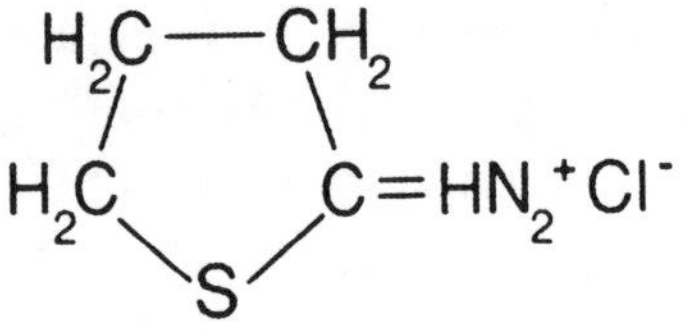

Fig. 1. Chemical structure of Traut's reagent (iminothiolane).

during synthesis, as discussed above. Otherwise, there are two reagents available for introduction of a thiol. One is 2-iminothiolane, also known as Traut's Reagent (Fig. 1). The other is *N*-hydroxysuccinimidyl 3-(2-pyridyldithio)propionate (Fig. 2). Refer to Section 5. for details of these reactions. Either one will introduce a free thiol onto an amino group. With 2-iminothiolane, the spacer moiety introduced between the modified side chain and the new thiol is three carbons (about 6.8 Å) long. SPDP introduces an activated thiol group (the 2-pyridyldithio moiety) onto the amino group. In this case, the linker is four carbons (about 8.1 Å) long, with retention of the positive charge.

2. Materials

2.1. Citraconylation

1. Citraconic anhydride.
2. Peptide: about 5 mg, with free amino groups. Calculate number of moles of amino groups present in peptide sample before reaction.
3. Buffer, pH 8–9. HEPES, 50 m*M*, is excellent. Avoid amine-containing buffers.
4. 1*N* NaOH.
5. Acetic acid, 1%.
6. Dialysis tubing or gel-filtration column.

2.2. Glutaraldehyde

1. Glutaraldehyde, 20 m*M* in water, freshly prepared: Although specially purified grades are available, technical-grade glutaraldehyde (stored at room temperature) works just as well.
2. Peptide: 6 mg, bearing one or two moieties available for reaction (N-terminal amino group, lysine ε amino group, cysteine thiol group, and so forth). Check for adequate solubility in buffer.

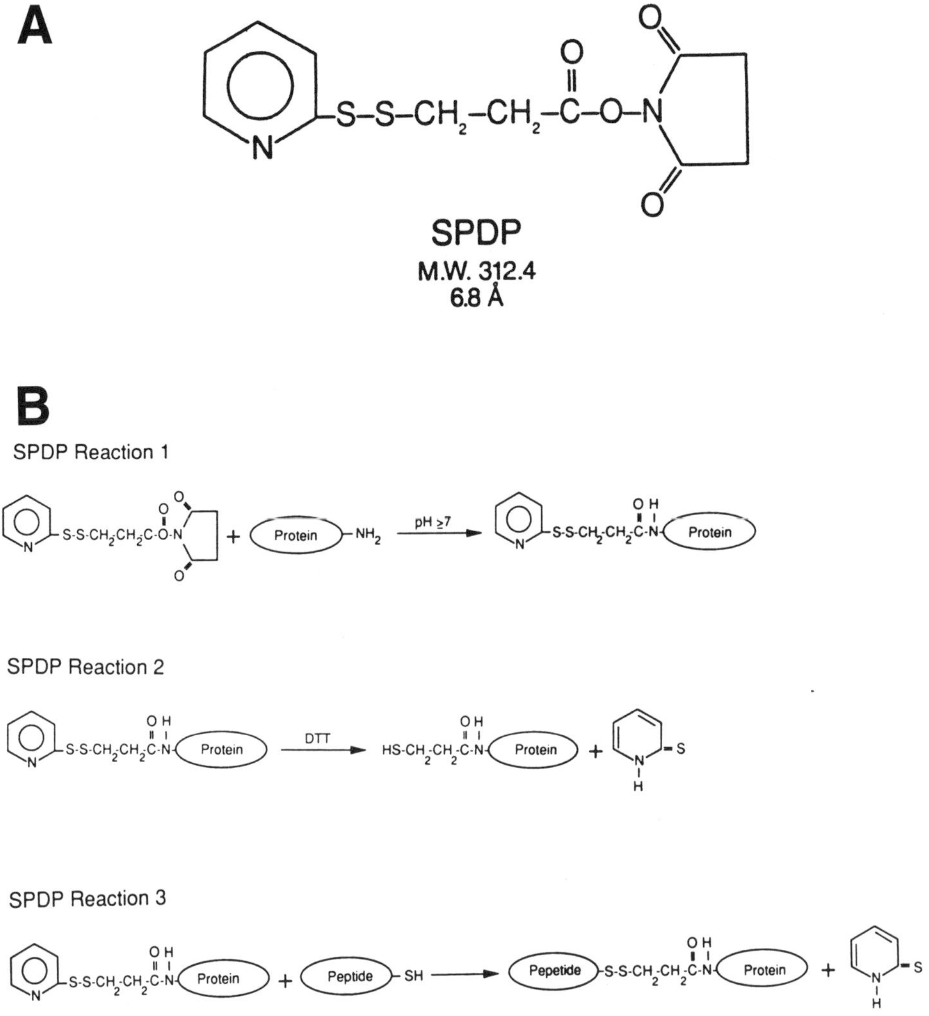

Fig. 2. (**A**) Chemical structure and (**B**) reactions of *N*-hydroxysuccinimidyl 3-(2-pyridyldithio)propionate (SPDP).

3. Carrier protein, e.g., KLH or ovalbumin: 6 mg.
4. Suitable buffer: PBS is usually quite effective. Alternatively, use phosphate buffer, pH 7–8, without saline, or use borate, pH 7–8, but avoid Tris and other amine-containing species. Also, you may add a little methanol to improve solubility of the peptide, but be careful not to precipitate the carrier protein. If pH is raised above 8, the glutaraldehyde itself may precipitate.
5. Optional: sodium borohydride or glycine.

2.3. Carbodiimide

1. 1-ethyl-3-(dimethylaminopropyl)carbodiimide (EDAC), 25 mg: The carbodiimide reagent must be used fresh or stored desiccated and frozen.
2. Peptide: 5 mg, bearing one or two free carboxyl moieties available for reaction. Amino groups may be temporarily blocked by citraconylation. Check to confirm adequate solubility in coupling solvent.
3. Carrier protein, e.g., KLH or ovalbumin: 5 mg.
4. Coupling solvent: PBS or 0.05*M* NaCl is usually effective. Alternatively, use phosphate buffer, pH 7.0–8.0, without saline, or use borate, pH 7.0–8.0, but avoid Tris and other amine-containing species. An excellent buffering solvent for the two-step reaction is 20 m*M* TES, pH 6.5. Also, you may add a little methanol or DMF to improve solubility of the peptide, but be careful not to precipitate the carrier protein, and avoid using buffers containing carboxylates (such as acetate) or phosphate.
5. Optional: ice, glacial acetic acid, dialysis tubing.

2.4. M-Maleimidobenzoyl-N-Hydroxysuccinimide Ester (MBS)

1. MBS, 10 mg/mL in DMF or DMSO. This solution may be sealed and stored at –20°C for several months.
2. Peptide: 5 mg, bearing one or two free thiol groups. Reduce with appropriate reagents before conjugation. Check to confirm reduction and to confirm adequate solubility in coupling solvent.
3. Carrier protein, e.g., KLH or ovalbumin: 5 mg.
4. Coupling solvent for step 1: 0.01*M* phosphate buffer, pH 6.0. You may add a little methanol or DMF to improve solubility of the peptide, but be careful not to precipitate the carrier protein.
5. Gel-filtration column (e.g., Sephadex G-25) equilibrated in coupling solvent.
6. Coupling solvent for step 2: PBS or phosphate buffer, pH 6.0–7.0, without saline. Add guanidine hydrochloride as necessary to facilitate solubility.

2.5. Thiol Alkylation

1. Peptide: 5 mg, with bromoacetylated N-terminus and no free thiol.
2. Carrier: 5 mg.
3. Reaction buffer: PBS is effective, but reaction proceeds faster in 0.1*M* $NaHCO_3$. Degas the buffer thoroughly to avoid air oxidation of free thiols.
4. Tributylphosphine: A solution 0.7*M* in methanol.

2.6. Bisdiazobenzidine

1. Benzidine: Caution: benzidine, its salts, and (to a lesser extent) BDB are known to be potent carcinogens. Use appropriate precautions when handling them.

2. 4*N* HCl.
3. Ice.
4. Sodium nitrite.
5. Peptide: 5 mg, bearing tyrosine.
6. Carrier: 5 mg.
7. Coupling solvent: PBS, phosphate buffer, pH 8–9, without saline, or borate. Do not use Tris.
8. 0.1*N* NaOH.

2.7. N-Hydroxysuccinimidyl 3-(2-Pyridyldithio)-Propionate (SPDP)

2.7.1. Method 1

1. Peptide: 5 mg, bearing one or two free thiol groups. Reduce with appropriate reagents before conjugation. Check to confirm reduction and to confirm adequate solubility in coupling solvent.
2. Carrier: 5 mg, bearing amino groups.
3. Coupling buffer: PBS works well. Alternatively, use practically any buffer, pH 5–9, but avoid Tris and other amine-containing species. Also, you may add a little methanol to improve solubility of the peptide, but be careful not to precipitate the carrier protein. Degas the buffer to avoid air oxidation of free thiols.
4. SPDP: 20 m*M* in ethanol. This solution may be sealed and stored at –20°C for several months.
5. Gel-filtration column equilibrated in buffer.
6. Optional (for method 2, below), dithiothreitol (DTT) and sodium acetate buffer, 0.1*M*, pH 4.5.

2.7.2. Method 2

1. Peptide: 5 mg, bearing one or two amino groups. Check to confirm adequate solubility in coupling solvent.
2. Carrier: 5 mg, bearing amino groups.
3. Coupling buffer: PBS works well. Alternatively, use phosphate buffer, pH 7.0–8.0, without saline, or use borate, pH 7.0–8.0, but avoid Tris and other amine-containing species. Also, you may add a little methanol to improve solubility of the peptide, but be careful not to precipitate the carrier protein. Degas the buffer to avoid air oxidation of free thiols.
4. SPDP: 20 m*M* in ethanol. This solution may be sealed and stored at –20°C for several months.
5. Sodium acetate buffer, 0.1*M*, pH 4.5.
6. Dithiothreitol (DTT).
7. Two gel-filtration columns, equilibrated in coupling buffer. At least one of these should be small cutoff, e.g., Sephadex G-10, for exclusion of the peptide.

2.8. Immunization

1. Two rabbits, preferably young (about 1 kg) females.
2. Peptide immunogen: 10 mg peptide or 20 mg conjugate, a solution or suspension in 10 mL PBS. Store this solution frozen.
3. Freund's complete adjuvant (FCA), 2 mL.
4. Freund's incomplete adjuvant, 8 mL.
5. Sterile syringes and a union to join their fittings.
6. Sterile hypodermic needles.

3. Methods

3.1. Citraconylation

In some cases, a peptide may contain several lysines with their side-chain ε amino groups available for conjugation. Effective conjugation may be performed indiscriminately through these amino groups by using limited amounts of coupling agent in order to achieve only one or two conjugating bonds per peptide. However, as previously discussed, this is not the preferred topology for effective antigen presentation by the peptide hapten.

As an alternative, the ε amino groups may be reversibly blocked by citraconylation with citraconic anhydride (Fig. 3). After conjugation is completed, decitraconylation may be facilitated by brief exposure to acetic acid. It has been reported that citraconylation also reversibly blocks thiol- and alkylhydroxyl-containing amino acids. Apparently aryl hydroxyls are not affected *(9)*.

1. Dissolve peptide in 5 mL buffer, and then add 5- or 10-fold molar excess citraconic anhydride. Keep pH adjusted to 8–9 via addition of 1*N* NaOH. When pH stabilizes, add 20% more citraconic anhydride and continue stirring. Monitor progress of citraconylation via pH: when pH remains stable after last addition of citraconic anhydride you have added enough. Allow it to react for an additional 15–30 min, and then conjugate.
2. Removal of the citraconic groups after conjugation is done at low pH. Simple dialysis vs 1% acetic acid for a few hours is effective. Alternatively, acetic acid may be added to the final conjugation mixture until pH drops below 4. After a 1-h incubation, the conjugate may be separated from the citraconic acid via gel filtration.

3.2. Glutaraldehyde

Glutaraldehyde (Fig. 4) couples through amino groups, especially lysine ε amino and peptide α amino groups. Cysteine sulfhydryls are also reactive, and tyrosine and histidine participate to a lesser extent. Because

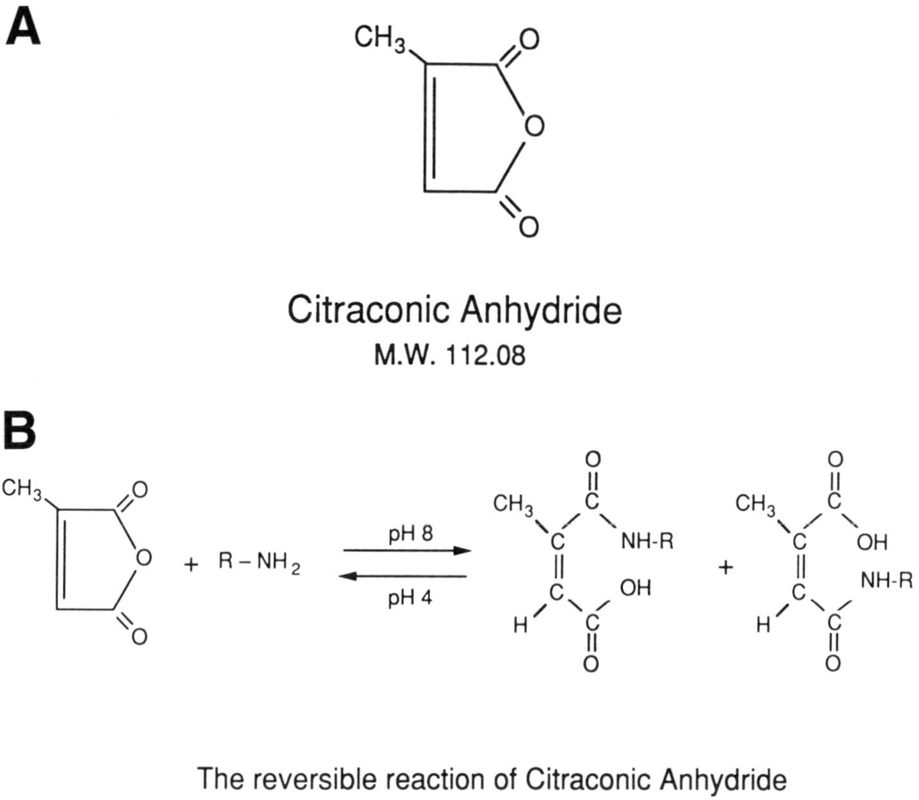

Fig. 3. Chemical structure of citriconic anhydride and reversible reactions used to block free amino groups.

of this reaction selectivity, the glutaraldehyde method is especially appropriate for peptides containing no reactive groups or one lysine at either the N- or C-terminus. This technique is excellent because of its simplicity, speed, and effectiveness. However, the reaction mechanisms and subsequent chemical linker structures remain poorly characterized. The reaction probably proceeds via a glutaraldehyde-lysine adduct Schiff base. In the absence of sodium borohydride to reduce this moiety, a rearrangement occurs leading to a nonhydrolyzable bond. UV spectra suggest that a quaternary pyridinium structure is formed involving four glutaraldehyde molecules per lysine amino *(11)*. Otherwise, a Michael addition or aldimine condensation may occur *(12)*. The following is a recipe for the conjugation after Baron and Baltimore *(13)*.

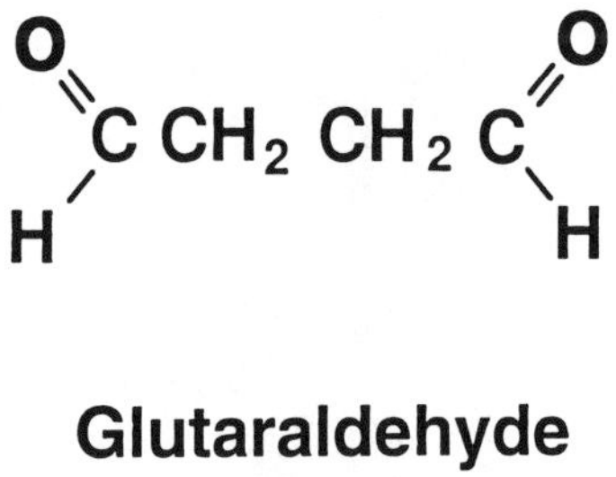

Fig. 4. Chemical structure of glutaraldehyde.

1. Weigh out 6 mg carrier and 6 mg peptide.
2. Dissolve them together in 2–3 mL buffer.
3. Add 1 mL glutaraldehyde solution, dropwise, mixing with a magnetic stir bar.
4. Continue mixing, and allow to react for 1 h at room temperature. The formation of some yellowish or milky precipitate is likely. This does not affect immunogenicity (*see* Chapter 11).
5. If desired, add sodium borohydride to 10 mg/mL. This will reduce the Schiff base to an amine bond before rearrangement occurs. In my hands, addition of borohydride does not improve immunogenicity, although it does improve solubility of the end product. Another common option is addition of glycine to 10 m*M* as a scavenger for unreacted glutaraldehyde. Again, in our hands addition of glycine does not improve immunogenicity of the conjugate, although it does improve solubility.
6. If you want to assume that the coupling went well, then the conjugate is ready to inject the animals, now. On the other hand, if you want to determine the level of substitution of the conjugation to see if it is near 20:1 peptide:carrier, then you must separate the conjugate from the free peptide. For glutaraldehyde conjugation, this is best performed via dialysis vs PBS.
7. If you prefer a soluble immunogen, you may perform the separation via gel filtration. For example, if the gel filtration is performed in PBS on Sephadex G-25, then the conjugate will elute in the void volume. The purified conjugate may then be subjected to analysis as described in Section 2.

3.3. Carbodiimide

One of the most commonly used carbodiimide reagents is the water-soluble reagent EDAC (*see* Fig. 5). Like other carbodiimides, EDAC couples an amino group to a carboxyl group (although side reactions involving cysteine sulfhydryl and tyrosine aryl hydroxyl are also reported). Usually the carboxyl group is the C-terminal of the peptide, and the

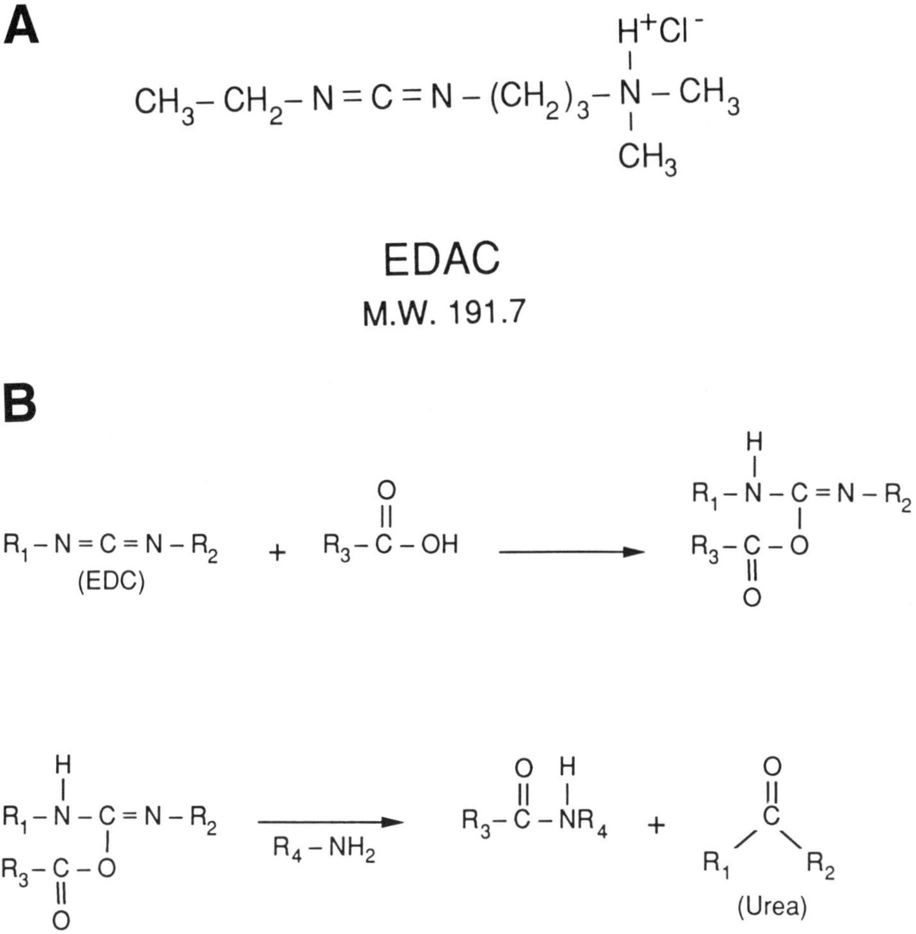

Fig. 5. (**A**) Chemical structure of EDAC and (**B**) reactions.

amino group is an ε amino group on a lysine residue contained in the carrier protein. Therefore, peptides with more than one carboxyl group (i.e., those peptides bearing aspartate or glutamate residues) may not be appropriate, and peptides should be quite free of residual acetic acid from synthesis. Also, peptide amino groups at the N-terminus as well as any lysine residues may be temporarily blocked via citraconylation (*see* Section 3.). On the other hand, a two-step reaction is possible, in which the carrier donates the carboxyl group, and the peptide provides the reactive amino group. In this case, preblocking with citriconylation is not required, although peptides with more than one amino group (i.e., those peptides bearing lysine residues) may react at more than one site. This may affect immunogenicity. Also, the reaction should be performed on ice to minimize rearrangement of the *O*-acyl intermediate to an unreactive *N*-acyl urea *(14)*.

This method is very versatile, and creates a well-characterized and stable amide bond. It provides no linker, and so avoids possible generation of neo-antigens. However, the method is somewhat susceptible to formation of intramolecular crosslinks. This may reduce the efficiency of the conjugation reaction, although it probably does not reduce immunogenicity of the conjugate. The following are recipes adapted from Bauminger and Wilchek *(14)*.

1. For the one-step reaction, dissolve 5 mg peptide in 1 mL solvent.
2. While stirring gently with a magnetic stir bar, add 25 mg EDAC.
3. Adjust pH to 4.0–5.0 by addition of 0.1*N* HCl.
4. Continue stirring, allowing to react 10 min at room temperature.
5. Dissolve 5 mg carrier in 0.5 mL solvent and add to the peptide solution.
6. Continue stirring, and allow to react for two more hours at room temperature.
7. For the two-step protocol, dissolve 5 mg carrier protein in 1 mL solvent.
8. Adjust pH to 4.0–5.0 by addition of 0.1*N* HCl.
9. Add 7 mg EDAC, stirring gently for 30 min on ice with a magnetic stir bar.
10. Add a drop of glacial acetic acid to quench the reaction.
11. Dialyze vs coupling solvent overnight.
12. Dissolve 5 mg peptide in 1 mL solvent.
13. Combine peptide solution with activated carrier solution. Allow to react 60 min, stirring on ice.
14. For either protocol, if you want to assume that the coupling went well, then the conjugate is ready to inject the animals. On the other hand, if you want to determine the level of substitution of the conjugate, then you must separate it from the free peptide. This may be done by gel filtration, as described above, or via final dialysis. The purified conjugate may then be subjected to analysis as described in Section 2.

3.4. M-*Maleimidobenzoyl-N-Hydroxysuccinimide Ester (MBS)*

MBS is a heterobifunctional agent that links a thiol group on the peptide with an amino group on the carrier protein at neutral pH (Fig. 6). Side reactions are generally not observed. The chemistry and immunology for this reagent are well characterized, providing a stable bond. Fur-

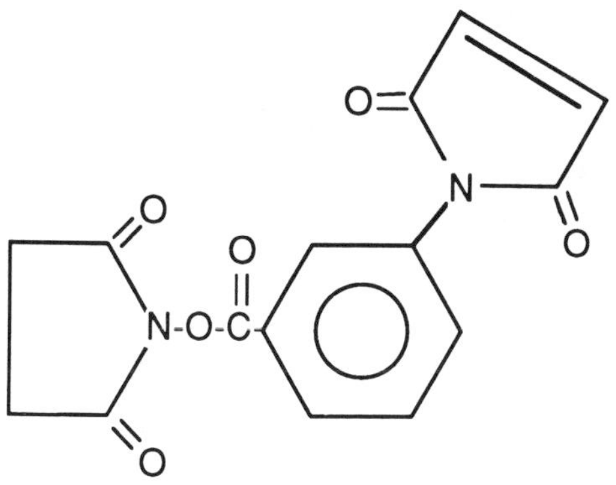

Fig. 6. Chemical structure of *m*-maleimidobenzoyl-*N*-hydroxysuccinimide ester.

thermore, because the reaction is carried out in two steps, intramolecular crosslinking is minimized. Although there exist a number of possible side reactions involving other amino acids, such as histidine, the desired reaction is quite fast, and problems owing to side reactivity are rare. Note that the linker, a benzyl moiety, is not suitable for use in humans because of probable toxicity. The following recipe is an adaptation of Liu et al. *(15)* and Lerner et al. *(16)*.

1. Dissolve 5 mg KLH in 0.5 mL solvent 1. Add 0.1 mL of the MBS solution. Stir gently with a magnetic stir bar, and allow to react for 30 min at room temperature.
2. Separate the MBS-activated KLH from free MBS via gel filtration. The protein should elute in the void volume, possibly appearing turbid. Otherwise, separation may be confirmed by measuring A_{280} of the fractions. The first peak will be the activated KLH, whereas the second peak will comprise unreacted MBS. Pool the MBS-KLH fractions.
3. Dissolve 5 mg peptide in 1 mL solvent 2. Combine the peptide solution with the MBS-KLH solution. Allow to react with gentle stirring for 3 h at room temperature.

If you want to assume that the coupling went well, then the conjugate is ready to inject the animals now. On the other hand, if you want to determine the level of substitution of the conjugate, then you must separate it from the free peptide. This may be done by gel filtration or via dialysis. The purified conjugate may then be subjected to analysis as described in Section 2.

Sulfo-MBS, the *N*-hydroxysulfosuccinimide ester of *m*-maleimidobenzoic acid, is a water-soluble analog of MBS, now commercially available. It is purported to have the same reactivity as MBS, but without the requirement for predissolving in organic solvent. Otherwise, use the same reagent proportions for conjugation reactions with sulfo-MBS.

3.5. Thiol Alkylation

1. Dissolve carrier in 1 mL buffer.
2. Add 0.2 mL tributylphosphine solution. Allow to react at room temperature for 30 min.
3. Add peptide, stirring for another 30 min.

The conjugate is now ready to inject into animals. If you want to determine the level of substitution of the conjugation, then you must separate the conjugate from the free peptide via dialysis, e.g., vs PBS. If you prefer a soluble immunogen, the separation may be performed via gel filtration. The purified conjugate may then be subjected to analysis as described in Section 2.

3.6. Bisdiazobenzidine

1. To 100 mg benzidine, add 1 mL 4*N* HCl, stirring vigorously to dissolve. Add 19 mL water, and chill on ice.
2. While stirring, add 70 mg $NaNO_2$. Allow to react for 1 h on ice in the dark. Aliquot and freeze. BDB, prepared thus from benzidine, may be stored at –70°C for over a year with retention of full coupling reactivity (Fig. 7).
3. Dissolve 5 mg peptide and 5 mg carrier in 5 mL solvent, and chill on ice. Add 0.1 mL thawed BDB solution, stirring gently with a magnetic stir bar. Allow to react for 30 min to 2 h on ice, maintaining the pH near 9 by addition of 0.1*N* NaOH.
4. The conjugate will turn yellow to orange-brown as the reaction proceeds. With longer reaction times, the formation of some precipitate is likely. However, this does not affect immunogenicity (*see* Chapter 11).

The conjugate is ready to inject into animals now. On the other hand, if you want to determine the efficiency of the conjugation reaction to confirm near 20:1 peptide:carrier, then you should dialyze it vs PBS. If you prefer a soluble immunogen, you may perform the separation via gel

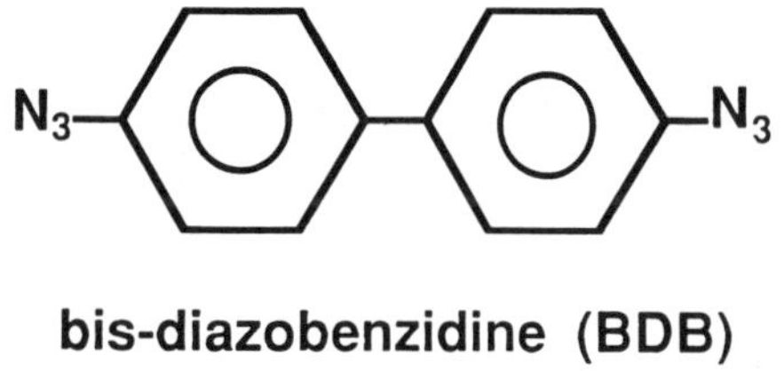

Fig. 7. Chemical structure of bisdiazobenzidine.

filtration. The purified conjugate may then be subjected to quantitative analysis as described in Section 2.

3.7. N-Hydroxysuccinimidyl 3-(2-Pyridyldithio)-Propionate (SPDP)

3.7.1. Method 1

1. Dissolve the carrier in 1 mL buffer. Add 0.2 mL SPDP. Allow to react for 1 h at room temperature. Separate the activated carrier from the leftover SPDP by gel filtration.
2. Dissolve the peptide in 2 mL buffer. Add the activated carrier solution, and allow reaction at room temperature for 30 min, or until the absorbance at 343 nm stops increasing.

Method 2 couples two molecules through their amino groups. Both molecules are first derivitized with SPDP to produce active thiols, as above. Then the pyridine-2-thione is released from one of the molecules via reduction leaving a free thiol. Finally, the free thiol is reacted with the active thiol from the other molecule. The procedure outlined below reduces the carrier, thus utilizing the activated thiol on the peptide molecule.

3.7.2. Method 2

1. Dissolve the peptide in 1 mL buffer. Add 0.2 mL SPDP. Allow to react for 1 h at room temperature. Separate the activated peptide from the leftover SPDP by gel filtration.
2. Dissolve the carrier in 1 mL buffer. Add 0.2 mL SPDP. Allow to react for 1 h at room temperature. Acidify by addition of acetic acid to make the pH 4.0–5.0. Add DTT to make the final solution 25 m*M*. Allow 1 h at room temperature for complete reduction. Separate the thiolated carrier from the other reagents by gel filtration.
3. Combine the two solutions, and allow reaction at room temperature for 30 min or until the absorbance at 343 nm stops increasing.

With either coupling method, the conjugate is ready to inject into animals now. If you want to quantitate the level of substitution of the conjugation, then you must separate the conjugate from the free peptide via dialysis, e.g., vs PBS. If you prefer a soluble immunogen, the separation may be performed via gel filtration. The purified conjugate may then be subjected to analysis as described in Section 2.

3.8. Immunization

1. Collect at least 5 mL normal (preimmune) serum from each animal before it is immunized. Store it at –70°C. The preimmune serum is an important reagent for subsequent experiments. A portion of it will be used as a control in every experiment using the immune serum.
2. Put 2 mL FCA into one syringe and 2 mL peptide solution into the other one. Couple the two syringes together tightly. Force the mixture back and forth through the small orifices between the syringes. This generally causes complete emulsion within a few minutes.
3. Inject each animal with 2 mL emulsion, as follows: Inject 0.5 mL into each of two sc sites on the animal's lower back or rump area. Inject another 0.5 mL emulsion into each of two im sites in the middle of the posterior of the animal's thigh.
4. One month later, make a second group of injections. The immunogen should be prepared as before, but use Freund's incomplete adjuvant for this and any subsequent group of injections. If abscesses are noted (especially at the sc sites), make the injection at least 1 cm away from these wound sites.
5. The following month, and every month subsequently, make test bleeds of about 5 mL from each animal. Draw blood from the marginal ear vein. This causes a minimum of discomfort for the animal. For convenience, make another round of injections (in incomplete adjuvant) on the same day as the test bleeds. Or for maximum titer, bleed 1 wk after injection.
6. By the third or fourth injection, the animals should begin to produce high-titer antibodies. This should be confirmed via ELISA results. Antibody-producing animals should be bled 20–25 mL every 2 wk or 40–50 mL/mo, depending on the mass of the animal. When antibody titer drops off, the animal should be retired.

4. The Three Most Popular Coupling Reagents

Literally dozens of different peptide- and protein-coupling agents are commercially available, but most of these bear one or two of only a few active groups. These common moieties include: *N*-hydroxysuccinimide, maleimide, 2,2'-pyridyldisulfide, and haloacetate (*see* Fig. 8).

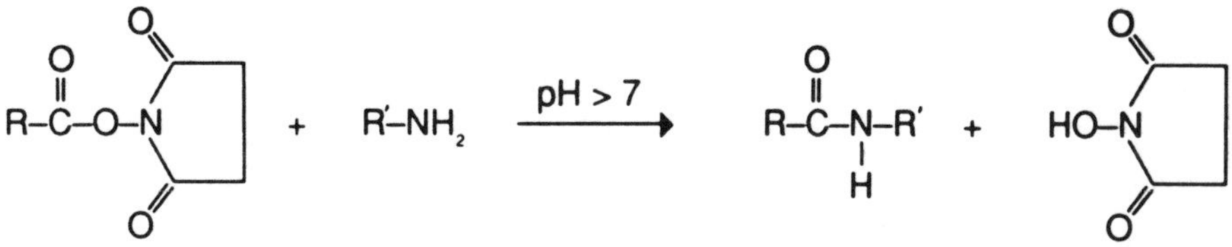

NHS Ester Reaction Scheme

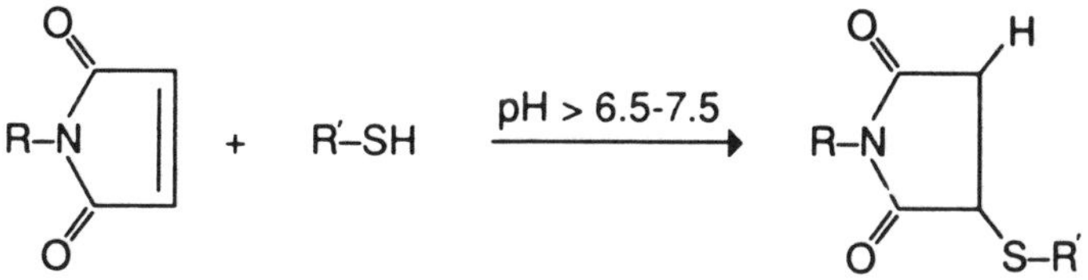

Maleimide Reaction Scheme

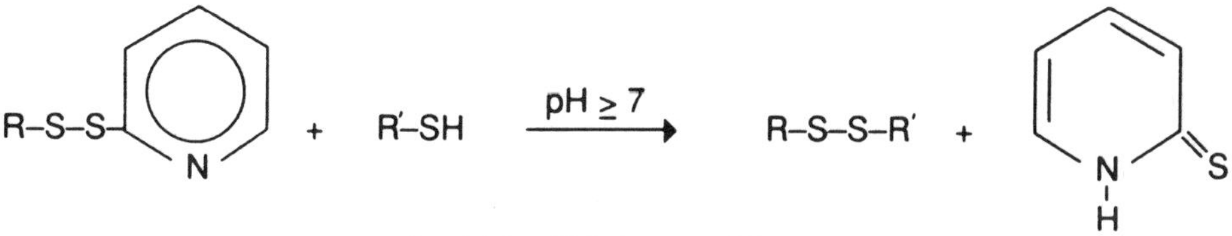

Pyridyl Disulfide Reaction Scheme

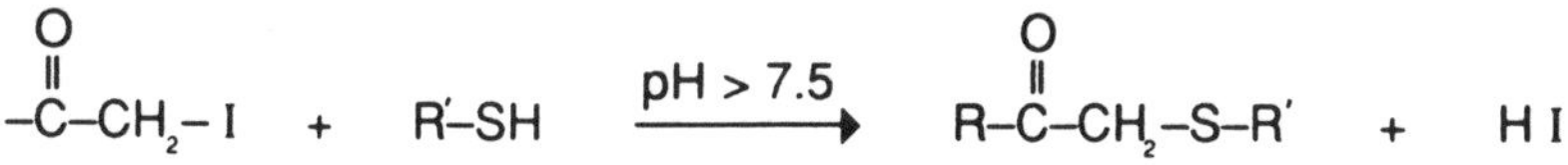

Active Halogen Reaction Scheme

Fig. 8. Structures and reactions of common conjugating reagents: *N*-hydroxysuccinimide, maleimide, 2-pyridyldisulfide, and haloacetate.

Photoreactive agents, which may also be used for peptide conjugation, usually bear arylazides. Investigators interested in studies of varying linker length, varying hydrophobicity, or varying chemistry will find most of these reagents available from Pierce (Rockford, IL). On the other hand, investigators wishing simply to conjugate their peptides and generate antisera may follow one or two of the procedures that follow.

I have chosen the three methods of glutaraldehyde, carbodiimide (EDAC), and *m*-maleimidobenzoyl-*N*-hydroxysuccinimide ester (MBS) because they are popular, inexpensive, and effective. Occasionally, for immunological reasons that are not well understood, any given coupling scheme may be successful at producing a conjugate, but unsuccessful at generating an antibody *(10)*. Also occasionally, the conjugate may produce antibodies reactive against the peptide, but not the parent protein from which the peptide sequence was taken. In these cases, alternate conjugation schemes may be effective. If the sequence is short or not predicted to be highly immunogenic (*see* Chapter 11), it is a good idea to spend a little extra energy and possibly save some time by performing the conjugation with two or even all three of these techniques, and then immunizing with all of them.

5. Three Other Protein-Coupling Chemistries

5.1. Thiol Alkylation

This method is used to couple an amino group to a free thiol. However, unlike MBS (*see* Section 3.4.), it introduces no toxic linker group and is extremely suitable for quantitation of conjugation efficiency *(17)*. To achieve this conjugation, a bromoacetyl moiety is introduced into the peptide at the N-terminus via acetylation before cleavage *(18)*. This may be performed using a carbodiimide with bromoacetic acid. Alternatively, the bromoacetyl group may be introduced in other places in the peptide via the side chain of *N*-ε-bromoacetyl-β-alanyl-lysine *(19)* in a *t*Boc synthesis (*see* Chapter 6, *PSP*). For the conjugation reaction, the carrier protein must provide the thiols in the form of cysteines (after suitable reduction). HBr is eliminated from the reaction intermediate, with formation of a stable thioether.

On hydrolysis in acid for amino acid analysis, the conjugate releases one molecule of carboxymethylcysteine for each peptide molecule bound via the thioether bond *(17)*. Quantitation of the carboxymethylcysteine is therefore quantitation of conjugation (*see* Section 2.).

5.2. Bisdiazobenzidine (BDB)

The second method is similar to glutaraldehyde conjugation in that it is a simple one-vessel reaction *(20)*. It couples principally via tyrosine aryl hydroxyl, but also via cysteine sulfhydryl, histidine imidazole, and lysine ε amino groups, and to a lesser extent, tryptophan indole and arginine guanidinium groups. Because it is fairly nonselective, it is a good choice for coupling of peptides that have performed poorly in previous conjugation reactions.

5.3. N-Hydroxysuccinimidyl 3-(2-Pyridyldithio)-Propionate (SPDP)

SPDP couples an amino group with a thiol *(21)*. However, it may also be used to conjugate two amino groups if both molecules are first derivitized with SPDP. Side reactions are generally not reported, and SPDP is reported to create no neo-antigens *(7)*. Other major advantages include the relative nontoxicity of the linker group and coupling under mild conditions. However, the conjugate link is through a disulfide bond. Hence, the conjugation is reversible under reducing conditions.

SPDP reacts with amino groups on either the peptide or the carrier protein. The active thiol thus produced can react with any free thiol, with the concomitant release of pyridine-2-thione. Otherwise, the pyridine-2-thione may be released through reduction. If desired, reduction may be effected with 25 m*M* DTT at pH 4.5 for 30 min to reduce the SPDP without reduction of protein disulfide bonds *(22)*. Liberation of the pyridine-2-thione chromophore may be monitored by measuring the change in absorbance of the solution at 343 nm ($\varepsilon = 8.1 \times 10^3$). Method 1 couples an amino group to a thiol. The amino on the first molecule is initially reacted with the SPDP. Then the thiol from the second molecule is allowed to react with the new activated thiol. Coupling occurs through formation of a disulfide bond. Note that the procedure outlined below in method 1 is for coupling such that the amino groups are provided by the carrier and the thiols are provided by the peptide. The converse conjugation may also be performed.

6. Miscellaneous Important Peptide-Coupling Phenomena

6.1. Solubility

Many synthetic peptides are poorly soluble in mild aqueous solutions. This characteristic can often be anticipated for longer peptides containing few charged and hydrophilic amino acids. This often does not cause problems in synthesis and characterization of the compounds, because polar organic solvents are effective in maintaining solubility of the peptides and other reactants. However, carrier proteins are frequently incompatible with organic solvents. For example, depending on the individual characteristics of the protein preparation used, KLH and BSA may not be soluble in >10% DMF or >70% DMSO. To maintain the solubility of the carrier proteins, therefore, most conjugations are performed at near-neutral pH in mild aqueous buffers.

For many difficult peptides, a little bit of DMF or DMSO may be enough to coax them into solution without interfering with the intended reaction. For example, conjugation with glutaraldehyde is unaffected by the presence of DMF, DMSO, or alcohols. On the other hand, DMSO can rapidly oxidize methionine, so it should be avoided with peptides bearing this amino acid. Another potential variable to improve the aqueous solubility of hydrophobic peptides is pH. This is especially true of peptides bearing no basic amino acids (arginine or lysine). Such weakly acidic hydrophobic peptides will often go into aqueous solution if the pH is raised. A pH as high as 8.0 is fine for most conjugation chemistries. Consult the theoretical titration curve of the peptide to determine the charge state of the molecule at various pHs.

Chaotropes, such as urea (at neutral pH or below) or guanidine, may be useful if they do not interfere with the conjugation chemistry. However, use of detergents to solubilize peptides is not recommended. They are often impossible to remove after the completion of conjugation, and they may be undesirable in the final (immunogen or assay antigen) preparation.

Many investigators have found that some immunogenic sequences are impossible to dissolve in aqueous buffers at any reasonable pH or temperature. These peptides have been resynthesized with the incorporation of two or three lysine residues at either end of the molecule. The increased number of positive charges on the molecule in neutral aqueous solution, in turn, increases the solubility of the sequence. Also, the new ε amino groups comprise conjugation sites that are far from the antigenic center of the molecule, and do not seem to perturb antigenicity.

6.2. Biotinylation

In order to facilitate binding of the peptide, for example, as a probe, a biotin molecule may be introduced. Avidin is a biotin-binding protein originally isolated from *Streptococcus*. Binding of biotin by avidin is so tight as to be essentially irreversible.

In order to allow coupling at one end of the peptide, biotinylation may be performed at N-terminal. This is easily accomplished by reaction of the side-chain-protected peptidyl resin with excess 2,4-dinitrophenyl-biotin before cleavage and purification. Alternatively, after cleavage, the biotin can be introduced at a thiol site by reaction with excess iodoacetylbiotin. This reaction is simply performed by reaction of the reduced peptide with a two- or threefold excess of iodoacetylbiotin in phosphate buffer, pH 8. Finally, biotin may also be introduced via standard solid-phase chemistry, with activation by HBTU, although solubility in DNF and NMP is limited.

6.3. Fatty Acylation

Before cleavage of the synthetic peptide from the resin, the N-terminus may be acylated by convenient solid-phase chemistry (*see* Chapter 8, *PSP*). This may be desired for incorporation of the peptide into liposomes for vaccine use. (*See* Section 8.) However, it results in greatly reduced aqueous solubility of the peptide. The reaction may be performed with the free fatty acid via standard carbodiimide coupling with or without HOBt. I routinely perform this reaction using powdered lauric or myristic acid on an automated peptide synthesizer.

6.4. Multiple-Antigen Peptides

Multiple-antigen peptides (MAPs) are complex branched molecules that are fairly easily synthesized *(23)*. The multiple branches are capable of presenting multiple antigens simultaneously, for example, both T- and B-Cell antigen *(24)*. They are simply prepared in a *t*Boc synthesis via incorporation of two or three cycles of bis-*t*Boc-Lysine. bis-Fmoc-Lysine may be used similarly in Fmoc syntheses. The MAP structure typically comprises four or eight identical branches of peptide attached to a branched polylysine core via a glycinylglycine spacer. The sterically buried C-terminus of the molecule remains accessible for chemical reactions in shorter (<12 amino acids, depending on size) MAPs.

6.5. Alternatives to Carrier Protein Conjugation

These include conjugation of B-cell epitope-bearing peptides to T-cell epitope-bearing peptides. Thus, immunogenesis may be achieved without the complication of a carrier protein, since such synthetic macromolecules will present both types of epitopes and exhibit a large immunostimulatory size. One method to generate such structures is colinear synthesis of B- and T-cell epitopes *(25)*. This may be done without intervening sequences or with spacer moieties, such as glycine residues. Otherwise, the B-cell and T-cell epitopes may be conjugated to each other, e.g., via MBS *(26)*, or even copolymerized into macromolecular size. For example, copolymerization has been effected via glutaraldehyde *(27)* or disulfide formation between cysteines on both ends of the molecules *(28)*. All of these methods have been successfully, but infrequently used.

7. Immunization Protocols

7.1. General

Immunizations may be performed according to a large number of protocols. These vary according to the choice of experimental animals, adjuvants, immunization schedule, and bleeding schedule. Each of these issues is discussed briefly below, although the choices are generally not critical, and variations, personal preferences, and opinions abound. However, a general scheme for effective immunization is presented. For a variety of alternate immunization schedules, refer to Weir *(20).*

7.2. Choice of Animals

A great variety of experimental animals have been exploited in the production of antibodies. Many studies require only the production of immune sera, containing polyclonal antibodies. In these cases, almost any animal on a valid protocol may be used. Females are most popular because their generally more docile personalities make for easier handling. Popular species include: mouse, rabbit, goat, and horse. Of course, larger animals require more expensive overhead. They also produce a larger volume of antiserum, ranging from a few microliters in a test bleed of a mouse to several liters in the exsanguination of an immune horse. In addition to these common experimental animals, special animals from practically any of various human disease models may be used.

Some studies require the production of monoclonal antibodies (MAb). MAb production requires the fusion of immune B-cells with immortal tissue-culture cells, and the subsequent culture of the hybridomas. For these protocols, specific strains of mice or hamsters are usually employed for immunization, although procedures are available for fusion with other animal immune cells.

For any protocol, it is wise to use at least two animals per immunogen. This will ensure against the failure of one of the animals owing to mishap or misfortune.

7.3. Injection Sites

Injections may be made in any of a number of sites: subcutaneously (sc), intramuscularly (im), intraperitoneally (ip), or intravenously (iv). Subcutaneous or im sites are usually preferred for early immunizations, but boosters are frequently given ip or iv.

Subcutaneous injections are probably most effective when made into the footpad for small experimental animals, such as rodents and rabbits. However, this is quite painful for the animals and should be avoided, whenever possible, for that reason. Subcutaneous injections at multiple sites in the back or rump region are also very effective for smaller animals, whereas larger ones do well with injections around the neck. Intramuscular injections are generally administered into the large thigh musculature of experimental animals.

Intraperitoneal injections must be made carefully into the abdominal cavity without piercing any vital organs. Nonetheless, ip boosters are popular for use with rodents, and sometimes also rabbits. Although they are difficult to administer in smaller animals, iv boosters may be given into the tail veins of the rodents, the marginal ear veins of rabbits, or the jugular veins of larger animals. Note that iv immunogens should be clarified and free of adjuvant.

7.4. Adjuvants

Adjuvants serve to form a physical depot for the immunogen as well as some nonspecific immune stimulation. The most popular adjuvant used is FCA. This is a preparation of paraffin oil containing a suspension of killed mycobacteria. Freund's incomplete adjuvant is simply the oil without the mycobacteria. FCA is extremely effective for priming, but booster injections are usually given in incomplete adjuvant in order to minimize formation of cysts and chronic inflammation. Even then significant irritation is common. For this reason, Freund's adjuvants are not suitable for use in humans. Emulsions in either FCA or incomplete adjuvant are not physically stable and should be prepared immediately before use. If necessary, they may be kept at 4°C overnight, but sterility is difficult to maintain, and contaminated preparations will result in health problems for the experimental animals. Freund's adjuvants are also unsuitable for iv immunizations.

One of the great strengths of liposomes is that they have been approved for use in human vaccines. These are simply lipid vesicles containing immunogens, such as peptides. Used as a vehicle for immunization, they are highly effective for presentation of peptides as antigen *(29)*. Indeed, they are occasionally able to overcome genetic restriction for B-cell epitopes *(30)*. They also reduce toxicity of certain immunogens *(29)*. In addition, they seem to be especially effective in inducing antibodies from peptides that are able to bind native proteins. They are prepared by evaporating a chloroform solution of the desired mixture of lipids, followed by resuspension in a PBS solution of peptide with vortexing. Another alternative immunogen vehicle is tripalmitoyl-glyceryl-cysteine-serine-serine, which may be synthesized at the N-terminus of the peptide *(31,32)*. This sequence is the immunologically active N-terminus of *Escherichia coli* lipoprotein. Similarly, dipalmityl-lysine conjugation was reported to be at least as effective for immunogenesis as KLH conjugation *(33)*. This moiety is also easy to introduce at the N-terminus of the peptide antigen via solid-phase synthesis. In studying different adjuvants, it is important to remember that the ability of any given adjuvant to produce antibodies of a particular specificity cannot be determined from its ability to produce antibodies of high titer *(30)*.

Some scientists are experimenting with peptidyl resins as immunogens. The Fmoc K-type polymer has been used with some success, suspended in PBS. For such uses, the peptide is deprotected and deblocked without cleaving it from the resin.

7.5. Timing on Injections and Bleeds

Some investigators insist that it is best to bleed animals 2 wk after each immunization in order to allow an immune response to the injected antigen. This belief is based on very early studies of the immune response to soluble antigen, and it probably applies to antigen preparations presented in solution. However, with the oily adjuvants, such as FCA, the antigen is injected in an emulsified depot that causes slow release. This allows successful bleeding at any convenient date, as long as it is at least 1 wk after the second injection (first boost).

Blood may be drawn from the tail veins of rodents, or the jugular veins of larger animals, using a heparinized needle. For rabbits, the marginal ear veins are usually used, either with a needle or by means of a small incision.

Test the sera for antibodies using an ELISA. (*See* Section 8.) Animals producing antibodies generally need not be further immunized, whereas those unproductive of specific antibodies may be further boosted. After 6 mo, unproductive animals may be retired; they will probably never produce antibodies. It is unfortunate that many animals may fall into this group. If fewer than half the animals make antibodies, the immunogen should probably be reformulated.

Once an animal is producing high levels of antibodies (titer > 10,000), it should be bled regularly. Twenty milliliters of blood may be drawn from a rabbit every 2 wk without consequences to the animal's health. After about 10 mo, even productive animals will show a significant decline in antibody production, and they may be "bled out" (exsanguinated) or boosted.

7.6. Treatment of Serum

The blood should be allowed to clot normally, for 10–20 min at room temperature. To maximize recovery of serum, the clot is then "wrung" by scraping it from the sides of the collection vessel to allow contraction of the thrombin fiber complex. After another 10–20 min, the clot is pelleted on a centrifuge and the serum decanted. Serum may be stored frozen for long periods of time: at least 6 mo at –20°C and several years at –70°C.

The easiest way to isolate the IgG from the serum, if this is desired, is via affinity chromatography. First precipitate the antibody protein by addition of ammonium sulfate to 50% saturation. Pellet the precipitate, dialyze it in PBS, and then load it onto an affinity column: either protein A or protein G. Only the antibody proteins will bind to the matrix. They are subsequently eluted with a low-pH buffer, such as glycine-HCl, pH 2.8. Maximal recovery of antibody binding activity requires immediate neutralization of the eluant, e.g., by the addition of Tris base. Antibody protein may alternatively be isolated via anion-exchange chromatography or affinity chromatography on immobilized antigen.

8. Basic ELISA

When it is necessary to determine the presence of specific antibodies in a solution, ELISA is one of the simplest methods to use. For example, it can easily and reliably detect small quantities (on the order of 1 µg/mL) of antibodies in hybridoma culture supernatant solution, ascites fluid, or immune serum. In hybridoma screening, often only the presence of the antibody is to be determined. On the other hand, in serum or ascites, a rough quantitation of antibody concentration and avidity can be made by determining the titer of the antibody solution. Titer is usually defined as the reciprocal of the dilution of antibody preparation that gives half the maximal response in the assay.

The ELISA is based on a series of molecular-binding reactions, taking place on a clear, flat-bottom, 96-well microtiter plate. First, the antigen is coated onto the plate. After blocking remaining sites for nonspecific binding, the antibody is allowed to bind to the immobilized antigen. This antibody protein is then probed by binding of an enzyme-conjugated second antibody. Finally, the second antibody is detected by means of a chromogenic substrate for the enzyme.

8.1. Notes on Procedure

1. Antigen loading: Many peptides will bind directly to microtiter plates without modification. Others, particularly small molecules <20 residues, require conjugation to an irrelevant carrier protein in order for them to stick. If a conjugate is used as an ELISA "capture" antigen, it should be made with a carrier protein other than the one chosen for the immunogen, to prevent false-positive reactions resulting from crossreactivity. The antigen solution should be 1–10 µg/mL (not counting carrier protein, if a conjugate is used) in any mild aqueous buffer, such as PBS, borate, or carbonate. Use 50 µL/well. Note that it may be necessary to test the antibody for binding to the parent protein as well as the peptide immunogen. The procedure for this is exactly the same, substituting the native protein antigen for the peptide in the first step, that of coating the plate with antigen. For this and subsequent incubations, more elevated temperatures (e.g., room temperature or 37°C) may be used successfully if microbial growth and proteolysis are suitably inhibited.
2. Washes: After each incubation, washes are performed to remove excess reagent. For the wash step, pipet 200 µL wash solution into the microtiter plate well, and then either flick it out into the laboratory sink or aspirate it, being careful not to disturb the bottom of the wells. Commercial apparatus ("microtiter plate washer" or "ELISA plate washer") is available for simplification and acceleration of the wash steps. Two or three wash steps are usually adequate.

   Note that in many buffers and reagent solutions used for ELISA, 0.1% Tween 20 is often added to the PBS. Tween 20 is a mild nonionic detergent that reduces nonspecific binding. Handling of Tween 20 solutions should be performed carefully so as to minimize aeration, since foaming will affect reproducibility.
3. Preparation of the blocking buffer: After the peptide antigen is attached to the plastic microtiter plate wells, remaining sites for nonspecific binding are blocked by treatment with a solution of irrelevant, inexpensive protein. Bovine serum albumin (BSA) is commonly used, unless it was utilized as a carrier protein in the immunogen conjugate. In that case, OVA or boiled casein may be substituted with equivalent results. The solution is made 1–2% in PBST. The blocker solution may be prepared in 1-L batches, filter-sterilized, and stored at 4°C for up to 2 wk.

4. First antibody: In general, make dilutions of the test antibody in PBST. However, if you are testing a hybridoma culture supernatant solution, you should probably use it undiluted. If you are testing a purified antibody, use at least 5 µg/mL. If you are testing a serum or ascites fluid (typically 100–1000 µg/mL specific antibody, and around 10 mg/mL irrelevant antibody), do serial dilutions to determine titer.

   At the very least, test the following 10-fold dilutions: $1/10^2$, $1/10^3$, $1/10^4$, and $1/10^5$. You may prefer to make half-log dilutions: $1/10^2$, $1/10^{2.5}$, $1/10^3$, $1/10^{3.5}$, $1/10^4$, and so forth, for greater resolution of the titer. On subsequent tests, you may wish to make two- or threefold serial dilutions or extend the range of dilutions tested. In any case, make enough of each antibody dilution to run the assay in duplicate or triplicate.
5. Preparation of the second antibody: The second antibody is conjugated to an enzyme. You may use second antibody preparations commercially available from many sources. These are usually conjugates of alkaline phosphatase (AP) or horseradish peroxidase (HRP). For the commercial reagents, the working concentration of the second antibody is specified by the manufacturer.

   In some cases, it may be necessary to prepare your own enzyme-conjugated second antibody. This is fairly easy to do, but it takes a long time. For example, to make an AP-conjugated goat antibody to recognize *Aotus* monkey antibodies in ELISAs, one group of investigators used the following protocol *(34)*. They first isolated several milligrams of nonimmune antibodies from the monkey serum by protein A affinity. They used most of this protein as an immunogen to raise antibodies in a goat. They next isolated several milligrams of goat immune antibody by protein A affinity. Then they coupled a few milligrams of the *Aotus* antibody immunogen to activated Sepharose. They used the immobilized *Aotus* antibody to affinity purify the goat antibody vs *Aotus* antibody. Then they conjugated the purified antibody to commercial alkaline phosphatase (AP) via glutaraldehyde. After diluting the conjugate in PBS with 0.2% NaN_3, they tested it at various concentrations to determine the appropriate working concentration for the reagent. The conjugate solution was stored frozen, avoiding refreezing. (This is a brief description not intended to be a detailed guide for the second antibody conjugation procedure. For such detailed information, refer to Weir *[20]*). Preparation of the substrate solution: For the nitrophenyl phosphate substrate, use 0.1*M* diethanolamine, pH 9.8, 0.01% $MgCl_2$, 0.02% NaN_3. The buffer may be prepared ahead and stored at 4°C for several months. However, it should be allowed to warm to room temperature before use. Immediately before use, dissolve the substrate *p*-nitrophenyl phosphate to a final concentration of 1 mg/mL. This substrate produces a yellow color. Absorbance should be read between 405 and 410 nm. Commercially available preprepared substrate solutions may be used with excellent results.

6. Development: In order to allow all the test wells of the ELISA to incubate for the same time, it is a good idea to use one of the special multiple pipets for addition of the substrate solution and stop solution. These tools pipet 8 or 12 wells at a time. In addition to synchronizing additions of reagent, these tools also cut down on time and work involved in the performance of the ELISA.

   To aid in visualization of color development, place the plates on a piece of white paper. Allow development to proceed until the positive reactions are fairly well colored. This usually takes 30–60 min. During development, avoid thermal gradients, which may be caused, for example, by drafts or sunlight. You may even cover the plates to ensure their isolation from environmental effects.

7. Termination of development: Do not allow development to proceed until the negative controls give a strong color reaction. Once the ELISA has developed some color (usually 10 min–2 h), add the stop solution. For the nitrophenyl phosphate substrate, use 0.01*M* EDTA as a stop solution. Chelation of the magnesium cofactor results in complete inhibition of the enzyme. For the HRP substrate, use 1% sodium dodecyl sulfate. In this case, denaturation of the enzyme results in its inhibition.

   Although the stop solution terminates enzymatic cleavage, the substrate is somewhat thermolabile, and the chromophore is somewhat photolabile. Therefore, read the plates on an automated microtiter plate reader within an hour. Also, when handling them, try to prevent your fingers from smudging the bottom of the ELISA plates before they are read.

   Times, volumes, and concentrations of reagents may require some adjustment in order to give good reproducible results. However, the general scheme outlined above is likely to give a reliable yes or no result on the first attempt.

8.2. Procedure

1. Coating with antigen: 100 µL of antigen solution are placed into each well of the microtiter plate and allowed to incubate overnight at 4°C. Put an irrelevant peptide into some of the wells to act as a negative control. Alternatively, put blocking solution (qv) in the negative control wells.
2. Wash: Wash one or two times with 200 µL PBST/well.
3. Blocking: 200 µL/well of a 2% solution of blocking protein are allowed to bind for 1 h at 4°C.
4. Wash: Perform three or four washes with 200 µL PBST/well, as before.

5. Test antibody: Pipet 50 µL antibody solution into each of the microtiter plate wells, with the duplicates side by side. Tap the side of the plate gently to mix and spread the solution, and allow to incubate for 1 h at 4°C.
6. Wash: Wash three times as before, with 200 µL PBST.
7. Second (enzyme-labeled) antibody: Pipet 50 µL of the second antibody solution into each well. Tap the side of the plate gently to mix and spread the solution, and allow to incubate for 1 h at 4°C.
8. Wash: Again, perform three washes with 200 µL PBST.
9. Substrate solution: Pipet 100 µL substrate solution into each well. Tap the side of the plate gently to mix and spread the solution.
10. Development: Allow to develop for 30–60 min at room temperature.
11. Terminate development: Terminate color development by addition of 100 µL stop solution to each well. The stop solution is added to the substrate solution already in the well. It is not usually necessary to tap the plate to mix these solutions.
12. Read the plates on an automated ELISA plate reader.

References

1. Tanaka, T., Slamon, D. J., and Line, M. J. (1985) Efficient generation of antibodies to oncogene proteins by using synthetic peptide antigens. *Proc. Natl. Acad. Sci. USA* **82,** 3400–3404.
2. Tam, J. P. and Zavala, F. (1989) Multiple antigen peptides. *J. Immunol. Meth.* **124,** 53–61.
3. VanRegenmortel, M. H. V., Briand, J. P., Muller, S., and Plaue, S. (1988) *Laboratory Techniques in Biochemistry and Molecular Biology,* vol 19 (Burdon, R. H. and Van Knippenberg, P. H., eds.), Elsevier, Amsterdam.
4. Dryberg, T. and Oldstone, M. B. A. (1986) Peptides as antigens. *J. Exp. Med.* **164,** 1344–1349.
5. Ponsati, B., Giraldt, E., and Andreu, D. (1989) A synthetic strategy for simultaneous purification-conjugation of antigenic peptides. *Analytical Biochem.* **181,** 389–395.
6. Satterthwait, A. C., Arrhenius, T., Hagopian, R. A., Zavala, F., Nussenzweig, V., and Lerner, R. A. (1988) Conformational restriction of peptidyl immunogens with covalent replacements for the hydrogen bond. *Vaccine* **6,** 99–103.
7. Peeters, J. M., Hazendonk, T. G., Beuvery, E. C., and Tesser, G. I. (1989) Comparison of four bifunctional reagents for coupling peptides to proteins and the effect of the three moieties on the immunogenicity of the conjugates. *J. Immunol. Methods,* **120,** 133–143.
8. Ruegg, U. T. and Rudinger, J. (1977) Reductive cleavage of cystine disulfides with tributylphosphine. *Meth. Enzymol,* **47,** 111–116.
9. Atassi, M. Z. and Habeeb, A. F. S. A. (1972) Reactions of proteins with citraconic anhydride. *Meth. Enzymol.* **25,** 546.
10. Schaaper, W. M. M., Lankohof, H., Pujik, W. C., and Meleon, R. H. (1989) Manipulation of antipeptide immune response by varying the coupling of the peptide with the carrier protein. *Mol. Immunol.* **26,** 81–86.

11. Reichlin, M. (1980) Use of glutaraldehyde as a coupling agent for proteins and peptides. *Meth. Enzymol.* **70,** 159–165.
12. Kirkeby, S., Jakobsen, P., and Moe, D. (1987) Glutaraldehyde—pure and impure. A spectroscopic investigation of two commercial glutaraldehyde solutions and their reaction products with amino acids. *Analyt. Lett.* **20(2),** 303–315.
13. Baron, M. H. and Baltimore, D. (1982) Antibodies against the chemically synthesized genome-linked protein of poliovirus react with native virus-specific proteins. *Cell* **28,** 395–404.
14. Bauminger, S. and Wilchek, M. (1980) The use of carbodiimides in the preparation of immunizing conjugates. *Meth. Enzymol.* **70,** 151–159.
15. Liu, F.-T., Zinnecker, M., Hamaoka, T., and Katz, D. H. (1979) New procedures for preparation and isolation of conjugates of proteins and a synthetic copolymer of D-amino acids and immunochemical characterization of such conjugates. *Biochem.* **18(4),** 690–697.
16. Lerner, R. A., Green, N., Alexander, H., Liu, F.-T., Sutcliffe, J. G., and Shinnick, T. M. (1981) Chemically synthesized peptides predicted from the nucleotide sequence of the Hepatitis B virus genome elicit antibodies reactive with the native envelope protein of Dane particles. *Proc. Natl. Acad. Sci. USA* **78,** 3403–3407.
17. Kolodny, N. and Robey, F. A. (1990) Conjugation of synthetic peptides to proteins: quantitation from *S*-carboxymethylcysteine released upon acid hydrolysis. *Anal. Biochem.* **187,** 136–140.
18. Lindler, W. and Robey, F. A. (1987) Automated synthesis and use of *N*-chloroacetyl-modified peptides for the preparation of synthetic peptide polymers and peptide-protein immunogens. *Int. J. Peptide Protein Res.* **30,** 794–800.
19. Inman, J. K., Highet, P. F., Kolodny, N., and Robey, F. A. (1991) Synthesis of *N*-alpha-(*tert*-butoxycarbonyl)-*N*-epsilon-[(*N*-bromoacetyl)-beta-alanyl]-L-lysine: its use in peptide synthesis for placing a bromoacetyl cross-linking function at any desired sequence position. *Bioconjug. Chem.* **2,** 458–463.
20. Weir, D. M. (ed.) (1986) *Handbook of Experimental Immunology,* vol. 1, Blackwell Scientific, Oxford, p. 20.14.
21. Gordon, R. D., Fieles, W. E., Schotland, D. L., Hogue-Angeletti, R., and Barchi, R. L. (1987) Topographical localization of the C-terminal region of the voltage-dependent sodium channel from *Electrophorus Electricus* using antibodies raised against a synthetic peptide. *Proc. Natl. Acad. Sci. USA* **84,** 308–312.
22. Carlsson, J., Drevin, H., and Axen, R. (1978) Protein thiolation and reversible protein-protein conjugation. *Biochem. J.* **173,** 723–737.
23. Tam, J. P. (1988) Synthetic peptide vaccine design: synthesis and properties of a high-density multiple antigenic peptide system. *Proc. Natl. Acad. Sci. USA* **85,** 5409–5413.
24. Tam, J. P. and Lu, Y. A. (1989) Vaccine engineering: enhancement of immunogenicity of synthetic peptide vaccines for Hepatitis in chemically defined models consisting of T- and B-cell epitopes. *Proc. Natl. Acad. Sci. USA* **86,** 9084–9088.

25. Borras-Cuesta, F., Petit-Camurdan, A., and Fedon, Y. (1987) Engineering of immunogenic peptides by co-linear synthesis of determinants recognized by B and T cells. *Eur. J. Immunol.* **17,** 1213–1215, also Borras-Cuesta, F., Fedon, Y., and Petit-Camurdan, A. (1988) Enhancement of peptide immunogenicity by linear polymerization. *Eur. J. Immunol.* **18,** 199–202.
26. Good, M. F., Maloy, W. L., Lunde, M. N., Margalit, H., Cornette, J. L., Smith, G. L., Moss, B., Miller, L. H., and Berzofsky, J. A. (1987) Construction of a synthetic immunogen: use of a new T-helper epitope on malaria circumsporozoite protein. *Science* **235,** 1059–1062.
27. LeClerc, C., Przewlocki, G., Schutze, M. P., and Chedid, L. (1987) A synthetic vaccine constructed by copolymerization of B and T cell determinants. *Eur. J. Immunol.* **17,** 269–273.
28. Patarroyo, M. E., Amador, R., Clavijo, P., Moreno, A., Guzman, F., Romero, P., Tascon, R., Franco, A., Murillo, L. A., Ponton, G., and Trujillo, G. (1988) A synthetic vaccine protects humans against challenge with asexual blood stages of *Plasmodium falciparum* malaria. *Nature* **332,** 158–161.
29. Alving, C. R., Richards, R. L., Moss, J., Alving, L. I., Clements, J. D., Shiba, T., Kotani, S., Wirtz, R. A., and Hockmeyer, W. T. (1986) Effectiveness of liposomes as potential carriers of vaccines: applications to cholera toxin and human malaria sporozoite antigen. *Vaccine* **4,** 166–172.
30. Hui, G. S. N., Chang, S. P., Gibson, H., Hashimoto, A., Hashiro, C., Barr, P. J., and Kotani, S. (1991) Influence of adjuvants on the antibody specificity to the *Plasmodium falciparum* major merozoite surface protein, gp195. *J. Immunol.* **147,** 3935–3941.
31. Deres, K., Schild, H., Weissmuller, K. H., and Jung, G. (1989) In vivo priming of virus-specific cytotoxic T lymphocytes with synthetic lipopeptide vaccine. *Nature* **342,** 561–564.
32. Weissmuller, K. G., Jung, G., and Hess, G. (1989) Novel low-molecular-weight synthetic vaccine against foot-and-mouth disease containing a potent B-Cell and macrophage activator. *Vaccine* **7,** 29–33.
33. Hopp, T. P. (1984) Immunogenicity of a synthetic HBsAg peptide: enhancement by conjugation to a fatty acid carrier. *Molecular Immunol.* **21,** 13–16.
34. Lyon, J. A., Geller, R. H., Haynes, J. D., Chulay, J. D., and Weber, J. L. (1986) Epitope map and processing scheme for the 195,000 dalton surface glycoprotein of *Plasmodium falciparum* merozoite deduced from cloned overlapping segments of the gene. *Proc. Natl. Acad. Sci. USA* **83,** 2989–2993.

CHAPTER 11

Epitope Prediction Methods

J. Mark Carter

1. Immunology Paradigm

1.1. Antigen Processing and Presentation

Before beginning this discussion, a brief review of immunology is required. The following is a very abbreviated overview of the paradigm as it pertains to the subject of synthetic peptides as antigens. In the vigorous field of immunology, theories of antigen presentation and cell regulation, in particular, are especially dynamic. The following discussion was accurate at the time of publication.

An epitope may be defined as the entity recognized and specifically bound by an immune cell through a specific antigen receptor molecule. There are two types of cells capable of recognizing epitopes: T-lymphocytes and B-lymphocytes. There are also two generally accepted classes of protein epitopes that may be modeled as peptides. Some epitopes comprise a single, short, continuous peptide, derived directly from the parent protein sequence. These are usually called "continuous epitopes." Most T-cell epitopes appear to be continuous epitopes. The other class contains epitopes comprising assemblages of amino acids from distant regions of the protein primary structure, brought together by folding of the chain. These are called "discontinuous epitopes." Most B-cell epitopes appear to be of the discontinuous class. There are three types of T-lymphocytes (or T-cells), distinguished via the presence of surface protein markers. One is the T-helper cells (defined by the presence of the protein CD4), which interact with and upregulate the activity of many other immune cells, increasing or sustaining an

From: *Methods in Molecular Biology, Vol. 36 Peptide Analysis Protocols*
Edited by: B. M. Dunn and M. W. Pennington Copyright ©1994 Humana Press Inc., Totowa, NJ

immune response. Another type is the T-suppressor cells (it is unclear whether this type of cell bears any CD4 or CD8 proteins), which similarly downregulate the immune response. The third type is the cytotoxic or killer T-cells (bearing the marker protein CD8), which destroy cells recognized as nonself. T-cells are embryologically produced in the thymus (hence "T"-cells), migrating out into the body very early in development.

Although the different life stages of B-cells are also identifiable via surface markers, there is only one type of B-cell. Naive B-cells first bind antigens via surface immunoglobulin. So primed, they may undergo blastogenesis (rapid division, also called clonal expansion) and begin production of soluble antibodies, when subsequently signaled by helper T-cells. Eventually, most B-cell clones will differentiate into plasma cells, although some will become memory cells. Plasma cells are large and extraordinarily efficient antibody factories. Memory cells take up long-term residence, thus providing for a rapid, high-level secondary response to the antigens that is termed "anamnestic." The B-cell life cycle is attended by the migration of the B-cells from their origin in the bone marrow throughout the lymphatic system.

The most current accepted model for presentation of foreign antigens involves recognition of the antigen in the context of self-MHC (Major HistoCompatibility antigen). The antigen-presenting cell (or APC) is generally either a B-cell or a macrophage-type cell. The APC binds pieces of processed antigen on its surface by means of MHC protein.

Successful stimulation of antibody production requires two complex occasions in cell–cell communication. The first episode comprises the three simultaneous binding reactions culminating in antigen presentation to the T-cell. One of these three reactions is the aforementioned binding of the antigen to the MHC protein on the surface of the APC. In addition, the antigen must be recognized by a specific antigen receptor on the surface of the T-cell. Finally, MHC on the surface of the APC must be recognized by the CD4-MHC receptor on the T-cell. (For foreign antigens, class II MHC molecules are important for proper recognition, whereas class I molecules are generally involved in self-recognition.) In the second cell–cell interaction, after the T-cell is presented with an antigen, it may stimulate an antigen-primed B-cell to divide and differentiate. The active T-cell signals the B-cell by secretion of soluble mediators, interleukins (IL). Figure 1 depicts a summary of all these interactions.

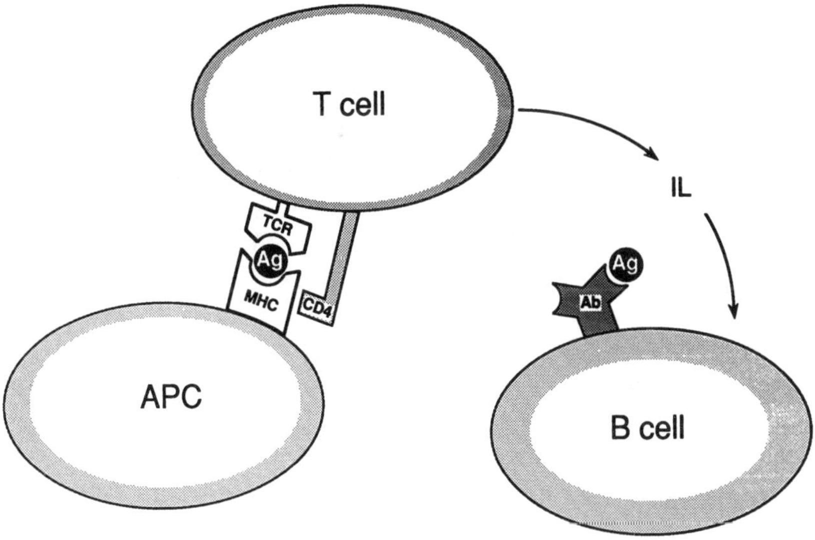

Fig. 1. Summary diagram of antigen presentation. This diagram is grossly simplified. Shown are the antigen-presenting cell (APC) presenting the antigen (Ag) by means of its major histocompatibility receptor protein (MHC) to the T-cell. The antigen actually bears two functional groups (not shown) each capable of binding either the T-cell or the B-cell antigen receptor. The T-cell recognizes its portion of the antigen only in context of the MHC. The T-cell then releases soluble mediators, such as interleukins (IL), that signal the B-cell (which has been primed by binding of its own respective portion of the antigen to its surface immunoglobulin) to begin to produce soluble antibodies.

The T-cell antigen receptor is an integral membrane protein bearing a long shallow cleft that binds its specific antigen somewhat like a hot dog bun wrapping around a wiener. The antigen in the cleft assumes a helical conformation that often shows amphipathicity *(1)*. About four of the amino acid side chains comprising the hydrophobic face of this helix make contact with the cleft of the receptor. Further details on antigen binding by the T-cell are presented in Section 3. below.

The B-cell antigen receptor, on the other hand, is simply a specialized antibody molecule. It binds its specific antigen in a highly shape-dependent manner. Binding observed often includes burial of a surface-oriented amino acid side chain from the antigen deep in the binding cleft of the antibody, as well as extensive surface contacts. Details on antigen binding by the B-cell are presented in Section 2. The intact antigen molecule, on presentation without processing, generally bears many B-cell antigens interspersed with a number of T-cell antigens. In contrast, for a proteolytically processed antigen, many of the processed products are peptides that often bear sites capable of binding to the B- and T-cell receptors in a manner analogous to the binding of the corresponding sites of the intact antigen protein. Investigators may prepare these same peptides via synthetic chemistry. Indeed, the delineation of B- and T-cell antigen is usually performed using synthetic peptides. Thus, when amino acid sequence is used to predict antigenicity, synthesis may be used to confirm it.

One common goal in peptide synthesis is generation of an antibody that is capable of binding to a native protein antigen. Because production of antibody requires the action of B-cells, such peptide immunogens should be planned to include known or predicted B-cell epitopes. However, generation of antibodies also requires effective T-cell-mediated antigen presentation and immune processing. It therefore requires the presence of both a T-cell epitope as well as a B-cell epitope on the immunogen molecule. This is the rationale behind conjugation of small peptide immunogens to protein carrier molecules (qv).

2. B-Cell Epitopes

2.1. Nature of the Antigen–Antibody Complex

Many antigen–antibody complexes have been studied via X-ray crystallography *(2–6)*. In these complexes, the antigen is typically bound very tightly by means of a number of specific interactions. Because B-cell antigens are often presented by the B-cells themselves, in the form of intact protein antigens, the surface regions of the antigenic proteins are often involved in the B-cell epitopes.

Obviously, a great deal is known about the secondary and tertiary structures of proteins whose X-ray crystal structure is known. Generally, antibody-binding sites lie on the solvent-accessible surface of these structures. Many protein B-cell epitopes have also been mapped via peptide-binding experiments. Most of these epitopes also lie at the surface of the native protein. For these reasons, effective methods for prediction of B-cell epitopes are often based on prediction of surface-accessible regions.

Data from an extensive study of proven B-cell epitopes reveal that:

1. All B-cell epitopes comprise eight or fewer usually noncontiguous amino acids;
2. The typical B-cell epitope contains 5 ± 1.3 residues that make contact with the antibody; and

3. A typical B-cell epitope contains 4 ± 1.2 residues that are essential for binding to the antibody *(7)*.

It has also been noted that most well-characterized B-cell epitopes actually comprise a mix of both hydrophilic and hydrophobic amino acid residues.

During the antibody–antigen binding reaction, it is supposed that initial hydrophilic interactions occur between surface-located amino acids in the antigen with the binding site on the antibody. These include hydrogen-bonding and electrostatic interactions *(8)*. These interactions lead to local denaturation of the antigen *(9)* allowing access of the antibody to buried regions of the antigen, where more hydrophobic amino acids may be located. This local denaturation occurs without significant change in the global structure of the antigen *(3)*. Hydrophobic interactions, thus facilitated, allow for tighter and more shape-dependent binding of the antigen–antibody complex. On the other hand, protein/peptide amino acid residues buried by the antibody-binding reaction are not necessarily involved in the binding. They may simply be buried by it *(10)*.

2.2. General B-Cell Epitope Prediction Procedure

For epitope prediction, the amino acid sequence of a protein is studied in short overlapping regions of about 3–12 amino acids (seven is most common) referred to as "windows." The score for any given measured or generated parameter in a window is the average score for the amino acids comprising the window. In all the different antigen prediction methods discussed below, the windows with the highest average scores are predicted to be the most antigenic. An algorithm for such calculations may be represented in equation form as follows, where *u* represents the scored value of the parameter examined for the particular amino acid at position *n* in the protein sequence, with a window size of *w*:

$$U_{i + w/2} = \sum_{i = n}^{n + w} u_i/w \qquad (1)$$

The right side of the equation describes the summation of the values for each of the amino acids in the window, divided by the number of amino acids in the window. The nomenclature of the left side of the equation simply indicates that the result obtained is assigned to the amino acid residue lying in the center of the window. Table 1 shows the values of the parameters (u_i) for each of the 20 common amino acids according to each of the methods discussed.

Table 1
Values for B-Cell Epitope Prediction[a]

	Hopp, 1981	Welling, 1985	Chou, 1978
Alanine	–0.5	0.115	0.66
Cysteine	–1.0	–0.120	1.19
Aspartic acid	3.0	0.065	1.46
Glutamic acid	3.0	–0.071	0.74
Phenylalanine	–2.5	–0.141	0.60
Glycine	0.0	–0.184	1.56
Histidine	–0.5	0.312	0.95
Isoleucine	–1.8	–0.292	0.47
Lysine	3.0	0.206	1.01
Leucine	–1.8	0.075	0.59
Methionine	–1.3	–0.385	0.60
Asparagine	0.2	–0.077	1.56
Proline	0.0	–0.053	1.52
Glutamine	0.2	–0.011	0.98
Arginine	3.0	0.058	0.95
Serine	0.3	–0.026	1.43
Threonine	–0.4	–0.045	0.96
Valine	–1.5	–0.013	0.50
Tryptophan	3.4	–0.114	0.96
Tyrosine	–2.3	0.013	1.14

[a] Higher values represent increased likelihood in epitope prediction.

Data resulting from these mathematical manipulations are usually displayed in a graph with the predictive parameter represented on the ordinate. By indicating the values lying above the mean plus two standard deviations, significant peaks and valleys may be identified. An example is shown in Fig. 2.

2.3. Hydrophilicity

Historically, the first methods used for prediction of protein B-cell antigens were based on hydrophilicity *(11)*. In this type of study, the 20 common amino acids were partitioned and quantitated in a biphasic aqueous/organic solvent mixture. Preference for the aqueous solvent was scaled as hydrophilicity. As predicted, it was noted that the most hydrophilic amino acids generally bear charged moieties: aspartate and glutamate (at physiological pH) bear negatively charged carboxyl

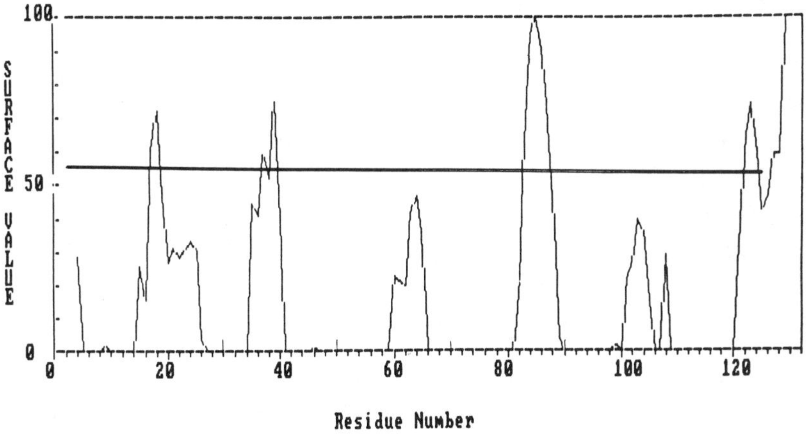

Fig. 2. Epitope prediction plot. In this example, the *X*-axis represents the hypothetical sequence, and the *Y*-axis represents the value of the epitope prediction parameter (normalized). The values shown are only those that lie above the mean value of the predictive parameter for the entire sequence. The solid horizontal line indicates the position of the second standard deviation. Signals above this line are predicted to be effective as B-cell antigen.

groups, whereas lysine bears a cationic ε amine, and arginine bears a guanidinium. Specifically, the hydrophilic amino acids generally include aspartate, glutamate, glycine, lysine, asparagine, glutamine, arginine, serine, and threonine, and often histidine. Conversely, the most hydrophobic amino acids generally bear aliphatic or aromatic side chains, whereas hydrophobicity increases with the size of the side group. The hydrophobic amino acids include alanine, cystine, phenylalanine, isoleucine, leucine, methionine, proline, valine, tryptophan, and tyrosine. Spans of the protein primary structure containing primarily hydrophilic amino acids are supposed to lie on the surface of the molecule, where they can interact with the aqueous solvent.

The hydrophilicity method is only about 50% accurate for monomeric globular proteins. That is to say that only about 50% of hydrophilic predicted surface regions of this type of protein will actually invoke a B-cell response in a typical experiment. The method is also only about 50% effective, meaning that only about 50% of known epitopes are predicted. It is noteworthy that the nonpredicted epitopes contain significant hydrophobicity.

2.4. Antigenicity

This method was developed based on a data base derived from antigen mapping experiments. The technique determined the statistical likelihood for each of the amino acids to occur in a known antibody-combining site *(12)*. Because few proteins had been studied by thorough antigen mapping, this method suffered from development from a necessarily limited and flawed data base. Although it seems like a good idea, it never enjoyed much popularity.

2.5. Predicted β Bends

As antigen structure became better understood, later methods for prediction of B-cell epitopes based on secondary structure were published. The known surface regions on larger proteins often fall into either β bends or Ω loops *(13)*. Such surface-lying regions are likely to be accessible for antibody binding in the native protein. Methods for predicting bends in protein secondary structure in soluble, monomeric, globular proteins *(14,15)* are sometimes successful in predicting B-cell antigen. However, again only about half of predicted β bends typically generate measurable B-cell responses.

It is also important to note that many proteins are not monomeric, soluble, or globular. In these cases, a more careful and detailed structural analysis is required. Interfacial and intramembranous regions can be predicted to exist among regions of exclusively hydrophobic amino acids. Hence, modeling can help to determine exposed areas of these proteins. Fibrous proteins, on the other hand, are generally exposed to antibody binding at positions all along their length. However, producing a synthetic peptide capable of attaining a fibrous conformation may not be a trivial task *(16)*.

2.6. Surface Accessibility

A fourth method has become popular more recently. It utilizes surface accessibility derived from "rolling hydration sphere" algorithms *(17)*. This method requires use of an established X-ray crystal structure. It gives an indication of the accessibility of the particular amino acid residues of a given structure to a hypothetical sphere, the diameter equal to that of a water molecule (1.4 Å), which is rolled over the surface of the molecule. Amino acids comprising regions of the protein that are contacted by such a sphere are called "surface accessible." Surface accessibility cannot be predicted from amino acid sequence alone, because it is very much a structure-dependent property of a molecule.

2.7. Thermal Mobility

In this method, X-ray data of a known structure are used to score each of the amino acid residues based on their positional uncertainty in the structure, also known as thermal mobility or "B factor" *(18)*. This is assumed to be a measure of structural "floppiness." The regions of a protein predicted to exhibit the greatest thermal mobility are predicted to comprise B-cell epitopes. Like surface accessibility, thermal mobility also cannot be predicted from amino acid sequence. However, it is useful to note that thermally mobile regions often occur at intron/exon boundaries at the genome level of eukaryotic protein sequences *(19)*.

2.8. Antigenic Index

When comparing the many different techniques above, one cannot help but be struck by the level of agreement between them. Most regions identified by one method are identified by one or more others as well. Because of this, the method many scientists prefer is a combination of all these methods called "antigenic index." This technique basically adds up the scores from each of the preceding methods and presents them normalized. Synthetic peptides based on the regions of the protein with the highest antigenic index have elicited antibody reacting with the native protein in about 60% of cases.

Computer software packages are available for performing all these (and many other) peptide and protein structure analyses. My personal favorites are PC gene for PC compatibles and GCG for the VAX *(20)*.

2.9. N- and C-Termini

In studying epitope maps of proteins of known structure, it is also apparent that the termini of the protein are also highly likely to be B-cell antigens. Most B-cell prediction methods also favor the termini, especially the N-terminus. For this reason, especially in the absence of computer support, the N- and C-termini are excellent first guesses for the synthesis of antigenic peptides *(21)*.

3. T-Cell Epitope Prediction Methods

3.1. T-Cell Receptor Assay

Limited data are available regarding the X-ray crystal structure of an antigen in complex with its T-cell receptor. However, a number of studies of T-cell antigenicity have been performed using the T-cell mitogenic response. In these experiments, polyclonal T-cells in culture are exposed to a series of peptides. Any peptide bearing a T-cell epitope will stimulate the particular T-cell binding it to undergo mitogenesis (rapid division similar to blastogenesis). This is detected by means of incorporation of radiolabeled nucleotides from the culture broth into the daughter cell nuclei. Results of such experiments have led to the recent development of methods for the prediction of T-cell epitopes, analogous to the techniques for the prediction of B-cell (antibody-binding) epitopes.

3.2. Amphipathic Helix

One of the first of these methods was developed even before the publication of the structure of the T-cell receptor was known *(22)*. In this method, amphipathicity of helical regions of proteins and peptides are predicted by means of the periodicity of hydrophilic and hydropathic amino acids in the sequence *(1)* (*see* Fig. 3).

Specifically, a periodicity of 3.4 amino acids (from the known structure of α helices) for alternating hydrophobic and hydrophilic amino acid residues is recognized by computer programs performing this algorithm *(23)*. An example of such a sequence is HHhhHHhHHhhHHh, where H = hydrophobic and h = hydrophilic. For the definition of hydrophobic and hydrophilic amino acids, *see* Section 2.

3.3. Hydrophobic Strip

Based on the binding of the hydrophobic face of the amphipathic helix to the T-cell receptor, another successful method for T-cell epitope prediction concentrates on the presence of a hydrophobic strip along one face of an α helix, without requiring hydrophilicity of the other face *(24)*. An example of this type sequence is HHxxHHxHHxxHHx, where x = any amino acid.

3.4. Helical Stability

Based on minimization of energy of solution conformation via computer simulation, the helical stability method stresses the importance of the thermodynamic stability of the predicted α helix of a potential T-cell antigen rather than its amphipathicity *(25)*. A detailed description of this method is beyond the scope of this volume. Briefly, it comprises complex calculations of free energy in order to determine the propensity of a peptide sequence to remain in a helical conformation when exposed to an amphipathic environment, such as the antigen-binding cleft of the T-cell receptor.

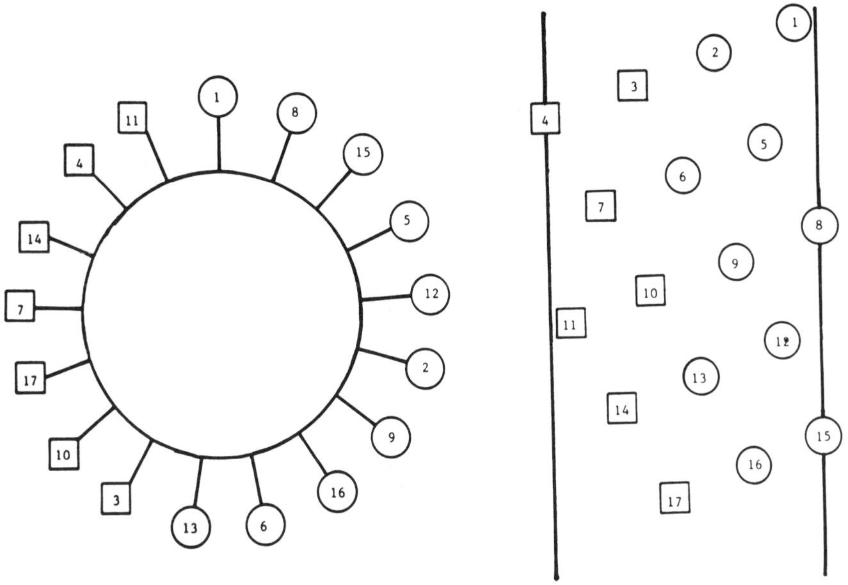

Fig. 3. Amphipathic helix representations. The figures show the results of representations of the sequence: HHhhHHhHHhhHHhHHh. **(A)** Helical wheel. **(B)** Helical net.

3.5. Information Theory-Based Pattern

Based on an analysis of known T-cell epitopes via information theory, a more rigorous structural motif for prediction has been developed *(15,26)*. In this method, predicted T-cell epitopes start with either glycine or a charged amino acid. This is followed by two or three consecutive hydrophobic amino acids, and then either a charged or polar residue. In a representation similar to those above, this type of sequence might appear as GHHH!hHH!, where G = glycine (or charged) and ! = a charged (or polar) amino acid. Obviously, this method agrees with the amphipathicity prediction methods outlined above.

3.6. Comparison of T-Cell Epitope Prediction Methods

A comparison of the accuracy and effectiveness of the published methods of Berzofsky *(22)*, Rothbard *(26)*, and Stille *(24)* found generally good agreement among the results. Agreement with experimentally determined T-cell antigenicity was about 40–70%, depending on the criteria used. This comparison duly noted the differences among these methods, but it suggested a lack of significant superiority of any one method over another *(27)*.

In general, in designing the synthesis of a T-cell epitope peptide, a stable amphipathic helix should be attempted. Helicity can be reasonably well predicted by the old and celebrated method of Chou and Fasman *(28,29)*, whereas amphipathicity can be ascertained by means of Edmundson's wheel model *(30)* or a helical net *(31)*. Both of these types of helical representations are depicted in Fig. 3. If computer support is available, many programs may be used to predict and analyze peptide structure as an amphipathic α helix *(20)*.

References

1. Berzofsky, J. A. (1985) Intrinsic and extrinsic factors in protein antigenic structure. *Science* **229,** 932–940.
2. Amzel, L. M., Poljak, R. J., Varga, J. M., and Richards, F. F. (1974) The three dimensional structure of a combining region-ligand complex of immunoglobulin NEW at 3.5 Å resolution. *Proc. Natl. Acad. Sci. USA* **71,** 1427–1430.
3. Amit, A. G., Mariuzza, R. A., Phillips, S. E. V., and Poljak, R. J. (1986) Three dimensional structure of an antigen–antibody complex at 2.8 Å resolution. *Science* **233,** 747–753.
4. Colman, P. M., Laver, W. G., Varghese, J. N., Baker, A. T., Tulloch, P. A., Air, G. M., and Webster, R. G. (1987) Three-dimensional structure of a complex of antibody with influenza virus neuraminidase. *Nature* **326,** 358–363.
5. Padlan, E. A., Silverton, E. W., Sheriff, S., Cohen, G. H., Smith-Gill, S. J., and Davies, D. R. (1989) Structure of an antibody–antigen complex: crystal structure of the HyHEL-10 Fab-lysozyme complex. *Proc. Natl. Acad. Sci. USA* **86,** 5938–5942.
6. Stanfield, R. L., Fieser, T. M., Lerner, R. A., and Wilson, I. A. (1990) Crystal structures of an antibody to a peptide and its complex with peptide antigen at 2.8. Å. *Science* **248,** 712–719.
7. Saul, A. J. and Geysen, H. M. (1990) Identification of epitopes through peptide technology, in *New Generation Vaccines* (Woodrow, G. C. and Levine, M. M., eds.), Dekker, New York, pp. 117–126.
8. Getzoff, E. D., Tainer, J. A., Lerner, R. A., and Geysen, H. M. (1988) The chemistry and mechanism of antibody binding to protein antigens. *Adv. Immunol.* **43,** 1–98 (review).
9. Sheriff, S., Silverton, E. W., Padlan, E. A., and Cohen, G. H. (1987) Three-dimensional structure of an antigen–antibody complex. *Proc. Natl. Acad. Sci. USA* **84,** 8075–8079.
10. Tainer, J. A., Deal, C. D., Geysen, H. M., Roberts, V. A., and Getzoff, E. D. (1991) Defining antibody–antigen recognition: towards engineered antibodies and epitopes. *Intern. Rev. Immunol.* **7,** 165–188.

11. Hopp, T. P. and Woods, K. R. (1981) Prediction of protein antigenic determinants from amino acid sequences. *Proc. Natl. Acad. Sci. USA* **78,** 3824–3828.
12. Welling, G. W., Weijer, W. J., van der Zee, R., and Welling-Wester, S. (1985) Prediction of sequential antigenic regions in proteins. *FEBS Lett.* **188,** 215–218.
13. Leszczynski, J. F. and Rose, G. D. (1986) Loops in globular proteins: a novel category of secondary structure. *Science* **234,** 849–855.
14. Chou, P. Y. and Fasman, G. D. (1977) Beta-turns in proteins. *J. Mol. Biol.* **115,** 135–175.
15. Gibrat, J. F., Garnier, J. O., and Robson, B. (1987) Further developments of protein secondary structure prediction using information theory. *J. Mol. Biol.* **198,** 425–443.
16. Ockenhouse, C. F., Deal, C. D., and Carter, J. M. (1991) A collagen multiple antigen peptide (MAP) binds CD36 and modulates CD36 specific epitope expression. *Twelfth American Peptide Symposium* Poster 545.
17. Novotny, J. and Haber, E. (1986) Static accessibility model of protein antigenicity: the case of scorpion neurotoxin. *Biochemistry* **25,** 6748–6754.
18. Tainer, J. A., Getzoff, E. D., Alexander, H., Houghton, R. A., Olsen, A. J., Lerner, R. A., and Hendrickson, W. A. (1984) The reactivity of anti-peptide antibody is a function of the atomic mobility of sites in a protein. *Nature* **312,** 127–133.
19. Tainer, J. A., Getzoff, E. D., Paterson, Y., Olsen, A. J., and Lerner, R. A. (1985) The atomic mobility component of protein antigenicity. *Ann. Rev. Immunol.* **3,** 501–535.
20. Devereux, J., Haberli, P., and Smithies, O. (1984) A comprehensive set of sequence analysis programs for the VAX. *Nucl. Acids Res.* **12,** 387–401.
21. Palfreyman, J. W., Aitcheson, T. C., and Taylor, P. (1984) Guidelines for the production of polypeptide specific antisera using small synthetic oligopeptides as immunogens. *J. Immunol. Methods.* **75,** 383–393.
22. DeLisi, C. and Berzofsky, J. A. (1985) T-cell antigenic sites tend to be amphipathic structures. *Proc. Natl. Acad. Sci. USA* **82,** 7048–7052.
23. Margalit, H., Spouge, J. L., Cornette, J. L. , Cease, K. B., DeLisi, C., and Berzofsky, J. A. (1987) Prediction of immunodominant helper T-cell antigenic sites from the primary sequence. *J. Immunol.* **138,** 2213–2229.
24. Stille, C. J., Thomas, L. J., Reyes, V. E., and Humphreys, R. E. (1987) Hydrophobic strip-of-helix algorithm for selection of T cell-presented peptides. *Mol. Immunol.* **24(10),** 1021–1027.
25. Nauss, J. L., Reid, R. H., and Boedeker, E. C. (1991) Helical stability as a method of predicting peptide T-cell epitopes, in *Proceedings of the Twelfth American Peptide Symposium,* Escom B. V., Leiden, Netherlands, pp. 855,856.
26. Rothbard, J. B. and Taylor, W. R. (1988) A sequence pattern common to T cell epitopes. *EMBO J.* **7,** 93–100.
27. Reyes, V. E., Fowlie, E. J., Lu, S., Phillips, L., Chin, L. T., Humphreys, R. E., and Lew, R. A. (1990) Comparison of three related methods to select T cell-presented sequences of protein antigens. *Mol. Immunol.* **27,** 1021–1027.
28. Chou, P. Y. and Fasman, G. D. (1974) Prediction of protein conformation. *Biochem.* **13,** 222–245.
29. Chou, P. Y. and Fasman, G. D. (1978) Prediction of the secondary structure of proteins from their amino acid sequence. *Adv. Enzymol.* **47,** 45–57.

30. Schiffer, M. and Edmundson, A. B. (1967) Use of helical wheels to represent the structures of proteins and to identify segments with helical potential. *Biophys. J.* **7,** 121–135.
31. Taylor, J. W., Miller, R. J., and Kaiser, E. T. (1982) Structural characterization of beta-endorphin through the design, synthesis, and study of model peptides. *Mol. Pharamacol.* **22,** 657–666.

CHAPTER 12

Epitope Mapping of a Protein Using the Geysen *(PEPSCAN)* Procedure

J. Mark Carter

1. Introduction

1.1. General

Immunity to many diseases is dependent on the ability of the host's antibodies to recognize foreign antigens, such as surface proteins or toxins, and bind them tightly and specifically. This binding is an important aspect of the immune response, and it is often required for subsequent immune processes that ultimately result in re-establishment of a disease-free state.

One of the toughest problems encountered in vaccine development is that of delineating the antibody response to a protein antigen. Whereas the overall response to an antigen may involve various molecular species of antibodies, each antibody molecule can bind specifically to one unique part of the antigen referred to as that antibody's epitope. Often only a subset of these epitopes is involved in blocking a protein's function, clearing of infectious organisms, or other steps in an effective immune response.

The *PEPSCAN* procedure, developed by Mario Geysen and marketed by Chiron Mimotopes (Victoria, Australia), is a variation of solid-phase peptide synthesis. It comprises the synthesis and immunochemical assay of hundreds of peptides covalently linked to plastic pins. This technology represents a major advance in the epitope mapping of protein antigens because of its ability to create the large numbers of overlapping peptides necessary for complete epitope mapping *(1)*.

From: *Methods in Molecular Biology, Vol. 36: Peptide Analysis Protocols*
Edited by: B. M. Dunn and M. W. Pennington

Fmoc-$NHCH_2CH_2CO$-$NHCH_2CH_2CH_2CH_2CH_2CH_2NH$-$COCH_2$polymer-polyethylene

Fig. 1. Structure of the prederivatized polyethylene pins commercially available for peptide synthesis.

Currently, the plastic pins for attachment of peptides are commercially available from Chiron Mimetopes. They are now prepared according to a modification of the method originally published *(1)*. First, the polyethylene matrix of the rods is grafted with an acrylic-like polymer. Then, the free carboxylic acid moiety on the polymer is amidated with one end of 1,6-diaminohexane. The other end of the amine is further derivitized with one of a number of linking moieties. Finally, the linker is acylated with Fmoc-β-alanine, yielding the structure shown in Fig. 1.

1.2. Issue of Reproducibility

One of the limitations plaguing early application of the *PEPSCAN* technique was poor reproducibility in the substitution level of these derivatized pins. In our tests of pins from the same lot at the Walter Reed Institute of Research (WRAIR), we found substitution levels ranging from 6–26 nmol NH_2/pin, with a mean of 12 nmol and an SD of 3 nmol. The issue here is not that this variability might prevent accumulation of worthwhile data. Rather, because of this limitation, all results from *PEPSCAN* must usually be considered qualitative. For confirmation, such results may be double-checked via synthesis and immunoassay of peptides via classical solid-phase methods.

Recent advances in pin design incorporate increased surface area as well as improved level and stability of pin derivitization. As the new pins become more widely available, the peptide pin methodology itself is expected to become more widely accepted and more generally utilized.

The pins for peptide synthesis are arranged in 8 × 12 arrays on 9-mm centers, like commercially available microtiter plates. This geometry allows for familiarity and simplification in subsequent enzyme-linked immunosorbent assay (ELISA) for the detection of antibody reactivity. Many laboratory technicians are already quite familiar with standard ELISA assays, and only minor modifications to this procedure are necessary to perform a *PEPSCAN* ELISA. Furthermore, automated microtiter plate readers are widely available for rapid determination of absorbance data in assays performed with these 96-well plates.

1.3. Computer-Automated Amino Acid Indexer

Other than variable substitution level in the pins, the most significant problem in *PEPSCAN* is the logistics of the simultaneous synthesis of several hundred peptides. Clearly computer support is required, but even a computer-generated synthesis schedule, such as the output from the software distributed by Chiron with the pins, leaves possible a large margin of error. The person performing the synthesis must manually transpose amino acid locations from the hard copy list to the microtiter wells. In our own laboratory, this procedure takes about 4 h to fill 10 microtiter plates, and it results in approx 3% error rate.

In order to address this problem, we have developed a computer-driven device that locates and identifies each of the different wells, and indicates their respective amino acid derivative requirements via illumination with LEDs. Using this computer-driven amino acid indexer, the time in filling 10 microtiter plates, for simultaneous synthesis of 960 peptides, is reduced sixfold to 40 min, and error becomes undetectable. The device is thoroughly described in VanAlbert et al. *(2)*, and it is commercially distributed by CRACO (Vienna, VA).

1.4. Linear vs Conformational Epitopes

PEPSCAN is particularly effective in the detection of linear (continuous) epitopes. Unfortunately, however, most antibodies are probably directed against discontinuous epitopes *(3–5)*. This fact becomes especially important when *PEPSCAN* is used to study the specificity of monoclonal antibodies (MAbs). In many cases, the results of such experiments are weak and equivocal. Nonetheless, Geysen has suggested that binding of antibodies to discontinuous epitopes (such as are reported for most MAbs) may be detected on peptide pins, at least in some instances. Such binding typically involves two or three discontinuous regions of the protein sequence that fold into a discrete conformation on the solvent-accessible surface of the native structure of the antigen. Theoretically, the antibodies should bind, although much more weakly, to each of these regions when presented separately. In fact, we have observed many sets of data suggesting this conclusion, but the binding so detected is often not statistically significant above background (nonspecific) binding.

Excellent results are generally obtainable using immune serum as a source of antibody. It is probably true that serum raised against a native protein antigen will contain only a limited subset of antibodies reactive to linear epitopes presented by the peptide pins *(6)*. However, there is usually such a large variety of reactivities represented by a polyclonal serum that a fair number of linear epitopes can be readily demonstrated by binding to the pins. Antisera raised against peptide antigens and peptide conjugates tend to contain a greater proportion of antibodies that are reactive to linear epitopes because of their more limited conformational freedom. Consequently, this type of immune serum generally gives the highest level of detected binding on *PEPSCAN*.

2. Materials

2.1. Synthesis

1. Prederivitized polyethylene pins and polyethylene microtiter plates.
2. *N,N,*-dimethylformamide (DMF).
3. Methanol.
4. PIP solution: 20% piperidine in DMF. (*See* Note 2).
5. Amino acid solution: 30 m*M* amino acid derivative and 30 m*M* 1-hydroxybenzotriazole (HOBt) in DMF. Derivatives used are 9-fluorenomethoxycarbonyl (Fmoc) amino acids as *O*-pentafluorophenol or *O*-dihydrobenzatriazine esters.
6. Dichloromethane (DCM)—**Note: This solvent is a suspected carcinogen.**
7. Basified DCM: 5% diisopropylethylamine in DCM. Prepare fresh within 1 h of use.
8. Acetylation cocktail: 5% acetic anhydride and 1% diisopropylethylamine (DIEA) in DMF. Prepare fresh immediately before use.
9. Deblocking cocktail: 2.5% phenol and 2.5% 1,2-dithioethane in trifluoroacetic acid. Prepare fresh within 1 h of use. **Note:** This reagent is extremely corrosive, and it smells absolutely terrible. Wear appropriate protective devices, and use it in a fume hood.
10. Deionized water.
11. Silica gel desiccant.
12. Plastic baths and sealable bags.

2.2. Disruption

1. Sonicator: *See* Note 10 about choice of ultrasonic cleaner.
2. Disruption buffer: 1% reagent-grade sodium dodecylsulfate, 0.1% 2-mercaptoethanol, and 0.1*M* sodium phosphate, pH 7.2.
3. Explosion-proof heating bath, filled with boiling methanol.
4. Silica gel.
5. Sealable bags, tongs.

2.3. ELISA Analysis

1. Peptide pins.
2. Phosphate-buffered saline (PBS): 150 m*M* NaCl and 25 m*M* phosphate, pH 7.4. Prepare in 1-L batches, filter-sterilize, and store at 4°C. PBS keeps for about 2 or 3 wk. For indefinite storage, add 2 g/L sodium azide.
3. PBS/Tween 20 (PBST): Phosphate-buffered saline (PBS, as above) with 0.1% Tween 20. Prepare in 1-L batches (*see* Notes 13 and 18).
4. Blocking solution (*see* Note 14): Use a commercial ELISA blocker solution or one of the following two solutions. Prepare in 1-L batches, filter-sterilize, and store at 4°C. Solution keeps for about 2 or 3 wk. For extended storage (up to 8 wk), add 2 g/L sodium azide.
   a. 1% Bovine serum albumin (BSA) and 1% chicken ovalbumin (OVA) dissolved in PBST.
   b. 2% Casein in PBST: Boil 20 g casein in 100 mL 1*N* NaOH until completely dissolved. Adjust pH to 7.4 by addition of HCl. Add PBST to make final volume 1 L.
5. Test antibody solution. For serum or ascites fluid, the concentration used should be the same as that which gives a good, strong signal on a standard ELISA. If a standard ELISA titer is not available, then use a dilution of 1/500. For a purified antibody, use 1–10 µg/mL.
6. Second antibody solution (*see* Notes 19 and 20): The working concentration of the second antibody is usually specified by the manufacturer. Make the dilution in PBST.
7. Substrate solution: 0.1*M* diethanolamine, pH 9.8, with 0.01% $MgCl_2$ and 0.02% NaN_3. This buffer may be stored at 4°C for several months. However, it should be allowed to warm to room temperature before use. Immediately before use, dissolve *p*-nitrophenyl phosphate (the substrate) to a final concentration of 1 mg/mL
8. Plastic boxes with tight-fitting lids (e.g., Tupperware).

3. Methods

3.1. Synthesis

Historically, Cambridge Research Biologicals (Cheshire, UK) distributed a recipe for synthesis of peptides via Fmoc chemistry. In it, the prederivatized polyethylene pins are deprotected, washed, neutralized, washed, and amino acylated repeatedly until peptides of the desired length are completed. These peptides are then N^{α}-acetylated, side-chain deblocked, and washed once more. Finally, the peptide pins are subjected to ultrasonic disruption before ELISAs are performed. All reactions are performed at room temperature in a fume hood. Details of the method, as well as a few modifications we and others have suggested, are described below.

1. Deprotection (removal of the N^{α}-Fmoc group): Pins are first pre-equilibrated in DMF baths for 5 min. Perform deprotection batchwise in polyethylene boxes (*see* Note 1) poured to a depth of about 2.5 cm with PIP solution for 1 h (*see* Note 2).
2. Wash: Deprotection is followed by washes in DMF (two washes, 5 min) and then methanol (three washes, 3 min each) (*see* Note 3). Then the blocks are allowed to air-dry in a fume hood for at least 1 h (*see* Note 4).
3. Coupling (amino acylation): After pre-equilibration in DMF bath 5 min, pins are amino acylated individually with 100 µL/well in the polyethylene microtiter plates. The plates bearing the peptide pins are carefully oriented and lowered so that the pins are inserted into their respective wells. In order to reduce evaporation and contamination, the reaction is allowed to proceed overnight inside a sealed ziplock bag (*see* Note 5).
4. Wash: Following the amino acylation reaction, pins are again washed with DMF (two washes, 5 min each) and methanol (four washes, 3 min each), and then air-dried again for at least 1 h.
5. Elongation: Deprotection, washing, amino acylation, and washing are repeated until peptides of the desired length are produced. After the last amino acid is coupled, final deprotection, washing, and air-drying are performed as above (*see* Notes 6 and 7).
6. N^{α}-acetylation: After pre-equilibration of the peptide pins in DMF bath for 5 min, α amino groups on the peptides are acetylated in polyethylene microtiter plates with 100 µL/well solution of the acetylation cocktail for 90 min.
7. Wash: Pins are then washed with DMF (two washes, 5 min each), methanol (four washes, 3 min each), and air-dried again for 1 h.
8. Deblocking: Blocking groups are removed from the peptide amino acid side chains by incubation of the pins for 4 h in 2.5-cm deep baths of deblocking cocktail.
9. Wash: The pins are then washed in baths of DCM (two washes, 2 min each), basified DCM (two washes, 2 min each), DCM (5 min), allowed to air-dry (1 h), then washed in deionized water (2 min), methanol (overnight), and air-dried. Finally, the peptide pins are dried over silica gel in ziplock bags overnight. The finished blocks of peptide pins are stored at –20°C over silica gel in ziplock bags.
10. Disruption: Before they will perform properly in ELISA assays, the peptide pins must be disrupted, (*see* Section 3.2.).

3.2. Disruption

After peptide synthesis is complete, ELISAs are typically unsuccessful without prior ultrasonic "disruption." In order to make the peptides on the pins accessible to antibody binding, high-power ultrasonic treatment at elevated temperature is absolutely necessary.

1. The sonicator (*see* Notes 8 and 9) is filled with the disruption buffer and allowed to heat to 60–70°C. The polyethylene blocks bearing peptide pins are floated in the buffer, with the pins pointing downward. The sonicator is then operated for 20 min.
2. Pins are removed from the sonicator with tongs and rinsed briefly, but thoroughly in 60–70°C water.
3. At this point, the pins may be used immediately for ELISA. If they are to be stored for more than a few minutes, they should be boiled in methanol for 2 min, air-dried for at least an hour in the fume hood, and finally stored in ziplock bags at –20°C over silica gel (*see* Notes 10 and 11).

3.3. ELISA Analysis

A typical ELISA has five main steps (plus washes): First, the antigen is allowed to bind to the microtiter plate wells overnight in a dilute solution with PBS (*see* Note 13). Next, the excess antigen solution may be removed, or even a brief rinse performed, before a "blocking" solution is added. After an hour or two, the blocking solution is removed, and the test or "first" antibody is added (*see* Note 14). The first antibody is usually allowed 1 or 2 h to bind the antigen on the plates. After a series of thorough washes, an enzyme-conjugated "second" antibody is added and allowed to bind to the first antibody. After an hour or two, another thorough wash is made, and then a chromogenic substrate solution for the enzyme is added and allowed to develop for a few minutes to 2 h. Finally, the results are read on an automated microtiter plate reader that generally stores the values for absorbance for each well in a computer file. (A general recipe for this type of basic ELISA appears in Chapter 10, Section 8).

The peptide pin ELISA is performed very much like a typical ELISA. Persons experienced in the latter generally have little trouble with the technical aspects of pin ELISAs. There is, indeed, but one major difference between the two. In a typical ELISA, the first step comprises the adsorption of the antigen onto the bottom of a microtiter plate. This molecule acts as a solid-phase "capture" antigen for the subsequent binding of antibodies. Contrarily, with peptide pins, the peptide antigen remains covalently linked to the solid-phase support pin at all times.

This means that a peptide pin ELISA has only four steps. Briefly, the pins are "blocked" with a suitable buffer, they are subjected to binding of a first antibody, they are probed with an enzyme-labeled second antibody, and they are developed with a substrate. Each of these steps, as well as washing between them, is detailed below.

1. Blocking: Into each well of a microtiter plate, pipet 200 μL of blocking solution. Insert the pins and incubate 1 h at room temperature.
2. Test antibody: Pipet 175-μL test antibody solution into each well of a microtiter plate. Insert the pins and leave to incubate overnight at 4°C, rocking gently on a platform (*see* Notes 15 and 16).
3. Wash: Pour PBST (*see* Note 17) into a clean plastic box so that the level of liquid comes at least halfway up the pins when the blocks of pins are inserted with their pins downward. Put the box with wash buffer and pins on a rotating platform for 10 min. Discard the used wash buffer down the sink. Repeat for a total of three washes.
4. Second antibody: Pipet 150 μL second antibody solution (*see* Notes 17, 19 and 20) into each well of a microtiter plate, insert pins, and allow 1–2 h for binding.
5. Second wash: After the second antibody, make another thorough wash to remove excess enzyme conjugate reagent. Again, three washes of 10 min each are sufficient.
6. Substrate: Pipet 125 μL substrate solution into each well of a good-quality ELISA plate. Before inserting them into the plate, carefully orient the pins so that the numbered edges of the plates correspond with the numbered edges of the block containing the peptide pins. This will prevent confusion when the plates are being read after development.
7. Development: Allow development to proceed until the positive reactions are well colored, usually 30–60 min (*see* Note 21). Stop the development by removing the pins. Do not allow development to proceed until the negative peptides give a strong color reaction (*see* Note 22). After development is complete, remove the pins and rinse them immediately in water.
8. Plate reader: Read the plates on an automated microtiter plate reader within an hour.
9. Disruption: Disrupt as soon as possible. If this cannot be done within a couple of hours, store the pins overnight in a methanol bath. Do not let any of the ELISA reagents dry onto the pins.

3.4. Data Interpretation

3.4.1. Epitope Analysis

For each plate, individually, subtract the mean of the lowest 10–25% of absorbance readings. This is background. This is facilitated by means of a spreadsheet program. In lieu of any officially established criterion for differentiating between positive and negative reactions, positive responses are identified through the judgment of the experimenter. In general, the highest responses will be scored as positive reactions, whereas most sequences will be unreactive. Peptides with intermediate reactivity are often borne out as positive or negative after a repeat of the ELISA experiment.

There are several different combinations of antibody and peptides commonly used in the peptide pin system. Each combination may be expected to give different results, although they will all generally allow the same conclusions to be drawn.

3.4.2. Polyclonal Antibody Epitopes

One of the most common applications of this system is epitope mapping of a full protein sequence of overlapping octamers, where the immunogen was the intact native protein (or even an entire organism). ELISA reactivity of such a polyclonal immune serum typically gives several peaks, each corresponding to an epitope. Frequently there is one relatively strong immunodominant epitope that stands out among the others (*see* Fig. 2).

3.4.3. Epitope Overlaps

Each of the peaks of epitope recognition will typically span several pins and, therefore, several overlapping peptides. The minimal region of recognition is the sequence contained in all the recognized peptides of a given epitope (*see* Fig. 3). This vital information is only accessible through synthesis of many overlapping peptides.

3.4.4. Antipeptide Antibodies

You may wish to map the fine specificity of a serum raised against a synthetic peptide immunogen. These experiments have generated the strongest ELISA signals we have seen. However, the results are often complicated by strong reactivity to two (or more!) closely neighboring epitopes. This gives a broad peak, so that it is difficult to tell where one epitope ends and another one begins. Of course, this is more of a problem with sera raised against larger peptides (30+ amino acids) as immunogens.

3.4.5. Monoclonal Antibody Epitopes

Most MAbs are raised with an intact protein as immunogen. The limited reactivity of these MAbs to peptide pins emphasizes the paradigm of underrepresentation of linear epitopes among the general population of

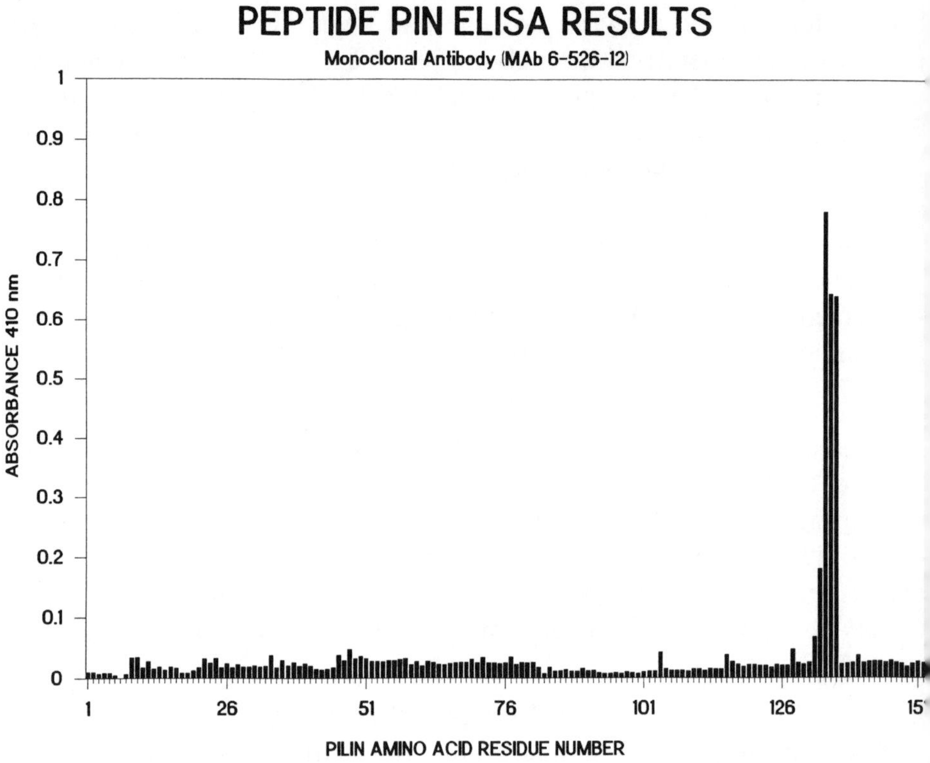

Fig. 2. Typical peptide pin ELISA. ELISA was performed according to standard methods presented in the text. Immune rabbit serum was used to probe pins bearing overlapping 8-mer peptides comprising a bacterial protein sequence. The *X*-axis of the figure represents the position in the protein sequence, and the *Y*-axis indicates absorbance (i.e., ELISA reactivity). Obviously, one epitope reacts much more strongly than the remainder of the protein. This phenomenon is referred to as immunodominance.

antibodies. Only about one out of every eight of the MAbs reacts strongly to any of the peptides on pins. Frequently, MAbs give somewhat ambiguous results with two or three peaks detected. This may seem like an artifact, since an MAb should only have one target sequence, but Geysen has suggested that these multiple regions of recognition suggest the location of the noncontiguous regions that would fold together in the native protein to give the conformational epitope for these MAbs. We have also seen gross crossreactivity of MAbs to several peptides with related sequences on pins. In our limited experiences with MAbs raised against peptide immunogens, we have seen only linear epitopes represented.

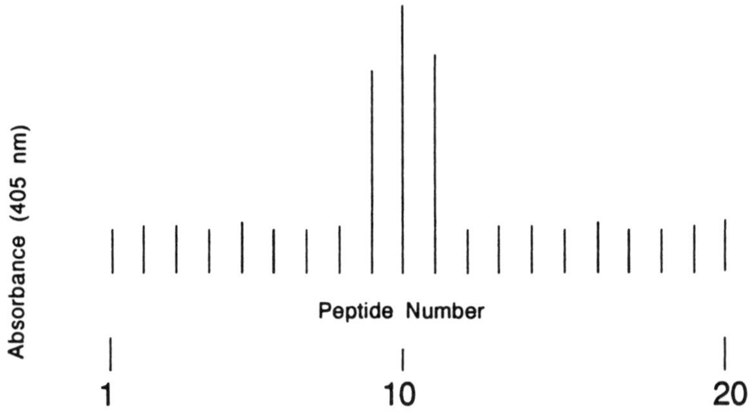

peptide number	residue number of amino acids in peptide
peptide #9:	9 10 11 12 13 14 15 16
peptide #10:	9 10 11 12 13 14 15 16 17
peptide #11:	11 12 13 14 15 16 17 18
common region:	11 12 13 14 15 16

Fig. 3. Schematic of overlapping peptides representing a linear epitope. Peptides 1–20 are overlapping 8-mer peptides. The figure shows their reactivity with an antibody preparation. Peptides 9, 10, and 11 are reactive. The amino acid sequence in common to these three peptides is 11–16. This sequence represents the epitope of the test antibody.

3.4.6. Human Serum Epitopes

Immunologists who regularly work with human sera are generally familiar with its idiosyncracies. Chief among these is a remarkably high background reaction in ELISAs. This is presumably the result of the broad sensitization of humans, owing to diversity of exposure experiences, as well as a large amount of low-specificity antibody in the naive state. Monkey sera exhibit these same problems, although to a lesser degree.

A brief treatment at 56°C will kill most disease organisms and viruses, as well as neutralizing complement and many other serum proteases. However, heat treatment increases nonspecific "stickiness" of the serum. We have found that an ELISA blocking solution based on 2% casein is very effective in reducing this background reactivity.

3.4.7. Differential Responses

Different species of immune animals will often react to different sets of epitopes in a given immunogen. Different individuals within a species often react differently, especially if they are "outbred" (not genetically homogenous). Even with inbred strains, differences will arise because of heterogeneity in the animals' immune history and in injection technique.

3.5. Other Applications

3.5.1. Mimetopes

One application now widely touted by Geysen is the construction of mimetopes. These are artificial epitopes comprising peptides containing natural and nonnatural amino acids in nonnative sequences *(7,8)*. Mimetopes can attain conformations in assays that have the same binding characteristics of naturally occurring conformational epitopes. It seems probable that mimetopes may also be able to elicit antibodies with affinity for naturally occurring conformational and even nonprotein (e.g., carbohydrate) epitopes *(9)*. For this reason, they are promising candidates for future vaccines.

3.5.2. Cleavable Pins

Another approach utilizes the chemical spacer built onto the peptides. This is the nonpeptide moiety that attaches the peptides to the plastic support pins. Incorporation of an acid-labile amino acid sequence (Asp-Pro) at this position in the peptide facilitates acidolytic cleavage from the pin after synthesis is completed. This results in generation of a large number of soluble peptides although in limited quantities. This technique has proven useful in studies demonstrating T-lymphocyte epitope specificity through mitogenesis assays *(10)*. More recently, pins bearing a water-cleavable chemical link have become commercially available from Chiron *(11,12)*. Depending on the linker and respective cleavage chemistry, the new pins can be used to generate peptides with C-terminal free carboxy acids, carboxamides, or diketopiperazines.

One fairly simple variation is the use of proteins other than antibodies to probe the peptide pin arrays. This approach is promising for structure–function studies on biological receptor molecules. Another example takes advantage of the reversibility of binding of antibodies to peptides in the typical ELISA application of the pins. By eluting the bound antibodies from the individual peptides, it is possible to affinity purify small quantities of antibody. The amounts of antibody protein isolated from each pin by this technique are vanishingly small, but sufficient to be detected by means of binding to Western blots.

4. Notes

1. Before deprotection, pins are first pre-equilibrated in DMF baths for 5 min. This step probably reduces nonspecific attachment of the piperidine molecule to the polyethylene pin matrix. Other chemists have increased the piperidine concentration in order to shorten the time for deprotection *(12)*.
2. At WRAIR, the original piperidine solution is reused every day (each deprotection cycle) for the entire synthesis. This is possible because the reaction of piperidine with the Fmoc groups on the peptide pins is stoichiometric: each mole of Fmoc removed requires only 1 mol of piperidine. This means that very little of the piperidine is actually consumed in one use of the reagent solution. With this in mind, and partly because of the difficulty and expense of obtaining and storing large quantities of piperidine (which is a controlled substance), we investigated the possibility of reusing the 20% piperidine solution. We found that, after 13 daily uses for deprotection of the peptide pins, the solution was as effective as it was when freshly prepared. This was in spite of the observation that, after 4 d, it began to develop a white crystalline material. We presumed that this material was a piperidine formate salt resulting from hydrolysis of the DMF solvent (by atmospheric water in the presence of the piperidine base as a catalyst). When we repeated the experiment, storing the piperidine/DMF reagent over molecular sieves to keep it anhydrous, the crystalline material did not form, and the reagent maintained its clarity as well as its efficacy.
3. When washing, we found that it was necessary to rinse both sides of the pin blocks in order to remove contamination resulting from splashes and condensation of solvents and reagent, which otherwise accumulate on their undersides.
4. From the residual odor remaining on the pins, it is apparent that the piperidine is not completely removed by the organic solvent washes alone, as described above. Because piperidine is a base, its presence is easily confirmed by testing the pH of an aqueous solution of the final wash solvent. Worried that residual piperidine will affect the peptide syntheses, we now typically add a 5-min wash in 1% acetic acid (freshly prepared) in DMF. This is intended to neutralize the piperidine base and reduce its affinity for the polyethylene. After washing with this modified protocol, we prepared a 50% solution of the final methanol wash in water, and found that its pH was neutral, indicating essentially complete removal of piperidine. Note that this acid-wash step is not appropriate for chemistry modification incorporating *in situ* activation of the amino acid derivatives.

5. If you elevate the concentration of amino acid derivative and HOBt in the amino acylation solution to 100 m*M*, you may reduce the time for acylation to 4 h. This allows two or even three amino acids to be added to the growing peptides in a 24-h period. Some have suggested the use of BOP or HBTU equimolar with the amino acid derivative, to accelerate the coupling reaction (*see also* Chapter 1, *PSP*). Still others have substituted dimethylacetamide for the DMF solvent in the coupling step, claiming that this improves solvation efficiency of the pin matrix *(13)*. Finally, Geysen himself now uses Fmoc amino acid derivatives as free carboxy acids, activating *in situ* with dicyclohexylcarbodiimide.
6. In order to expedite and improve the accuracy of the placement of amino acid esters in the appropriate microtiter plate wells, an automated indexer is used in our laboratory *(2)*. Driven by menu software on a PC-type computer, this device indicates the appropriate wells for each of the amino acids for the synthesis *(14)*. Chiron is expected to market a similar device.
7. The laboratory at WRAIR typically makes peptides 6–12 amino acids in length. Shorter molecules may not have a measurable affinity for the test antibody, whereas longer molecules will probably contain little of the full-length peptide because of the limited efficiency of the non-sequence-optimized coupling chemistry.
8. We used an ultrasonic cleaner instrument manufactured by Blackstone and rated for 500 W at 25 kHz. This sonicator has an electrical heater and a thermostat that we operate at 70°C. Our several attempts using less-powerful sonicators were ineffective, resulting in high-background signals in the ELISA and residual protein on the pins, as detected via amino acid analysis. Similarly, poor results were obtained when the bath temperature was allowed to drop below 60°C.
9. Our sonicator has a vol of 20 L, so it can fit eight blocks of peptide pins at once, floating in a single layer on the top. Although we add fresh 2-mercaptoethanol every day, we routinely reuse the disruption buffer 10 or 12 times, until it begins to darken.
10. To keep the silica gel from intimate contact with the pins, pouches may be made from paper towels, filled with a generous handful of indicator-grade silica, and then stapled shut. These silica pouches may be regenerated when necessary by baking overnight at 120°C.
11. To keep the silica gel from intimate contact with the pins during storage, pouches may be made from paper towels, filled with a generous handful of indicator-grade silica, and then stapled shut. These silica pouches may be regenerated when necessary by baking overnight in an oven at 120°C.
12. It is critical to avoid microbial contamination of the peptide pins. Amino acid analysis indicates that the peptides are rapidly destroyed by microbial

action. Indeed, pins left overnight in PBS at room temperature are thereby completely ruined. It is also probably best to prevent any of the ELISA solutions from drying onto the pins.

13. In all buffers and reagent solutions used for the pin ELISA, 0.1% Tween 20 is typically added. Tween 20 is a very mild nonionic detergent. It serves as a wetting agent, thereby improving reproducibility and helping to reduce nonspecific binding. Because Tween 20 is surface active, all pipeting should be performed carefully so as to minimize aeration, since foaming will affect reproducibility.
14. Two percent casein gives lower background for some antibody, such as human serum. Either of the blocking buffers described will keep for 2 or 3 wk if sterility is maintained. If desired, 0.2% NaN_3 may be added. This will increase the practical storage time for the reagent to several weeks, but 1 L is typically consumed in a few days of ELISA work. Blocking is generally performed for 1 h at room temperature, but if the solution contains 0.2% sodium azide, it may be left overnight in the refrigerator.
15. As an alternative to a rocking platform, we have used a rotating (orbital) platform for incubations. It seems to work just as well.
16. For the overnight incubation with test antibody, put the filled plates into a sealable plastic box lined with a moistened paper towel to maintain humidity and minimize evaporation. If this step is allowed longer than about 12 h, evaporation and condensation may nonetheless begin to affect reproducibility, especially for the pins closest to the edge of the plate. After the overnight incubation, the first antibody solution is usually discarded. However, we have occasionally pooled and reused this reagent up to four times without any discernible loss in signal-to-noise ratio. In these cases, we have added a single wash step between the blocking and first antibody to minimize dilution and contamination of the valuable test antibody solution. Of course, we also stored the antibody solution at 4°C.
17. For PBST, we purchase 10X PBS in liter bottles, and then add Tween 20 and sodium azide. We store this 10X stock solution in a carboy at room temperature for up to 2 wk. From the stock, we prepare 1X PBST for each day's use by diluting 1/10 with deionized water.
18. Remember that proper reactions for control peptides, if they are used, will probably require a different first antibody solution. They may also require a different second antibody solution.
19. For second antibody (enzyme conjugates), we have successfully used commercially available reagents from various sources, as well as our own conjugates. Although we have also used conjugates with alkaline phosphatase (AP) and horseradish peroxidase (HRP), we prefer the AP conjugates for maximum reproducibility with good sensitivity since the HRP substrate

contains peroxides that damage the peptides. We perform incubations for second antibody binding on the laboratory benchtop, but it does not hurt to do it in the refrigerator or to use a rocking platform. This incubation may also be performed overnight.

20. In some cases, it may be necessary to prepare your own enzyme-conjugated second antibody. To make an AP-conjugated goat antibody to recognize *Aotus* monkey antibody in ELISAs, we used the following protocol: Isolate several milligrams of nonimmune antibody from the monkey serum by protein A affinity. Use most of this protein as an immunogen to raise antibody in a goat. Isolate several milligrams of goat immune antibody by protein A affinity. Couple a few milligrams of the *Aotus* antibody immunogen to Sepharose. Use the immobilized *Aotus* antibody to affinity purify the goat antibody vs *Aotus* antibody. Conjugate the purified antibody to commercial AP via glutaraldehyde. Dilute in PBS with 0.2% NaN_3. Test the second antibody conjugate at various concentrations to determine the appropriate working concentration for the reagent. Store frozen, avoiding refreezing. For details, refer to Lyon and Haynes *(15)*.
21. To aid in visualization of color development, place the plate containing the substrate solution on a piece of white paper. If you are using more than one plate for the ELISA (which is likely), number them on their outer edges. During development, avoid thermal gradients. These may be caused, for example, by drafts or sunlight. We cover the pins to isolate them from environmental effects.
22. After development, removal of the pins stops the color generation catalyzed by the enzyme conjugate on the pins, but the substrate is thermolabile, and the chromophore is photolabile, so avoid unnecessary delays by setting up the reader while development is still taking place. Avoid touching the bottom of the ELISA plate before it is read. Do not discard the plates until you are certain that you have two legible copies (either "soft" or "hard" copies, according to your preference) of your data. If the signals are weak, you may return the pins to the plates for further development and read them again later.

References

1. Geysen, H. M., Meleon, R. H., and Barteling, S. J. (1984) Use of peptide synthesis to probe viral antigens for epitopes to a resolution of a single amino acid. *Proc. Natl. Acad. Sci. USA* **81,** 3998–4002.
2. VanAlbert, S., Lee, J., Lyon, J. A., and Carter, J. M. (1991) Amino acid indexer for synthesis of Geysen peptides, application US Patent # 5,243,540.
3. Barlow, D. J., Edwards, M. S., and Thornton, J. M. (1986) Continuous and discontinuous protein antigen determinants. *Nature* **322,** 747–748.

4. Amit, A. G., Mariuzza, R. A., Phillips, S. E. V., and Poljak, R. J. (1986) Three dimensional structure of an antigen–antibody complex at 2.8 Å resolution. *Science* **233,** 747–753.
5. Sheriff, S., Silverton, E. W., Padlan, E. A., and Cohen, G. H. (1987) Three-dimensional structure of an antigen–antibody complex. *Proc. Natl. Acad. Sci. USA* **84,** 8075–8079.
6. Thomas, A. W., Carter, J. M., and Lyon, J. A. (1994) Identification of biologically significant epitopes in *Plasmodium falciparum* gp195 protein and protective results of peptide vaccine trial in monkeys, submitted.
7. Geysen, H. M., Rodda, S. J., and Mason, T. J. (1986) A priori delineation of a peptide which mimics a discontinuous antigenic determinant. *Mol. Immunol.* **23,** 709–715.
8. Geysen, H. M., Rodda, S. J., and Mason, T. J. (1986) Synthetic peptides as antigens. *Ciba Foundation Symposium* **119,** 130–149.
9. Geysen, H. M., MacFarlan, R., Rodda, S. J., Tribbick, G., Mason, T. J., and Schoofs, P. G. (1987) in *Towards Better Carbohydrate Vaccines* (Bell, R. and Torrigiani, G., eds.), Wiley, Chinchester, pp. 103–118.
10. Van der Zee, R., van Eden, W., Meloen, R. H., Noordzij, A., and van Embden, J. D. A. (1989) Epitope mapping and characterisation of a T-cell epitote by the simultaneous synthesis of multiple peptides. *Eur. J. Immunol.* **19,** 43.
11. Bray, A., Maeji, N. J., and Geysen, H. M. (1990) The simultaneous multiple production of solution phase peptides; assessment of the Geysen method of simultaneous peptide synthesis. *Tetrahedron Lett.* **31,** 5811–5814.
12. Maeji, N. J., Bray, A. H., and Geysen, H. M. (1990) Multi-pin peptide synthesis strategy for T cell determinant analysis. *J. Immunol Methods* **134,** 23–33.
13. Arendt, A. and Hargrave, P. A. (1991) Optimization of peptide synthesis on polyethylene pins. *Twelfth American Peptide Symp.* Poster 269.
14. Carter, J. M., VanAlbert, S., Lee, J., Lyon, J. A., and Deal, C. D. (1992) An aid to peptide pin syntheses. *Biotechnology* **10,** 509–513.
15. Lyon, J. A. and Haynes, J. D. (1986) *Plasmodium falciparum* antigens synthesized by schizonts and stabilized at the merozoite surface when schizonts rupture in the presence of protease inhibitors. *J. Immunol.* **136,** 2245–2251.

CHAPTER 13

Analysis of Proteinase Specificity by Studies of Peptide Substrates

The Use of UV and Fluorescence Spectroscopy to Quantitate Rates of Enzymatic Cleavage

Ben M. Dunn, Paula E. Scarborough, Ruth Davenport, and Wieslaw Swietnicki

1. Introduction

Synthetic peptides are used for many purposes in chemistry and biology. Among these, one of the most profitable has been the exploration of the specificity of proteolytic enzymes through quantitative studies of the enzymatic cleavage of sets of peptide substrates with systematic changes in specific positions. Similar information can be obtained when the inhibition of enzymatic activity by sets of peptides or peptide derivatives is studied; however, substrate studies have one major advantage in that the position of cleavage of a substrate peptide will always report on the orientation of the peptide in the active site. Inhibition might occur through the binding of peptides to different regions of the active site or, in extreme cases, through binding outside the active site.

This consideration of the orientation of active site binding is especially significant when the activity of proteolytic enzymes is under study, since most of these enzymes have a large, extended active site crevice where binding occurs. Therefore, it is conceivable that the binding of a linear peptide might occur in several different ways through the active site, differing in the specific subsites occupied by the amino acids of the substrate and in the particular peptide bond that would be presented near the catalytic apparatus

From: *Methods in Molecular Biology, Vol. 36: Peptide Analysis Protocols*
Edited by: B. M. Dunn and M. W. Pennington Copyright ©1994 Humana Press Inc., Totowa, NJ

of the enzyme. This is often seen when a standard substrate, such as the oxidized B-chain of insulin, is utilized to examine the specificity of a newly discovered proteinase. Multiple bonds are often cleaved by the enzyme, providing some information on the preferences of that proteinase, but also revealing the multiple-binding interactions possible with such enzymes.

In the studies described in this chapter, we prefer to limit the cleavage of substrate peptides to a single peptide bond, thus simplifying enzyme kinetic studies. This has been accomplished in the first instance by fixing the P_1 and P_1' residues in a substrate series as those that satisfy the primary specificity of the enzymes under study. The large, hydrophobic, aromatic amino acids, Phe and *p*-nitrophenylalanine (Nph), are used as the P_1 and P_1' residues, respectively. The aspartic proteinase family of enzymes have large and very hydrophobic S_1 and S_1' pockets bracketing the catalytic apparatus. Thus, peptides with the —Phe*Nph— sequence typically bind to the active site of enzymes of this family with Phe in the S_1 pocket and with Nph in the S_1' pocket, leading to cleavage between these two residues only.

The residues in the peripheral positions, P_2–P_n, on the left-hand side of the point of cleavage, and P_2'–$P_{n'}$, on the right-hand side of the cleavage point, can then be varied over a fairly large range. The effects of the resulting structural variation on the kinetics of cleavage of the —Phe*Nph— bond report on the strength of interaction of the peripheral amino acids of the substrate in the other subsites of the active site cleft and on the influence that those interactions have on catalysis.

Quantitation of the rates of cleavage in such peptide series has been facilitated by the shift in UV absorbance that occurs when the —Phe*Nph— bond is cleaved. This shift in absorbance maximum, from 278–280 nm to 270–272 nm, is subtle, but large enough to be accurately quantitated by sensitive spectrophotometers. The balance of this chapter will describe the experimental protocol for efficient and accurate studies of cleavage rates of peptides of this type.

Two further points should be made before proceeding: First, any enzyme that can accommodate the Nph residue in the P_1' position could be studied by this method, although it might require a different residue in the P_1 position to achieve recognition by the enzyme. For example, we have studied HIV proteinase using substrates with a number of different hydrophobic amino acids in P_1 *(1)*. Also, Hofmann and Hodges utilized the sequence —Lys*Nph— in the preparation of excellent substrates for the proteinase from *Penicillium janthinellium (2)*.

Second, it must be understood that the —Phe*Nph— peptide bond is a normal peptide bond. The *p*-nitro substitution on the aromatic ring of Phe does **not** create an activated peptide bond. The rates of cleavage of —Phe*Phe— and —Phe*Nph— peptide bonds are nearly identical. Some confusion on this point has arisen because of the historical use of *p*-nitrophenyl anilides as substrates for chymotrypsin, for example. *p*-Nitrophenyl anilides, where the *p*-nitro group is in direct resonance with the nitrogen involved in the amide bond, are more highly reactive, and their kinetics of cleavage do reflect this activation. In the —Phe*Nph— peptide bond, such a direct resonance interaction does not occur, since the nitrophenyl group is separated from the nitrogen of the peptide bond by two saturated carbon atoms, effectively preventing resonance.

The preceding section has described our general approach to the study of proteinase specificity. The emphasis above has been on the use of substrates containing peptide bonds that have an Nph residue in the P_1' position, leading to a measurable change in absorbance properties on cleavage by an enzyme. However, many proteolytic enzymes of great interest are unable to accommodate the Nph residue in the P_1' position most likely because of a restricted space or a hydrophilic pocket provided in the S_1' region of the active site. In particular, enzymes present in the picornavirus family prefer a very small residue in the P1' position, typically Gly, but occasionally Ala or Ser.

In such cases, the quantitative study of enzyme catalysis is still possible, but a shift in strategy is required. In these cases, we prefer to use the internally quenched fluorescent substrate approach, described by Yaron et al. many years ago *(3)*. Improvements in chromophores have made this a reasonably convenient strategy, although several problems can arise (*see* Notes 1–3).

First, the addition of large, hydrophobic chromophores onto relatively small peptides can frequently cause major changes in the solubility or the binding of the peptides to proteins, including proteolytic enzymes. This can have deleterious consequences for the substrate properties of the peptide. Fortunately, this has proven to be rare, with the approx 50-fold reduction in rate of cleavage reported by Weidner and Dunn for poliovirus 3C proteinase-catalyzed cleavage of Ac-Arg-Cys[CPM]-Nle-Glu-Ala-Leu-Phe-Gln-Gly-Pro-Leu-Tyr-Lys[DABTC]-Asp, the classic example *(4)*.

The second, and unfortunately more prevalent, problem that arises with fluorescent substrates is that the fluorescence measurement is never as linear with concentration as the absorbance measurement. Therefore, the linear range of fluorescence response is invariably restricted to the range below 50-µmol substrate concentration. This usually prevents the determination of the apparent K_m parameter by variation of substrate concentration, and typically makes the determination of V_{max} values problematic as well. As a consequence, comparison between substrates is limited to the k_{cat}/K_m parameter, which, of course, is the preferred measure of enzyme specificity. Thus, the limitation on the concentration range does not cause insurmountable difficulty in exploring enzyme specificity. As described in the discussion, however, care must be taken in using this method.

To construct an internally quenched fluorescent substrate, one simply places appropriate chromophores at the ends of a peptide sequence that fits the specificity of the enzyme under study (*see* Section 3. for description of procedures). For example, the sequence Leu-Arg-Thr-Gln-Ser-Phe-Ser was found to be an acceptable substrate for the proteinase from the hepatitis A virus, HAV 3C proteinase, with cleavage occurring between the Gln and Ser residues *(5)*. Studies by Petithory and colleagues demonstrated that the P_2' Phe residue could be replaced with nearly any other amino acid *(6)*. Therefore, this residue was replaced with the same Nph residue used in our other studies, but in this case, it is being used as a quencher of fluorescence of a dimethylaminonaphthalenesulfonate (dansyl) group attached to the amino terminus, yielding dansyl-Leu-Arg-Thr-Gln-Ser-Nph-Ser. This peptide was still cleaved between the Gln and Ser residues at a rate nearly equivalent to that of the parent peptide. On hydrolysis of the Gln*Ser peptide bond, the separation between the dansyl group on one product and the Nph on the other product is increased, and the fluorescence quenching is diminished, leading to a direct measurement of the rate of cleavage.

In the case describe above, the increase in fluorescence on relief of quenching is approximately twofold. In other cases, including those reported by Matayoshi et al. *(7)*, where different pairs of chromophores are utilized, the increase in fluorescence can be as large as 50-fold. Obviously, the larger the fluorescence change on enzyme-catalyzed hydrolysis, the more sensitive the assay will be for the presence of the enzyme. Thus, the fluorophore-quencher pair that gives the largest change in readily measurable fluorescence, while still yielding a peptide that retains kinetic properties similar to the cleavage of the natural cleavage site peptide, should be employed.

2. Materials

1. The peptides used in these studies were all synthesized on an Applied Biosystems (ABI, Foster City, CA) 430A peptide synthesizer, using procedures described below, and reagents from ABI. N^{a}Boc-*p*-NO_2Phe (Boc-Nph) was purchased from Chemical Dynamics (South Plainfield, NJ).
2. Enzymes were either obtained from commercial sources (porcine pepsin, from Sigma, St. Louis, MO), from recombinant methods *(8)*, or purified from natural sources *(9)*. The enzyme, HAV 3C proteinase, was supplied by Malcolm and colleagues, at the Chiron Corporation, (Emeryville, CA).
3. Chromogenic activity assays employ a Hewlett-Packard (Sunnyvale, CA) 8452A diode array spectrophotometer equipped with a seven-place multicell transport thermostated to 37°C by a circulating water bath. Also required are a 37°C heating block and a vortex mixer.
4. Spectrophotofluorometric assays utilized an SLM 4800C system (SLM Instruments, Inc., Urbana, IL) equipped with a four-place multicell turret thermostated to 37°C by a circulating water bath. Also required is an IBM-compatible computer (PC/AT or higher) with a math coprocessor to control the spectrofluorimeter (*see* Note 4).

3. Methods

3.1. Synthesis of Peptide Series in Which One Residue Is Varied

The example given is for the preparation of the series Lys-Pro-P_3-Lys-Phe-Nph-Arg-Leu, where P_3 represents the amino acid to be varied.

1. Initiate a 0.5-mmol scale synthesis by adding Boc-Leu-Pam resin (ABI) to a clean reaction vessel and running the manufacturer's coupling procedures. For coupling of Boc-Nph, weigh 2 mmol of the reagent into a clean Phe cartridge and use the rboc 1d, cboc 1d, and aboc 1d programs for the reaction, concentration, and activation vessels, respectively. When entering the sequence into the synthetic program, use Phe for the sequence position where the Nph is to be added. (For the small-scale, rapid-cycle procedures, 0.1 mmol, use cycles rboc 21, cboc 23, and aboc 23). Continue the synthesis until the point at which variation is desired.

2. Dry the resin by running program end-dry twice. Weigh the dried resin on a balance that reads accurately to 0.1 mg, and split into five portions. Place each aliquot in a separate 0.1-mmol reaction vessel.
3. The synthesis is then continued until the end. Schematically, this would appear as follows (Scheme 1) for variation at the P3 position.
4. Cleave the peptides, following complete drying using the rdry21 cycle, by standard HF methods, (*see* Chapter 4, *PSP*). Analyze the resulting peptides by amino acid analysis, RP-HPLC (*see* Chapter 3), capillary electrophoresis (*see* Chapter 6), and mass spectroscopy (*see* Chapter 7), if available.

3.2. Synthesis of Fluorescent Substrate

The example given is for the preparation of DNS-Leu-Arg-Thr-Gln-Ser-Nph-Ser.

1. Initiate a 0.5-mmol synthesis by adding that amount of N^{α}Boc-Bzl-Ser-PAM resin to a clean, small-scale reaction vessel. Run the standard coupling programs of the manufacturer to add Boc Arg(Tos) in the first coupling step. Then add Boc-Nph using the procedure described above in Section 3.1.
2. Continue synthesis until the complete peptide is assembled, utilizing standard procedures of the manufacturer.
3. Deprotect the amino-terminal Leu residue by running program N-end of ABI, and neutralize the resulting NH_3^+ group by washing three times with 10% diisopropylethylamine (DIEA). Dry resin with end-dry cycle, and transfer resin to a clean polypropylene tube.
4. Weigh 115 mg of DNS-Cl into a clean polypropylene tube, dissolve with a 5/1 mixture of DCM and DMF, and add 200 µL of DIEA. Add this mixture to the tube containing the deprotected resin, cover the tube with aluminum foil, and allow the reaction to proceed for 1 h. Add 200 µL DIEA, and allow the coupling to proceed for an additional hour. Then repeat the neutralization and the reaction with a second portion of DNS-Cl to ensure complete coupling.
5. Transfer the slurry into a clean reaction vessel, wash the resin with DMF and DCM, and dry the resin first with air pressure. Then assemble the reaction vessel by adding a top, and complete the drying with the end-dry cycle on the synthesizer.
6. Cleave the dansylated peptide using standard HF conditions, as described in Chapter 4 of this vol. Verify the purity of the resulting peptide by RP-HPLC (*see* Chapter 3), and purify by HPLC, if needed.

3.3. Chromogenic Enzyme Activity Assays

Within one cycle of an assay, the decrease in absorbance in the range of 284–324 nm is monitored at 0.1-s intervals every 2 nm of the wavelength range over a total time of 0.5 s. The readings are averaged to give

Boc-Leu-Pam-resin

↓ add Boc-Arg

↓ add Boc *p*-nitroPhe

↓ add Boc Phe

↓ add Boc Lys

Boc-Lys-Phe-Nph-Arg-Leu-resin

↓ weigh, split into five portions

↓ add Boc Ala	↓ add Boc Asp	↓ Add Boc Leu	↓ add Boc-Arg	↓ add Boc-Ser
↓ add Boc-Pro	↓ add Boc-Pro	↓ add Boc-Pro	↓ add Boc-Pro	↓ add Boc-Pro
↓ add Boc-Lys	↓ add Boc-Lys	↓ add Boc-Lys	↓ add Boc-Lys	↓ add Boc-Lys
A.	B.	C.	D.	E.

cleave each separately to yield:

A. Lys-Pro-**Ala**-Lys-Phe*Nph-Arg-Leu

B. Lys-Pro-**Asp**-Lys-Phe*Nph-Arg-Leu

C. Lys-Pro-**Leu**-Lys-Phe*Nph-Arg-Leu

D. Lys-Pro-**Arg**-Lys-Phe*Nph-Arg-Leu

E. Lys-Pro-**Ser**-Lys-Phe*Nph-Arg-Leu

Scheme 1. Diagrammatic representation of the synthesis of the P_3 set of peptides.

one time-point. For each cuvet, the cycles are repeated with a cycle time of 17.2 s. Data are typically collected for 1000 s. The initial velocity is calculated from the slope during the linear phase of the reaction and plotted vs substrate concentration at the start of the reaction. These data are fitted to the standard Michaelis-Menten equation by Marquardt analysis to yield calculated values of V_{max} and apparent K_m or K_i. Enzyme preparations are titrated with a tightly binding inhibitor. Specific methods are described below, but may be altered according to the objectives of a specific experiment.

3.3.1. Stocks, Buffers, and Cuvets

1. Dissolve synthetic peptides with Nph incorporated at the P_1' site in sterile-filtered deionized water to a concentration of approx 10 mg/mL to make stock solutions.
2. Hydrolyze duplicate samples of stock solutions, and analyze by amino acid analysis to determine peptide concentration accurately.
3. Prepare dilution series, typically with 12 dilutions ranging from 62.5–3125 µ*M*, by mixing measured volumes of the stock solutions with sterile-filtered deionized water. When 20 µL of these dilutions are used in 250-µL assays, the peptide concentration range examined experimentally is 5–250 µ*M*. Store peptide solutions at 0–5°C.
4. Prepare 2X reaction buffer, usually at an ionic strength of 0.2*M* (except in the case of variable ionic strength experiments), by mixing, for example, 0.4*M* sodium acetate, 0.4*M* acetic acid (in the appropriate ratio to give the desired pH value, based on the Henderson/Hasselbach equation), 0.4*M* sodium chloride, and water. The total volume of 0.4*M* acetate plus water is equal to the total volume of 0.4*M* acetic acid plus 0.4*M* sodium chloride, ensuring that the final ionic strength is independent of the final pH. Also, the volume of acetate plus acetic acid is half of the total, yielding a final concentration of acetate species of 0.2*M*. Buffering components vary with desired pH: sodium formate for buffers of pH 3.0–4.4; sodium acetate for buffers of pH 4.4–5.4; and 2-[*N*-morpholino]ethanesulfonic acid (MES) for buffers of pH 5.6–6.8. Filter sterilize 2X buffers and deionized water with 0.2-µm filters before use. Store buffers at 0–5°C.
5. Prepare enzyme stock solutions at concentrations of ~1 mg/mL in precooled microfuge tubes with ice-cold, sterile-filtered deionized water or, for less-stable enzymes, an enzyme stock buffer. For chromogenic assays, make dilutions from enzyme stocks into precooled tubes with ice-cold, filter sterilized deionized water or enzyme dilution buffer such that, within a vol of ~1–5 µL, ~0.01–1 µg enzyme will be introduced into the 250-µL assay depending on optimal enzyme concentration determined (*see* Section 3.3.2.). Store 100–200 µL aliquots of enzyme dilutions at –20°C. It is important for the pH values of all enzymes solutions to be in a range such that the enzyme does not undergo self-degradation.

6. Dissolve inhibitors typically in dimethylsulfoxide (DMSO) at stock concentrations of 1 or 10 m*M*. If an inhibitor contains within its structure at least one amino acid, hydrolyze duplicate samples of stock solutions and analyze by amino acid analysis to determine inhibitor concentration accurately.
7. Make 10-fold serial dilutions of inhibitor stocks, also in DMSO. Store inhibitor solutions at 0–5°C. DMSO tolerance (<10% inhibition of control reaction rates) of a particular enzyme must be determined for inhibitor experiments. The concentration of DMSO allowable in kinetic experiments typically will not exceed ~4% DMSO or no more than 10 µL DMSO in a 250-µL reaction.
8. Clean quartz cuvets with detergent, 95% ethanol, deionized water, and then acetone. Prewarm cuvets, numbered 1–6 plus a Blank, in the thermostated seven-place cell holder of the spectrophotometer for at least 4 min before assay.

3.3.2. Optimal Enzyme Concentration Determination

Prior to K_m or K_i determination experiments, the enzyme concentration necessary in the assay is determined.

1. Measure into a microfuge tube (without cap) 125 µL 2X buffer, ~0.1 µg enzyme (should be in ~1–5 µL; for solution of unknown concentration use ~1–2 µL), and adjust the volume to 230 µL with sterile-filtered deionized water.
2. Mix the solution on a vortex mixer, and prewarm by floating in water-filled holes of a 37°C heating block for 4 min. During this time, add 20 µL of a 625 µ*M* dilution of a known good substrate to a second microfuge tube such that, on mixing, the substrate concentration in the assay is 50 µ*M*.
3. Place the second tube in the heating block to prewarm. At the end of the 4-min enzyme incubation time, add the 230-µL enzyme mixture to the 20-µL peptide solution, mix quickly on vortex mixer, and then pipet quickly into a prewarmed cuvet in the spectrophotometer cell holder.
4. Monitor the reaction for 5–15 min in order that the duration of initial velocity may be determined. Adjust the enzyme concentration for subsequent reactions such that initial velocities may be measured by a linear reaction trace over at least 100–150 s and up to ~10 min. If no decrease is observed in the first reaction, the following troubleshooting actions are suggested:
   a. After ~5 min, the reaction mixture should be scanned for the absorbance spectrum ~200–350 nm. If the peak of absorbance has shifted from 280 to 272 nm, peptide cleavage is indicated, and equilibrium of the reaction may have been reached before monitoring was begun. If so, an additional reaction may be done at 5- to 10-fold less enzyme in an attempt to capture the initial velocity phase of the reaction.

b. If the peak of absorbance has not shifted from 280 nm, an additional reaction may be done at 5- to 10-fold more enzyme to make sure there is enough enzyme to give detectable cleavage of peptide.

3.3.3. K_m Determinations

After the optimal enzyme concentration has been determined that will yield reactions in which initial velocities may be measured for at least 100 s and up to 10 min, K_m determination experiments may proceed.

1. Into a microfuge tube rack, place two rows of six microfuge tubes (without caps; numbered one to six) plus two additional tubes, one for a master mix of 2X buffer, water, and enzyme, and the other for mixing the blank (1X buffer).
2. The components of the master mix are calculated in the following manner:
   a. Calculations are made based on 6.25 × 250 µL reactions from which six reactions are prepared. Therefore, the total vol to be considered is 1562.5 µL.
   b. Half the total volume, 781.25 µL, is 2X buffer. The remaining volume is accounted for in substrate, enzyme, and water.
   c. Although not included in the master mix, the substrate volume is taken into account. Subtract 6.25 × 20 µL or 125 µL to leave 656.25 µL to be made up by enzyme and water.
   d. The enzyme amount in each assay reaction should typically range from 1–5 µL, so the volume accounted for by enzyme will range from 6.25–31.25 µL.
   e. Finally, adjust the remaining volume with filter sterilized deionized water.
3. Vortex the master mix containing 2X buffer, enzyme, and water thoroughly, and aliquot 230 µL into each of six numbered microfuge tubes. Prewarm the tubes in the heating block for 4 min. During this incubation, add 20 µL of six different concentrations of a peptide dilution series to the **other** set of six numbered tubes. Prewarm the tubes during the remainder of the enzyme incubation.
4. At the end of the 4 min, beginning with #1 through #6, add the 230-µL enzyme mixture to the 20-µL peptide solution, mix quickly with a vortex mixer, and pipet into the cuvet of the same number (*see* Note 5).
5. Monitor the reactions beginning immediately following the mixing of the final pair of tubes. Monitor reactions for up to 10–15 min. Fit a line tangent to the initial linear portion of the reaction curve such that the calculated slope yields a measure of the initial velocity.
6. Plot these data vs peptide concentration. Repeat the procedure with the remaining six concentrations of the peptide dilution series (usually odd-numbered dilutions are used in one run and the even-numbered dilutions in the second run, and then plotted together in order to detect abnormalities more easily in a given data set; *see* Fig. 1).

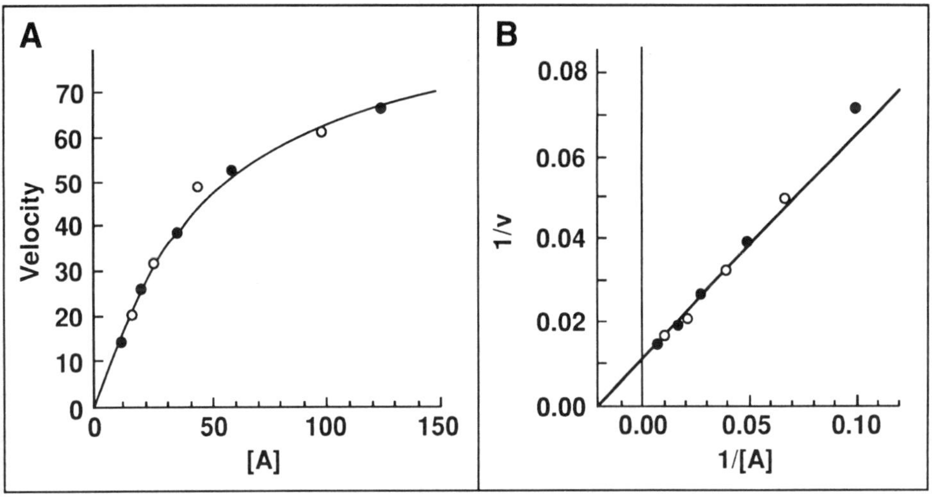

Fig. 1. Plots of v_i vs [S] for cleavage of a typical oligopeptide containing Nph in the P_1' position. **(A)** (○) Data obtained in the first set of determinations; (●) indicates data obtained in the second set of determinations. **(B)** Plot of the reciprocals of velocity ($1/v_i$) vs the reciprocals of the corresponding substrate concentration (1/[S]) with the slope yielding K_m/V_{max}, the intercept yielding $1/V_{max}$, and the ratio of those two yielding K_m.

7. Fit the data to the standard Michaelis-Menten equation by Marquardt analysis to yield calculated values of V_{max} and apparent K_m (Fig. 1).

$$v_i = V_{max}[S]/(K_m + [S]) \quad (1)$$

Two special cases should be noted here (Fig. 2). If a velocity vs peptide concentration plot is horizontally linear, the apparent K_m for the reaction may be lower than the peptide concentrations examined. If the velocity vs peptide concentration plot is diagonally linear within the peptide concentration range, the apparent K_m may be higher than the concentrations examined. Therefore, V/K values may be calculated.

3.3.4. K_i *Determinations*

K_i determination experiments are very similar to K_m determination experiments with one major exception: The inhibitor is prewarmed in the enzyme mixture for the same 4 min before mixing with substrate.

1. The first experiment is a determination of the percentage of inhibition of the initial rate of reaction in the presence of 4 μ*M* inhibitor when compared

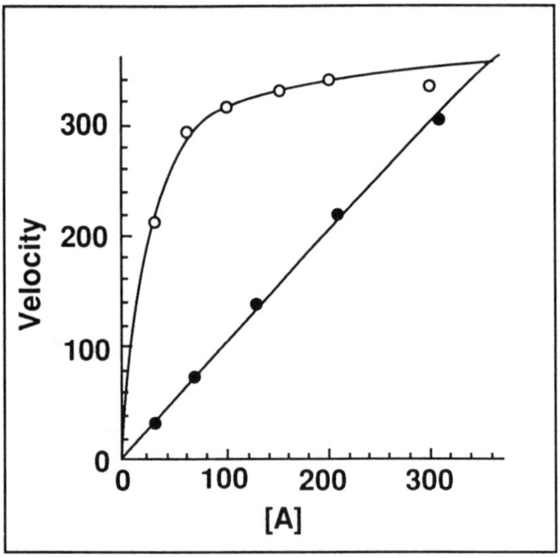

Fig. 2. Two special cases arising in studies of initial velocity as a function of substrate concentration: In one case, the measured velocity at all concentrations of substrate is about the same, indicating that the K_m value is much lower than any of the concentrations of substrate utilized; in the second case, the velocity increases linearly with increase in substrate concentration. The lack of curvature prevents the use of the Michaelis-Menten equation, as in Fig. 1, and only the ratio of V_{max}/K_m can be determined.

to a control reaction. Into one tube, measure 125 µL 2X buffer, 1 µL 1 m*M* inhibitor, and then enzyme and water to bring the vol up to 230 µL.
2. Vortex the contents, and then prewarm for 4 min.
3. Add the 230 µL to prewarmed 20 µL of the 625 µ*M* dilution of a known good substrate, and monitor the reaction in parallel with a control reaction minus inhibitor. The result will give an estimate of how potent the inhibitor is in the enzymatic reaction.

To determine the K_i of an inhibitor, the control apparent K_m of a good substrate is determined in the presence of the allowable concentration of DMSO. The DMSO is included in the master mix, thereby decreasing the volume of water to be added. From the rate vs peptide concentration plot, concentrations of substrate are selected for inhibitor experiments. The master mix is then made similar to that in K_m determinations with the exception that 6.25× inhibitor volume to be used per reaction is incorporated into the calculations, thereby decreasing the volume of

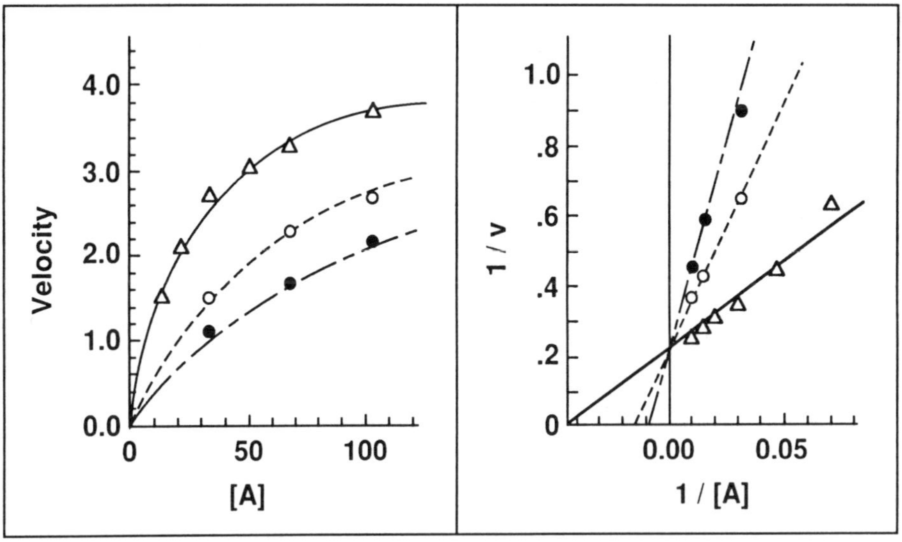

Fig. 3. Determination of K_i by analysis of the effect of two different inhibitor concentrations on the kinetics of cleavage of a good substrate.

water to be added. Also included in the master mix should be the volume of DMSO to bring the concentration up to that used in the control K_m determination.

1. Vortex the master mix, then aliquot 230 µL into tubes, and prewarm for 4 min.
2. Add 20 µL of the selected substrate concentrations to the second set of tubes and prewarm.
3. Begin the reactions with mixing, and then monitor for up to 15 min.
4. Repeat the procedure for several concentrations of inhibitor to examine fully the area under the control curve of rate vs peptide concentration. Calculate rate data, plot along with the control K_m data, and analyze to yield K_i (Fig. 3).

$$v = V^*[A]/\{K_a^*(1 + [I]/K_{is}) + [A]\} \quad (2)$$

3.3.5. Enzyme-Active Site Titrations

Active site titration reactions are prepared with fixed concentrations of enzyme and a known good substrate with variable concentrations of a tight binding inhibitor (K_i of around 1 nM or below).

1. In this case, add into each of six tubes increasing amounts of an inhibitor dilution and then DMSO to adjust the volume to the amount of allowable

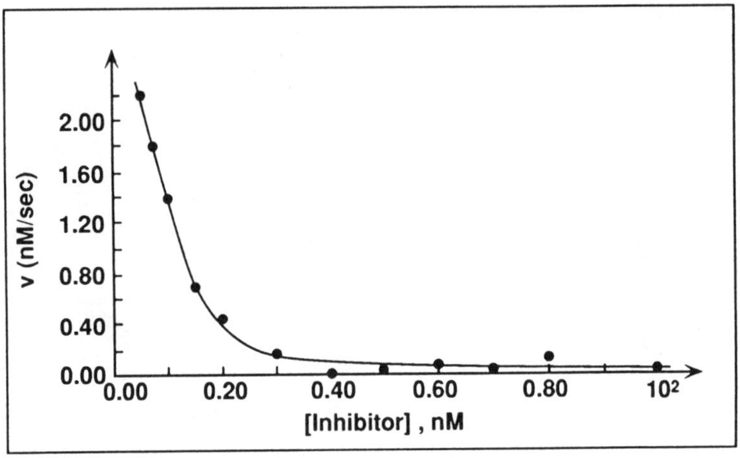

Fig. 4. Titration of enzyme activity with a tight-binding inhibitor.

DMSO. For example, into tubes #1–#6 add 0, 1, 1.5, 2, 3, and 4 µL of 10 n*M* inhibitor to examine the inhibition at reaction concentrations of 0, 0.04, 0.06, 0.08, 0.12, and 0.16 n*M* inhibitor, respectively.

2. The master mix then will contain only 2X buffer, enzyme, and water. Aliquot in 226-µL increments into tubes containing inhibitor, vortex, and then prewarm.
3. At the end of 4 min, add the enzyme/inhibitor mixtures to prewarmed tubes of substrate (typically for an assay concentration of 50 µ*M* substrate), mix, and monitor reactions.
4. Plot initial rates vs inhibitor concentration (*see* Fig. 4). Repeat the procedure with increasing concentrations of inhibitor until the rate of reaction is inhibited maximally. Analyze these data by the Henderson equation *(10)*, which accounts for multiple depletion of enzyme, to yield a measure of active site concentration.

$$v = 0.5 * v_0/[E]_t * \{(-1)*[I] - [E]_t + K_i * ([A]_t + K_a)/K_a\} + \text{SQRT}\,\{(\text{ABS}([I] - [E]_t + K_i * ([A]_t + K_a)/K_a))\text{^}2 + 4 * [E]_t * K_i * ([A]_t + K_a)/K_a)\} \quad (3)$$

3.4. Fluorescence Activity Assays

During one cycle of an assay, the increase in the emission at a given wavelength with a constant excitation (for example, emission at 555 nm with excitation at 346 nm for the DNS-L-R-T-Q*S-Nph-S peptide) is monitored for 2 s, and the readings are averaged to give one time-point.

For each of the four cells, the cycles are repeated every 16 s. Data are typically collected for 4000 s. Some specific methods are described below, but can be modified according to the needs of a particular enzyme/substrate pair.

3.4.1. Solutions and Cuvets

Because of the specific nature of fluorescence experiments, all solutions should be filter sterilized before use. Cuvets made of quartz, polystyrene, or methacrylate (Fisher Scientific, Pittsburgh, PA) should be dusted with compressed air and kept covered during the experiments.

1. Dissolve synthetic substrates in a suitable solvent (for example, sterile buffer or DMSO) to a concentration of approx 1 mg/mL to make a stock solution.
2. Hydrolyze a sample of a stock solution, and analyze for amino acid composition to determine peptide concentration accurately.
3. Prepare at least one 10-fold dilution of the stock substrate solution.
4. A typical buffer to assay hepatitis A virus 3C proteinase activity consists of 50 m*M* potassium phosphate with a pH of 7.5 and 0.2 m*M* EDTA. Prepare this buffer by dissolving 27.2 g of KH_2PO_4 and 76 mg of $Na_4EDTA \cdot H_2O$ in 950 mL of doubly distilled water and adjusting the pH with 1*N* KOH. Add water to 1000 mL. Store buffers at 0–5°C.
5. Thaw an aliquot of enzyme on ice. Prepare this sample not more than 15 min before use.
6. Whenever possible, use disposable cuvets. Some enzyme tend to adhere to glass very tightly and cannot be removed without a complicated washing procedure. Polystyrene cuvets are suitable for measurements above 340 nm, and the methacrylate cuvets can be used above 280 nm.

3.4.2. Optimal Substrate Concentration

1. Mix the reaction buffer and the substrate directly in the cuvet, and let incubate for 15 min at desired temperature. A typical total volume ranges from 0.7–1.4 mL.
2. Read the fluorescence.
3. Repeat the measurements from steps 1 and 2 with increasing amounts of substrate. The practical limit is about 0.04 m*M*.
4. Plot the fluorescence vs the concentration to determine the linear range of response (Fig. 5).
5. Choose a concentration within the linear portion of the range, and repeat step 1.
6. Add enzyme (not more than 20 µL/1400 µL of the total vol), and mix three times with a pipetor.

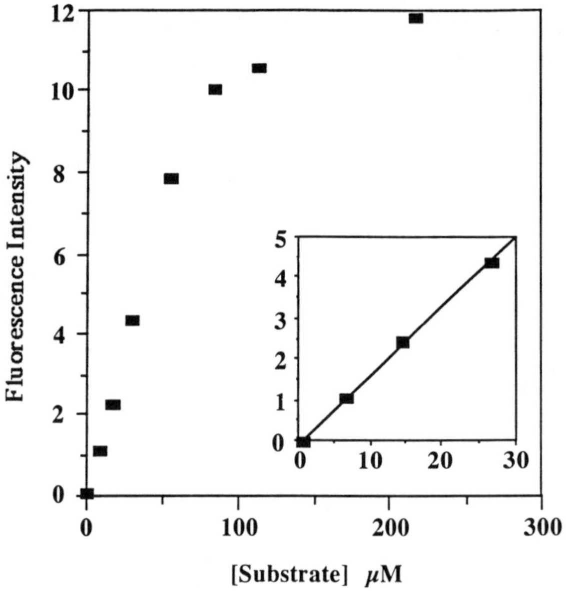

Fig. 5. Fluorescence response. Plot of observed fluorescence intensity vs concentration of fluorescent substrate. Deviation from linearity can be seen at concentrations above 100 μ*M*. Inset: Expansion of the region below 30-μ*M* fluorescent substrate concentration, demonstrating the linearity of fluorescence response in this range.

7. Monitor the fluorescence with time until the change with time is practically negligible.
8. Compare the final fluorescence value with the fluorescence at the end of the linear range of substrate concentration. If the final value is within the linear range, the chosen substrate concentration is appropriate. Otherwise, decrease the substrate concentration two times, and repeat the measurements from steps 6 and 7 until the final value is within the linear range.

3.4.3. Optimal Enzyme Concentration

Prior to rate constant determination, the enzyme concentration necessary in the assay must be determined.

1. Mix the reaction buffer and the substrate directly in the cuvet, and incubate for 15 min at the desired temperature. When doing measurements at 37°C, prewarm the buffers to room temperature before adding to the cuvet.
2. Add enzyme, and mix the solution with a pipet three times.
3. Measure the increase in fluorescence emission for about 2000 s.

4. Fit the data to the equation:

$$F = F_0 + (F_\infty - F_0)(1 - e^{-kt}) \quad (4)$$

where F_∞ is the final fluorescence, F_0 is the initial fluorescence, k is the rate constant, and *t* is the time, using a nonlinear data analysis program (for example, ENZFITTER, Elsevier-Biosoft, UK for IBM-compatibles or Kaleidagraph for a Macintosh computer). A reasonable fit requires at least 20 data points in the steep part of the curve and sufficient data points in the period of the reaction where it reaches at least 90% of the final fluorescence value.
5. When the steep part of the curve has <20 points, decrease the enzyme concentration two times and repeat from step 1. On the other hand, if the reaction cannot reach 90% of the final value within 2000 s (that is, the fluorescence is still increasing), increase the amount of enzyme two times and repeat from step 1.

4. Notes

1. To obtain meaningful results from fluorescence experiments, it is important to know the range of substrate concentration where change in fluorescence is proportional to change in substrate or product concentration (it is a good practice to verify it independently, e.g., by HPLC). This can be achieved in two ways. The first method assumes that a linear change in fluorescence with time corresponds to a linear change in substrate or product concentration. This approach may be used in the range where a total change in fluorescence owing to a complete digest by a protease is not linear with a substrate concentration. The second method relies on the assumption that the linearity of initial fluorescence with substrate concentration is a guarantee of a linearity of product concentration as long as the final fluorescence is within that $F_{init.}$ vs. $[S_0]$ linear range (Fig. 5). This method is usually hard to implement for substrates with 10 or more fold change on a complete cleavage by a proteinase (e.g., most substrates based on the Dabcyl-Edans pair).
2. Another problem connected with the fluorescence measurements is the order of reaction between the enzyme and the substrate. Most fluorescent substrates are not very soluble in water, and the useful measurement range might be for $[S_0] \leq [E]_{total}$. In that case, the observed change in fluorescence might be influenced by a competition between binding of the substrate to the enzyme and a cleavage of the scissile bond.
3. Most fluorescent substrates use bulky chromogenic groups that may influence the rate of cleavage by its size and proximity to the active site *(4)*. They will not be very useful in the substrate specificity studies and should be treated only as a general probe for enzyme activity.

4. The choice of instrument for fluorescence measurements is not as important as the choice of substrates. Typical spectral band widths for fluorescent substrates are about 50 nm, and even a simple band-pass filter may replace more sophisticated monochromators. For kinetic measurements, however, computer-controlled data collection and storage are of great advantage in extracting rates of cleavage. In this laboratory, kinetic measurements were carried out on an SLM 4800C spectrofluorometer from SLM Aminco interfaced with an IBM PC AT (with math coprocessor added) running the 4800 ver. 1.61 software. The data, after removing plus signs with a Windows A Write (Microsoft Co., US) word processing program, were accessed directly by Enzfitter (Elsevier Biosoft, UK) nonlinear data analysis software.
5. The process of initiating six reactions, as described in this chapter, inevitably means that some of the initial part of the reaction time-course will be lost during the time required to mix and transfer all reactions. For the highest accuracy, it is, therefore, advisable to use an amount of enzyme that will produce the longest possible linear phase to the reaction. Under these conditions, the loss of the first few percent of the reaction trace is not a deficit. Optimally, the linear phase can be extended over the typical 15-min time-course we utilize. The use of the seven-position cuvet transport permits data collection in a time-efficient manner, even under the conditions of slow reaction described here.

References

1. Richards, A. D., Phylip, L. H., Farmerie, W. G., Scarborough, P. E., Alvarez, A., Dunn, B. M., Hirel, Ph.-H., Konvalinka, J., Strop, P., Pavlickova, L., Kostka, V., and Kay, J. (1990) Sensitive, soluble chromogenic substrates for HIV-1 proteinase. *J. Biol. Chem.* **265,** 7733–7736.
2. Hofmann, T. and Hodges, R. S. (1982) A new chromophoric substrate for penicillopepsin and other fungal aspartic proteinases. *Biochem. J.* **203,** 603–610.
3. Yaron, A., Carmel, A., and Katchalski-Katzir, E. (1979) Intramolecularly quenched fluorogenic substrates for hydrolytic enzymes. *Anal. Biochem.* **95,** 228–235.
4. Weidner, J. R. and Dunn, B. M. (1991) Development of synthetic peptide substrates for the poliovirus 3C proteinase. *Arch. Biochem. Biophys.* **286,** 402–408.
5. Jewell, D. A., Swietnicki, W., Dunn, B. M., and Malcolm, B. A. (1992) Hepatitis A virus 3C proteinase substrate specificity. *Biochemistry* **31,** 7862–7869.
6. Petithory, J. R., Masiarz, F. R., Kirsch, J. F., Santi, D. F., and Malcolm, B. A. (1991) A rapid method for determination of endoproteinase substrate specificity: specificity of the 3C proteinase from Hepatitis A virus. *Proc. Natl. Acad. Sci. USA* **88,** 11,510–11,514.
7. Matayoshi, E. D., Wang, G. T., Krafft, G. A., and Erickson, J. (1990) Novel fluorogenic substrates for assaying retroviral proteases by resonance energy transfer. *Science* **247,** 954–958.

8. Conner, G. E., Udey, J. A., Pinto, C., and Sola, J. (1989) Nonhuman cells correctly sort and process the human lysosomal enzyme cathepsin D. *Biochemistry* **28,** 3530–3533.
9. Scarborough, P. E., Guruprasad, K., Topham, C., Richo, G. R., Conner, G. E., Blundell, T. L., and Dunn, B. M. (1992) Exploration of subsite binding specificity of human cathepsin D through kinetics and rule-based molecular modeling. *Protein Science* **2,** 264–272.
10. Henderson, P. J. F. (1972) A linear equation that describes the steady-state kinetics of enzymes and subcellular particles interacting with tightly-bound inhibitors. *Biochem J.* **127,** 321–333.

CHAPTER 14

Synthesis of Recombinant Peptides

Gino Van Heeke, Jay S. Stout, and Fred W. Wagner

1. Introduction

The development of recombinant methods to clone and express genes of one organism in another organism has been one of the greatest advances in biology. More recently, refined methodologies have been utilized to clone and express peptides of diverse complexities into a variety of organisms.

Theoretically, there is no limitation on the size or the sequence of the peptide that can be incorporated into a gene-expression system, but in practice, there are a number of factors limiting the size of the peptide that can be effectively expressed in good yields, especially in microorganisms. The principal problem in cloning and expressing peptides results from cellular mechanisms that degrade proteins and peptides *(1)*. The half-lives of peptides under 50 residues is generally very short *(2)*. Other subtle processes, such as limited proteolysis and deamidation reactions, can serve to alter the primary peptide structure and to render the peptide product useless.

To date, solid-phase methods for large- and small-scale synthesis of peptides have been adequate for virtually all production needs. However, the realization of the therapeutic value of peptides, such as calcitonin (32 residues, amidated) and growth hormone-releasing factor (44 residues, amidated), may increase future production requirements to tens of kilos per year. Although solid-phase methods may be able to meet such demands, in theory, the recombinant production of peptides at this

From: *Methods in Molecular Biology, Vol. 36: Peptide Analysis Protocols*
Edited by: B. M. Dunn and M. W. Pennington

level should be much less expensive, should yield products devoid of contaminating levels of D-amino acids, and should be free of rearrangement byproducts known to occur when synthetic peptides are deblocked or treated at extremes of pH. Moreover, when recombinant methods are developed to produce peptides in extremely large quantities, new uses for these molecules, so far unimagined, should develop.

1.1. Experimental Design of a Recombinant Peptide

1.1.1. General Considerations

Based on current technology, a recombinant microbial expression system for a peptide is valuable when the peptide possesses at least 15 residues and is to be produced in extremely high quantities or smaller quantities intermittently over a long period of time. Peptides (50 residues or less) for research purposes are much easier and less expensive to produce by organic chemical means.

Since heterologous peptides of <50 residues are generally degraded in cells, it has been common practice to clone them as a part of another "carrier" protein. This approach involves the genetic construction of a protein construct, commonly referred to as a fusion protein *(3)*.

The protein construct is composed of a carrier protein, and the peptide product connected by an interconnecting peptide sequence that allows the peptide product to be released from the carrier protein by chemical or enzymatic cleavage. Technically, the product peptide can be constructed either at the amino or carboxy terminus of the protein construct. Since most cleavage protocols are specific for the carboxyl side of a given amino acid residue (usually not a part of the desired peptide sequence), product peptides are usually cloned onto at the carboxy terminal end of the carrier protein.

1.1.2. Choosing the Protein Carrier

The protein construct mainly functions to protect the product peptide from proteolysis from the time it is synthesized in the cell until it is purified from cellular or adventitious proteases; however, it has other important functions. A carrier protein expressed at high levels ensures a comparable level of expression when the product peptide is included in its primary structure. It also may provide a means to purify the protein construct. Carrier proteins possessing a unique biological property can be purified by affinity purification methods that exploit this property.

The most common method has been to use a protein carrier for which there is a corresponding polyclonal or monoclonal antibody. As an example, the Flag™ technology utilizes immobilized monoclonal antibodies specific for a small peptide sequence cloned onto the amino terminus of the desired genetic product *(4)*. Another example is the use of protein A as the carrier protein *(5)*. Fusions with Protein A bind selectively to immobilized IgG antibodies. The protein construct can be eluted as a purified protein.

Immunoaffinity chromatography has several major disadvantages when used for purifying protein constructs. Large-scale immunoaffinity columns are expensive to synthesize, have low binding capacities to solid supports, have limited half-lives, are difficult to sterilize, and are easily contaminated. Antibodies also leak from the resin and contaminate the protein product. Antibody products can be contaminated by pathogenic viruses derived from the host organism *(6)*.

Another approach has been to use enzymes as carrier proteins that can be purified by conventional affinity chromatography using immobilized inhibitors or substrates *(7–9)*. For large-scale productions, most systems are inadequate because they use expensive ligands with low affinities for the carrier protein, which limits column binding capacity. The ligands are generally cellular metabolites or their analogs, and are difficult to keep sterile when immobilized. Most enzymes used as carrier proteins have relatively large K_i values ($>10^{-5}M$) for the corresponding inhibitors. The inhibitor cost makes large quantities of affinity resins (more than 10 kg) expensive to synthesize. Many of the enzymes used as carrier proteins have large molecular weights (β-galactosidase; 116,000 Dalton) *(7)*, reducing the percentage by weight of product peptide that constitutes the protein construct.

We have developed a protein construct based on the latter approach that uses the low-mol-wt enzyme, human carbonic anhydrase II (hCAII; 29,000 Dalton), as the carrier protein. This enzyme is inhibited by *p*-aminobenzene sulfonamide and many of its derivatives (K_i values on the order of 0.1 μ*M*) *(10,11)*. Several very effective affinity resins have been developed that selectively bind hCAII with capacities up to 20 mg hCAII/mL of resin matrix *(12,13)*. The ligand is inexpensive and is easily coupled to a variety of resins to yield chromatography supports that can be washed with either 0.1*M* HCl or NaOH, and can be stored in organic solvents indefinitely. We have repeatedly used such resins for up to 2.5 yr with little change in chromatographic properties. Eventually, the mechanical integrity of the resin fails, and the resolution and binding capacity begin to deteriorate.

1.1.3. Choosing the Interconnecting Peptide

The product peptide is separated from the carrier protein by selective cleavage of the interconnecting peptide. The cleavage is either performed by enzymatic means or by chemical means. A comprehensive list of these methods is given by Carter *(14)*. There is no single best cleavage method, and usually the method of choice depends on the sequence of the peptide being produced. Often when the protein construct is produced in *E. coli*, inclusion bodies (insoluble aggregates of the recombinant protein construct) are obtained. Inclusion bodies are notoriously insoluble and only dissolve in highly concentrated chaotropic solutions (5–7*M* guanidine-HCl). Once solubilized in guanidine-HCl, they are usually not soluble in aqueous solutions required for proteolytic cleavage of the interconnecting peptide. These protein constructs are best processed by chemical cleavage means.

As an example, a good interconnecting peptide sequence is AsnGly when the product peptide possesses a Gly residue at the amino terminus *(15,16)* and no other AsnGly sequences. After the fusion protein is isolated, the peptide can be cleaved in 5*M* guanidine-HCl containing 2*M* NH_2OH. We have used this linker for two different peptides with cleavage yields up to 90%. The method has the advantage of minimal deleterious effects on Asn and Gln amide residues. The disadvantage is that other peptides are susceptible to cleavage, but at much slower rates. It is recommended that the cleavage kinetics be determined for any cutting method used.

Proteases used to digest interconnecting peptides must possess limited specificities. Enterokinase and thrombin are good choices since they have specificity requirements for at least five amino acid residues. The V8 protease, specific for the carboxyl side of Glu, can be used for peptide products devoid of this residue *(17)*. Trypsin, on the other hand, is a poor choice since it will cleave most Lys—Xxx or Arg—Xxx bonds. A secondary consideration is the action of the protease on the carrier protein and the effects that its hydrolysis products may have on peptide purification.

In any of the methods, the amino acid recognition sequence must be constructed adjacent to the product peptide. Incubation of the pure protein construct with the appropriate enzyme (e.g., enterokinase) allows for the release of the product peptide. In the case of enterokinase, the recognition sequence is AspAspAspAspLys. Since it is unlikely that this sequence occurs in the carrier protein (it does not occur in hCAII), cleavage is selective and the product can often be obtained in a highly pure state simply by size-exclusion ultrafiltration. Enterokinase preparations are notoriously contaminated with trypsin and chymotrypsin. Thus, only preparations free of other proteases can be used successfully.

When the recognition sequence occurs in the carrier protein, it may be changed by site-specific mutagenesis. As an example, AsnGly sequences may be changed either to GlnGly or AsnAla sequences when the interconnecting peptide is cleaved by NH_2OH. Each of these changes involves a one-carbon change in structure and should have a minimal effect on the protein.

1.1.4. Providing for Carboxy Terminal Amidated Peptides

Many peptides of biological interest exist as α-carboxy terminal amides rather than free acids. This poses a significant problem since genetic codons do not exist for amino acid amides. Thus, amides must be constructed by posttranslational manipulations. Since peptides have such a diverse number of side-chain functional groups, posttranslational chemical amidation schemes are not feasible.

Two methods have been elucidated for the production of recombinant α-carboxy terminal peptide amides. The first of these involves the use of the enzymes that naturally amidate peptides *(18)*. Peptides to be amidated are expressed with an additional Gly residue at the carboxy terminal end. The enzymes glycine monooxygenase *(19)* and peptidylamidoglycolate lyase *(20)* remove the terminal Gly residue, but leave its amino group on the penultimate residue as the amide.

The second method employs the use of serine carboxypeptidases that are capable of catalyzing transpeptidation reactions *(21)*. The principle of this approach is to produce a peptide with a carboxy terminal residue susceptible to transpeptidation by a serine carboxypeptidase. This method has utility. However, it is limited by the specificity of the enzyme to the residues in both the P_1 and P_1' positions (nomenclature of Schechter and Berger *[22]*) of the peptide. Thus, using an enzyme such as carboxypeptidase Y (CPD-Y), a peptide ending with the sequence of $LeuPheNH_2$ could be synthesized from a mixture of $PheNH_2$, CPD-Y, and a peptide ending in LeuAla. However, peptides ending in Xxx-$ProNH_2$, Xxx-

$GluNH_2$ (or Asp), and Lys (or Arg)-$XxxNH_2$ cannot be made with CPD-Y *(23)*. The singular advantage of this method is the availability and relatively low cost of CPD-Y. Care must be taken to use affinity-purified CPD-Y *(24)* since it does not possess detectable amounts of endopeptidases.

In the case of peptides ending in $ProNH_2$, an alternate synthetic route has been devised by Henriksen et al. *(25)* utilizing CPD-Y and a novel L-α-amino acid amide, *o*-nitro-L-α-phenylglycine amide (*o*-NPGA). The substrate peptide ends in the sequence ProAla, and is transpeptidated in the presence of CPD-Y and *o*-NPGA at pH 6.5 to produce a peptide ending in the sequence Pro-*o*-NPGA. When exposed to light at wavelengths above 320 nm, the *o*-nitrophenyl group photolytically degrades, and the amino group of *o*-NPGA is left as the amide of Pro *(25)*. The technique was used successfully to amidate analogs of human calcitonin with yields in excess of 90% *(25)*.

2. Methods

2.1. Cloning and Expression of a Peptide: A Practical Example

We have cloned and expressed about 20 different peptides ranging in size from 8–300 amino acid residues using hCAII as the carrier protein. As a practical example, the cloning, expression, purification, enterokinase cleavage, and amidation of an eight-residue peptide of the sequence ThrAsnThrGlySerGlyThrPro is described. This sequence corresponds to the C-terminal region of salmon calcitonin. For amidation purposes, the DNA sequence designed to code for this peptide was adapted to code for an extra Ala at the C-terminal end of the peptide. However, the procedures and protocols presented below are generally applicable to other peptides as well.

2.1.1. Description of pBN

Plasmid pBN (*see* Fig. 1) is a bacterial expression plasmid that is capable of producing large amounts of a particular polypeptide fused to the C-terminus of hCAII. It contains an expression cassette based on the bacteriophage T7 promoter in a pUC plasmid-derived backbone *(26–28)*. The T7 promoter directs the transcription of a fusion gene consisting of two parts: the structural gene of hCAII lacking the C-terminal 2 codons, followed by the linker sequence GTC GAC GAC GAC GAT ATC, which encodes $ValAsp_4Ile$. The T7 promoter can only be transcribed by T7

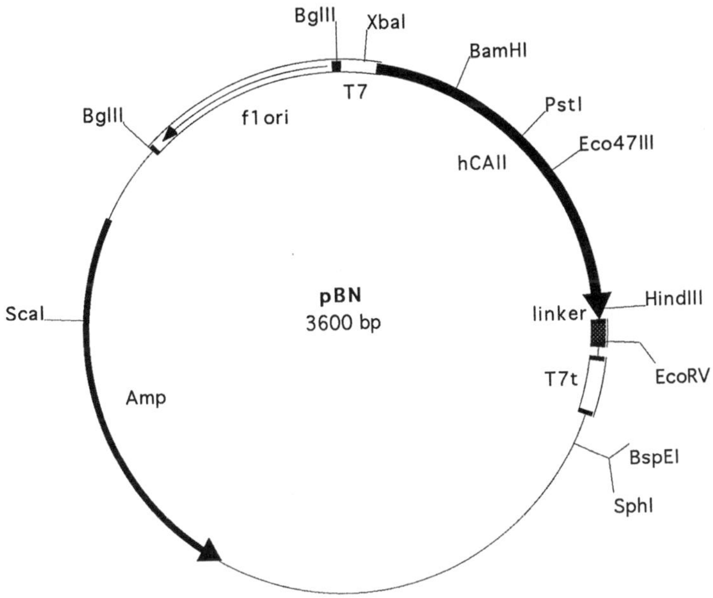

Fig. 1. Map of the pBN plasmid needed to construct and express fusion proteins with hCAII. The unique *Eco*RV restriction enzyme site is used to insert heterologous genes in the correct reading frame with the gene coding for hCAII. The linker sequence encodes Asp_4Ile, part of the recognition sequence for enterokinase. The genes coding for β-lactamase (confers ampicillin resistance) and hCAII are shown, as well as the f1 origin useful to produce single-stranded DNA for mutagenesis and sequencing purposes.

RNA polymerase, which itself is foreign to *E. coli.* The hCAII fusion protein can therefore only be overproduced in an *E. coli* host that also expresses T7 RNA polymerase. Three such systems have been described in the literature. T7 RNA polymerase can be encoded on a second compatible plasmid and under the control of a promoter functional in *E. coli* *(29)*. A lysogenic *E. coli* host containing a chromosomal version of the T7 RNA polymerase gene under the control of bacterial regulatory sequences, such as BL21 (DE3) or JM109 (DE3), is also available (from Novagen [Madison, WI] and Promega [Madison, WI], respectively) *(26)*. Alternatively, expression from the T7 promoter can be induced by infection with a λ bacteriophage derivative that carries the T7 RNA polymerase *(26)*. For practical purposes, the latter method is less desirable for large-scale production of fusion proteins.

The linker sequence contains a unique *Eco*RV (GAT/ATC; the "/" indicates site of cleavage) restriction enzyme site. *Eco*RV cleaves between the codons for the fourth Asp and the adjacent Ile. This allows a foreign gene to be inserted in the correct translational reading frame with hCAII.

Ideally, the peptide of interest is encoded by a synthetic gene. This allows the design of a DNA sequence using codons preferred by *E. coli* for abundantly expressed proteins *(30)*. In addition, restriction enzyme sites can be edited throughout this sequence for future mutagenesis and structure/function studies. Often, the peptide is short enough that its corresponding gene can be constructed with one single complementary oligonucleotide set.

Synthetic genes should be designed to code for an extra Lys codon immediately upstream of the gene. This Lys amino acid is encoded by an AAA or AAG codon, which enables the addition of a *Dra*I (TTT/AAA) or an *Afl*II (C/TTAAG) restriction enzyme site for further cloning purposes. This Lys codon, when inserted in the *Eco*RV site of pBN, will complete the coding region for an enterokinase recognition sequence (AspAspAspAspLys). In case a natural gene is to be inserted in pBN, a Lys codon has to be added to the 5' end of the gene by appropriate methods, such as PCR *(31)*.

2.1.2. Preparation of Linearized pBN

1. Digest 300 ng of pBN with 5 U of *Eco*RV in 10 m*M* Tris (pH 7.9)/50 m*M* NaCl/10 m*M* $MgCl_2$/1 m*M* DTT/100 µg/mL BSA at 37°C for 1 h.
2. Add 1–2 U calf intestinal phosphatase (Promega), and incubate at 37 and 50°C for 30 min at each temperature.
3. The vector is ready to be purified by standard phenol extraction procedures or gel-purification methods *(32)*. Alternatively, the reaction mixture can be separated on a low-gelling temperature agarose gel, and the vector fragment ligated to the insert DNA by an in-gel ligation procedure *(33)*.

2.1.3. Preparation of the Insert DNA Coding for the Nonapeptide

1. The following oligonucleotide: 5'-AAA AGA AAT ACT GGA TCC GGT ACC CCG GCT TAA GAT-3' and its complementary sequence 5'-ATC TTA CGA GGG GGT ACC GGA TCC AGT ATT TGT TTT-3' were designed and synthesized for the nine-residue peptide (NP).
2. Phosphorylate the oligonucleotides in a reaction mixture containing 10 m*M* Tris, pH 8.0, 10 m*M* $MgCl_2$, 5 m*M* DTT, 1.5 µ*M* of each oligonucleotide, 6 µ*M* ATP, and 10 U of T4 polynucleotide kinase (Promega) for 60 min at 37°C.

3. Combine both phosphorylated oligonucleotides. Heat at 95°C for 5 min, and cool slowly to room temperature to allow both strands to anneal.
4. Ligate the double-stranded phosphorylated oligonucleotide for 16 h at 16°C using T4 DNA Ligase (Promega) in a reaction mixture containing 30 m*M* Tris-HCl, pH 7.8, 10 m*M* $MgCl_2$, 10 m*M* DTT, and 1 m*M* ATP.
5. Transform *E. coli* DH5 with an aliquot of this ligation mixture, and select colonies on Luria broth agar plates containing ampicillin at a final concentration of 100 µg/mL *(32)*.
6. Purify plasmids from isolated colonies by a standard procedure, such as the alkaline lysis method *(32)*, and identify those plasmids containing the oligonucleotide insert by restriction enzyme analysis.

2.1.4. Expression of the hCAII-Nonapeptide Fusion Protein

1. Grow a bacterial culture of BL21(DE3)/pBN-NP or JM109 (DE3)/pBN-NP for 16–18 h at 37°C in Luria broth containing ampicillin at 50 µg/mL.
2. Dilute the culture 100-fold in fresh Luria broth plus ampicillin, and incubate at 37°C until the optical density measured at 550 nm reads 0.6–0.7.
3. Add isopropylthio-β-D-galactopyranoside and $ZnCl_2$ to a final concentration of 0.4 m*M* and 12.5 µ*M*, respectively. Continue incubation at 37°C for 3 h.
4. Harvest the cells by centrifugation at 4°C. Wash the cell pellet once in ice-cold 50 mM Tris-HCl, pH 7.6, and store the cell pellet at –20°C.

2.1.5. Affinity Purification of the hCAII-NP Fusion Protein (Laboratory Scale)

1. Thaw and resuspend the frozen bacterial pellet on ice in one-tenth of the original culture volume of ice-cold 50 m*M* Tris, 0.5 m*M* EDTA, and 0.5 m*M* EGTA, pH 7.8.
2. Add PMSF to a final concentration of 1 m*M*. Immediately sonicate three to four times for 30 s with 1-min intervals at 0–4°C.
3. Add $ZnCl_2$ to a final concentration of 50 µ*M*. Keep on ice for 10 min.
4. Spin the cell extract for 30 min at 45,000*g* at 4°C. Recover the supernatant fraction, and add 2 vol of ice-cold 50 m*M* Tris, 0.5 m*M* EDTA, and 0.5 m*M* EGTA, pH 7.8.
5. Adjust the pH of the lysate to 8.7 with solid Trizma base or with a 1*M* Tris-base solution.
6. Stir lysate for 16 h at 4°C with *p*-aminomethylbenzenesulfonamide-agarose resin (Sigma). Use 1 mL of resin for every 2 mL of lysate.
7. Transfer the affinity resin to an appropriately sized column, and collect the flow-through. Wash the resin with cold 0.1*M* Tris-SO_4, 0.2*M* K_2SO_4, and 0.5 m*M* EDTA, pH 9.0, until the absorbance of the effluent measured at 280 nm becomes negligible.

8. Wash the resin with 15 bed vol of cold 0.1*M* Tris-SO_4, 0.2*M* K_2SO_4, pH 7.0, or until the absorbance of the effluent measured at 280 nm returns to zero.
9. Elute the hCAII-fusion protein with 5 bed vol of cold 0.1*M* Tris-SO_4, 0.4*M* KSCN, and 0.5 m*M* EDTA, pH 6.8. Collect fractions of a volume equal to the resin bed volume. The hCAII-fusion protein usually elutes in fractions 2 and 3 (*see* Fig. 2).
10. Wash the resin with another 10 bed vol of elution buffer and 10 bed vol of 0.1*M* Tris-SO_4, 0.2*M* K_2SO_4, 0.5 m*M* EDTA, and 1 m*M* NaN_3, pH 7.5, and store at 4°C.
11. The eluted fusion protein is now ready for further concentration by standard procedures, such as ultrafiltration.

2.1.6. Enterokinase Cleavage of the hCAII-NP Fusion Protein

1. Dissolve the hCAII-NP fusion protein in 50 m*M* Tris, pH 8.0, and 1 m*M* $CaCl_2$.
2. Add enterokinase (highest grade available from Biozyme [San Diego, CA], SA 100,000 U/mg) at a 1/500 to 1/100 (w/w) ratio. Incubate the mixture for 16 h at 37°C.
3. Apply the entire reaction mixture to the affinity resin to remove the hCAII moiety and any remaining uncleaved fusion protein. Recover the peptide, along with the enterokinase, from the effluent (*see* Fig. 3).

2.1.7. Amidation

1. Dissolve recombinant nonapeptide, 10 mg, in 1 mL of 10 m*M* MOPS and 1 m*M* EDTA, pH 6.5, containing 0.5 mg CPD-Y and 0.1*M* *o*-NPGA. Incubate at 35°C for 2 h *(25)*.
2. Purify the transpeptidation product, ThrAsnThrGlySerGlyThrPro-*o*-NPGA, by HPLC using a Polysulfoethyl aspartamide column eluted with 65% acetonitrile/35% 10 m*M* TFA-TEA. Collect the peptide peak and lyophilize.
3. Dissolve the powder in 2.5 mL of a solution containing 1.25 mL of methanol and 1.25 mL of 60 m*M* $NaHSO_3$.
4. Adjust the pH to 9.5, and purge the solution with nitrogen. Then photolyze using a Xenon lamp filtered with a Pyrex™ glass filter.
5. Purify the peptide amide by HPLC as before.
6. Structural confirmation of the product is ascertained by conventional procedures.

3. Notes

1. The oligonucleotide sequence contains a 5' end AAA Lys codon followed by the octapeptide coding sequence. A *Bam*HI (G/GATCC) and *Kpn*I (GGTAC/C) restriction enzyme site (underlined; *see* Section 2.1.3.) are

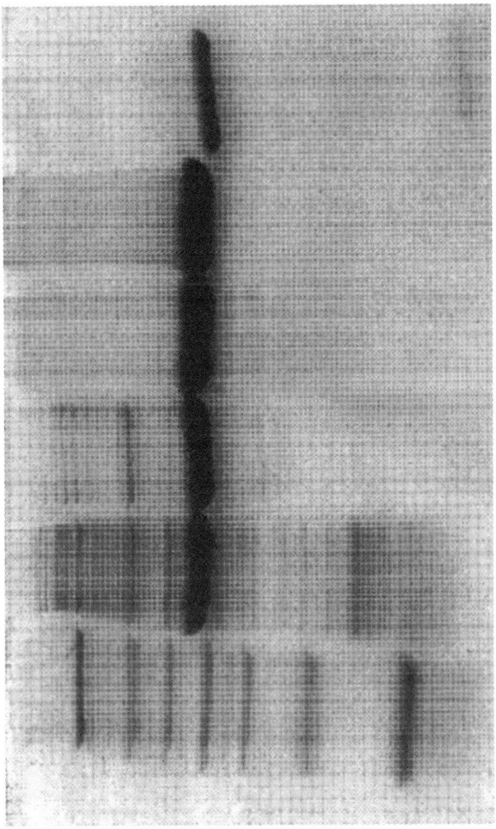

Fig. 2. SDS-PAGE fractionation of cell extract containing the hCAII-NP fusion protein at different stages during purification. Lane 1: Molecular-weight markers (α-lactalbumin:14,200; soybean trypsin inhibitor: 20,100; trypsinogen: 24,000; bovine carbonic anhydrase: 29,000; glyceraldehyde-3-phosphate dehydrogenase: 36,000; egg albumin: 45,000; bovine albumin: 66,000); lane 2: total cell extract of an *E. coli* BL21(DE3)/pBN-NP culture (10 µL), lane 3: total cell extract of an *E. coli* BL21(DE3)/pBN-NP culture (5 µL), lane 4: total cell extract of BL21(DE3)/pBN-NP after affinity chromatography (10 µL), lane 5: total cell extract of BL21(DE3)/pBN-NP after affinity chromatography (5 µL), lane 6: bovine carbonic anhydrase standard (Sigma). Samples were separated on a 12% SDS-denaturing polyacrylamide gel and proteins were visualized by Coomassie Brilliant Blue staining.

included to facilitate subsequent screening. The coding sequence is followed by a TAA stop codon and a GAT base sequence that restores the *Eco*RV site on insertion in the pBN vector.

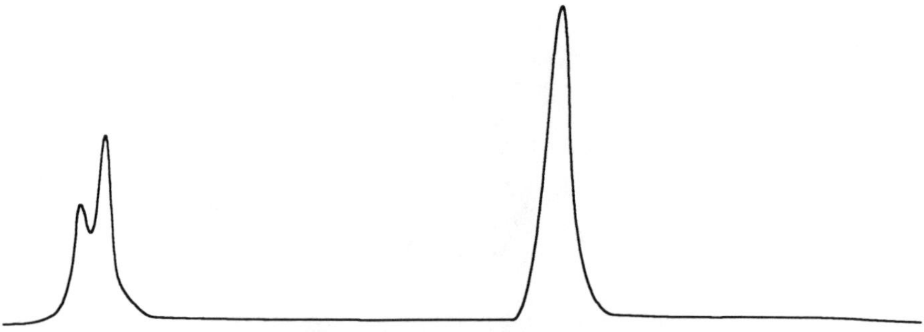

Fig. 3. HPLC chromatogram of hCAII-NP fusion protein cut with enterokinase after filtration through an Amicon 10K membrane. The sample was chromatographed on a polysulfoethyl aspartamide column eluted with 65% acetonitrile/35% 10 m*M* TFA-TEA. Besides the injection doublet peak at the left, the single peak on the right corresponds to the cleaved and recovered NP peptide.

2. The nonapeptide gene is designed to contain a *Kpn*I restriction enzyme site that is unique to the entire construct. Insertion of the fragment is monitored by digesting the plasmid with *Kpn*I. Clones containing the oligonucleotide can be resolved as a linear fragment by agarose gel electrophoresis. If such a diagnostic restriction enzyme site cannot be designed in the synthetic gene, an increase in the size of a fragment spanning the *Eco*RV region and bracketed by unique restriction enzyme sites in the pBN vector (e.g., *Pst*I and *Sph*I) will be apparent (*see* Fig. 1). Although these analyses confirm the presence of the peptide gene, its sequence should still be verified *(32)*.
3. All plasmid constructions should be performed in regular *E. coli* K-12 cloning hosts (such as HB101 or DH5). These hosts do not contain the T7 RNA polymerase gene and do not allow expression of the potentially toxic fusion protein. Once the plasmid has been characterized, it is used to transform the *E. coli* expression hosts available, i.e., *E. coli* BL21(DE3) or JM109(DE3), by standard protocols *(32)*.
4. Genes for peptides larger than about 20 amino acid residues are preferably constructed using multiple oligonucleotides. Detailed methods for the design and construction of synthetic genes have been described elsewhere *(34)*. In essence, the entire gene is synthesized as complementary oligonucleotide fragments, which, when annealed, should result in double-stranded fragments with short single-stranded protruding ends. Proper alignment of these fragments is ensured by the sequence of these protruding ends. They are then ligated together, and the entire gene is assembled in a separate cloning vector. The synthetic gene is designed to possess an engineered *Dra*I (TTT/AAA) or *Afl*II (C/TTAAG) restriction enzyme site

at the 5' end of the intact gene. Either one of these sites encompasses a Lys codon (AAA or AAG) necessary for restoring the enterokinase recognition site in pBN. Transfer of the gene fragment from the vector containing the synthetic gene to the pBN fragment by *Dra*I digestion is straightforward since this leaves a 5' end Lys codon (AAA) for blunt-end and in-frame ligation into the *Eco*RV site of pBN. Transfer of the gene fragment using an *Afl*II site leaves a 5' end four-base protruding sequence that is incompatible with the *Eco*RV site of pBN, and thus requires further manipulation. The sticky *Afl*II site can be converted to a blunt site by partially filling in the four-base overhang (TTAA) of the *Afl*II site with T4 DNA polymerase in the presence of dTTP only. This will result in a two-base 5' protruding overhang (TT) that is then removed by treatment with an exonuclease, such as mung bean nuclease *(32)*.

5. Nonsynthetic genes that do not contain a Lys codon as part of a restriction enzyme site immediately adjacent to the N-terminal peptide sequence need to be adapted by other methods, for instance PCR *(31)*.
6. When the cells are harvested, a small aliquot should be removed for analysis on a denaturing polyacrylamide gel prior to starting the purification procedure. This will not only confirm production of the fusion protein, but also give an estimate of how much is expressed and what fraction is soluble vs insoluble. An immunoblot of gel-fractionated cell extract using polyclonal antibodies against hCAII may be useful for a fine analysis of the temperature effect on solubility (*see* next note) and to determine the extent of intracellular degradation, if any.
7. The concentration of the protein construct in the soluble cell extract can also be quantitated by assaying an aliquot for its enzymatic activity using *p*-nitrophenylacetate *(35)*. The assay is performed using 3 m*M* substrate in 100 m*M* diethylmalate buffer, pH 7.5, and the hydrolysis of substrate is monitored at 400 nm. Although the specific activity of each protein construct is slightly different, the value of 8.5 m*M* substrate cleaved/min mg protein can be used to estimate the protein construct concentration.
8. It is possible that part or all of the expressed fusion protein is present in the cell as an inclusion body. Often, the proportion present in soluble form can be increased favorably by lowering the incubation temperature to 30 or 21°C just prior to induction *(36,37)*.
9. Some potential problems relating to plasmid stability have been described for the T7 expression system *(26)*. If these problems arise, they can usually be solved by substituting the gene conferring tetracyclin resistance for the gene conferring ampicillin resistance. Although this does not improve plasmid instability, it does provide a more effective selection of host cells containing the intact expression plasmid.

10. When the method is used for the first time or when a new fusion protein is purified, it is advisable to remove aliquots at different stages of the procedure to check where possible losses might occur.
11. The affinity resin should be washed with 10 vol of 0.1*M* Tris-SO_4, 0.2*M* K_2SO_4, and 0.5 m*M* EDTA, pH 7.5, prior to use to remove all traces of NaN_3 present in the storage buffer. NaN_3 competes with sulfonamide for the active site of hCAII.
12 Affinity resins can be synthesized in bulk quantity by the procedure of Osborne and Tashian *(12)*.
13. Concentrations of the protein construct for the nonapeptide could be measured using the molar absorptivity of the hCAII at 280 nm. For other protein constructs, the value can be adjusted by the Tyr or Trp content of the product peptide.
14. Because enterokinase has a high molecular weight, the peptide can be further purified by size-exclusion methods as well as conventional peptide chromatography protocols. Alternatively, for many purposes, further purification may not always be necessary.

References

1. Goldberg, A. L. and St. John, A. C. (1976) Intracellular protein degradation in mammalian and bacterial cells: part 2. *Ann. Rev. Biochem.* **45,** 747–803.
2. Goldberg, A. L. and Goff, S. A. (1986) The selective degradation of abnormal proteins in bacteria, in *Maximizing Gene Expression* (Reznikoff, W. and Gold, L., eds.), Butterworths, Stoneham, MA, pp. 287–314.
3. Sassenfeld, H. M. (1990) Engineering proteins for purification. *TIBTECH* **8,** 88–93.
4. Hopp, T. P., Prickett, K. S., Proce, V. L., Libby, R. T., March, C. J., Cerretti, D. P., Urdal, D. L., and Conlon, P. J. (1988) A short polypeptide marker sequence useful for recombinant protein identification and purification. *Bio/Technology* **6,** 1204–1210.
5. Moks T., Abrahmsen, L., Osterlof, B., Josephson, S., Ostling, M., Enfors, S.-O., Persson, I., Nilsson, B., and Uhlen, M. (1987) Large-scale affinity purification of human insulin-like growth factor I from culture medium of *Escherichia coli. Bio/ Technology* **5,** 379–382.
6. Bailon, P. and Roy, S. K. (1990) Recovery of recombinant proteins by immunoaffinity chromatography, in *Protein Purification: From Molecular Mechanisms to Large Scale Processes*, American Chemical Society, Washington, DC, pp. 150–167.
7. Ullmann, A. (1984) One step purification of hybrid proteins which have β-galactosidase activity. *Gene* **29,** 27–31.
8. Knott, J. A., Sullivan, C. A., and Weston, A. (1988) The isolation and characterization of human atrial natriuretic factor produced as a fusion protein in *Escherichia coli. Eur. J. Biochem.* **174,** 405–410.
9. Smith, D. B. and Johnson, K. S. (1988) Single-step purification of polypeptides expressed in *Escherichia coli* as fusions with glutathione *S*-transferase. *Gene* **67,** 31–40.

10. Maren, T. H. (1967) Carbonic anhydrase: chemistry, physiology and inhibition. *Physiol. Rev.* **47,** 595–781.
11. Lindskog, S., Henderson, L. E., Kannan, K. K., Liljas, A., Nyman, P. O., and Strandberg, B. (1971) Carbonic anhydrase, in *The Enzymes,* vol. V (Boyer, P. D., ed.), Academic, New York, pp. 587–665.
12. Osborne, W. R. A. and Tashian, R. E. (1975) An improved method for the purification of carbonic anhydrase isozymes by affinity chromatography. *Anal. Biochem.* **64,** 297–303.
13. Johansen, J. T. (1976) Isolation of human carbonic anhydrase B and C and apocarbonic anhydrase by affinity chromatography. *Carlsberg Res. Commun.* **41,** 73–80.
14. Carter, P. (1990) Site-specific proteolysis of fusion proteins, in *Protein Purification: From Molecular Mechanisms to Large Scale Processes,* American Chemical Society, Washington, DC, pp. 181–193.
15. de Geus, P., van den Bergh, C. J., Kuiper, O., Verheij, H. M., Hoekstra, W. P. M., and de Haas, G. H. (1987) Expression of porcine pancreatic phospholipase A2. Generation of active enzyme by sequence-specific cleavage of a hybrid protein from *Escherichia coli. Nucleic Acids Res.* **15,** 3743–3757.
16. Moks, T., Abrahmsén, L., Holmgren, E., Bilich, M., Olsson, A., Uhlén, M., Pohl, G., Sterky, C., Hultberg, H., Josephson, S., Holgren, A., Jörnvall, H., and Nilsson, B. (1987) Expression of human insulin-like growth factor I in bacteria: use of optimized gene fusion vectors to facilitate protein purification. *Biochemistry* **26,** 5239–5244.
17. Gearing, D. P., Nicola, N. A., Metcalf, D., Foote, S., Willson, T. A., Gough, N. M., and Williams, R. L. (1989) Production of leukemia inhibitory factor in *Escherichia coli* by a novel procedure and its use in maintaining embryonic stem cells in culture. *Bio/Technology* **7,** 1157–1161.
18. Kizer, J. S., Busby, W. H., Jr., Cottle, C., and Youngblood, W. W. (1984) Glycine directed peptide amidation presence in rat brain of 2 enzymes that convert pyro glutamylhistidylprolyl glycine into pyro glutamylhistidyl prolinamide TRH. *Proc. Natl. Acad. Sci. USA* **81,** 3228–3232.
19. Pekins, S. N., Husten, E. J., and Eipper, B. A. (1990) The 108-kDa peptidylglycine alpha-amidating monooxygenase precursor contains two separable enzymatic activities involved in peptide amidation. *Biochem. Biophys. Res. Commun.* **171,** 926–932.
20. Katopodis, A. G., Ping, D., and May, S. W. (1990) A novel enzyme from bovine neurointermediate pituitary catalyzes dealkylation of alpha hydroxyglycine derivatives thereby functioning sequentially with peptidylglycine alpha-amidating monooxygenase in peptide amidation. *Biochemistry* **29,** 6115–6120.
21. Breddam, K., Widmer, F., and Johansen, J. T. (1981) Carboxypeptidase Y catalyzed C-terminal modifications of peptides. *Carlsberg Res. Commun.* **46,** 121–128.
22. Schechter, I. and Berger, A. (1967) On the size of the active site of proteases. I. Papain. *Biochem. Biophys Res. Commun.* **27,** 157–162.
23. Breddam, K. and Sørensen, S. B. (1987) Isolation of carboxypeptidase III form malted barley by affinity chromatography. *Carlsberg Res. Commun.* **52,** 275–283.

24. Breddam, K. and Ottesen, M. (1987) Determination of C-terminal sequences by digestion with serine carboxypeptidases: the influence of enzymatic specificity. *Carlsberg Res. Commun.* **52,** 55–63.
25. Henriksen, D. B., Breddam, K., Møller, J., and Buchart, O. (1992) Peptide amidation by chemical protein engineering. A combination of enzymatic and photochemical synthesis. *J. Am Chem. Soc.* **114,** 1876,1877.
26. Studier, F. W., Rosenberg, A. H., Dunn, J. J., and Dubendorff, J. W. (1990) Use of T7 RNA polymerase to direct expression of cloned genes. *Methods Enzymol.* **185,** 60–88.
27. Tanhauser, S. M., Jewell, D. A., Tu, C. K., Silverman, D. N., and Laipis, P. J. (1992) A T7 expression vector optimized for site-directed mutagenesis using oligodeoxyribonucleotide cassettes. *Gene,* **117,** 113–117.
28. Studier, F. W., Davanloo, P., Rosenberg, A. H., Moffatt, B. A., and Dunn, J. J. (1990) US Patent # 4,952,496.
29. Tabor, S. and Richardson, C. C. (1985) A bacteriophage T7 RNA polymerase promoter system for controlled exclusive expression of specific genes. *Proc. Natl. Acad. Sci. USA* **82,** 1074–1078.
30. de Boer, H. A. and Kastelein, R. A. (1986) Biased codon usage: an exploration of its role in optimization of translation, in *Maximizing Gene Expression* (Reznikoff, W. and Gold, L., eds.), Butherworths, Stoneham, MA, pp. 225–285.
31. Innis, M. A., Gelfand, D. H., Sninsky, J. J., and White, T. J. (eds.) (1990) *PCR Protocols, a Guide to Methods and Applications.* Academic, San Diego.
32. Sambrook, J., Fritsch, E. F., and Maniatis, T. (1989) *Molecular Cloning, A Laboratory Manual,* Cold Spring Harbor Laboratory, Cold Spring Harbor, New York.
33. Sheng, S., Moraga, D. A., Van Heeke, G., and Schuster, S. M. (1992) High level expression of human asparagine synthetase and production of monoclonal antibodies for purification. *Protein Expression and Purification,* **3,** 337–346.
34. Brousseau, R., Wu, R., Sung, W., and Narang, S. A. (1987) Synthetic gene assembly, cloning, and expression, in *Synthesis and Applications of DNA and RNA* (Narang, S. A., ed.), Academic, San Diego, pp. 95–114.
35. Verpoorte, J. A., Metha, S., and Edsall, J. T. (1967) Esterase activities of human carbonic anhydrases B and C. *J. Biol. Chem.* **242,** 4221–4229.
36. Schein, C. H. and Noteborn, M. H. M. (1988) Formation of soluble recombinant proteins in *Escherichia coli* is favored at lower growth temperature. *Bio/Technology* **6,** 291–294.
37. Van Heeke, G. and Schuster, S. M. (1989) Expression of human asparagine synthetase in *E. coli. J. Biol. Chem.* **264,** 5503–5509.

CHAPTER 15

De Novo Design of Proteins

Template-Assembled Synthetic Proteins (TASP)

Gabriele Tuchscherer, Verena Steiner, Karl-Heinz Altmann, and Manfred Mutter

1. Introduction

More than three decades have passed since Anfinsen's classical experiment on Ribonuclease A unequivocally established that all the information required for a protein to adopt its native globular conformation is solely contained in its amino acid sequence *(1)*. Since then, dramatic advances have taken place in the methodology of peptide and protein chemistry, as well as molecular biology, that have led to a vastly improved understanding of the complex interplay among sequential, structural, and functional properties of natural proteins *(2–5)*. With the development of recombinant DNA techniques, it is now possible via gene cloning and expression in bacterial systems to isolate virtually any protein consisting of the 20 natural amino acids *(6)*. Even the incorporation of unusual amino acids is currently under investigation *(7)*.

At the same time, rapid progress in the development of various spectroscopic techniques (especially NMR spectroscopy [*8,9; see* Chapter 9] as well as X-ray crystallographic methods) has made possible the determination of even subtle differences in protein structure, thus providing powerful tools for the evaluation of the importance of any single amino acid residue within a given three-dimensional protein network (by studying appropriately designed mutants obtained via site-directed mutagenesis). Most notably, detailed studies on several natural proteins

From: *Methods in Molecular Biology, Vol. 36: Peptide Analysis Protocols*
Edited by: B. M. Dunn and M. W. Pennington

have demonstrated that only a limited number of modifications is tolerated before the structural and functional integrity of a protein is lost *(10)*.

Yet, despite the flood of data accumulated over the last decade, the underlying molecular mechanisms causing a particular polypeptide sequence to fold into a certain three-dimensional structure (and causing others not to fold at all!) still remain elusive. Consequently, the design of new proteins with tailor-made structural and functional properties, a goal that has fascinated protein and peptide chemists alike ever since the basic structural features of protein molecules were unraveled at the beginning of the century by Emil Fischer, still remains an overwhelmingly ambitious task. Recent successful attempts in this area *(11)* should be considered "single hits" rather than products of a generally applicable design strategy; furthermore, convincingly strong experimental evidence for the actual conformation of those *de novo* designed proteins is still unavailable.

It appears that the major obstacle in the construction of artificial proteins rests in the complexity of the folding pathway as well as the limited diversity of structural motifs in natural proteins, e.g., βαβ folding units or four-helix bundle arrangements. Among the surprisingly small number of recurring secondary structural motifs (α-helix, β-sheet, β-turn) *(12–14)*, the same structural type is adopted by many different sequences, and the formation of small globular folding units is not confined to a specific amino acid sequence ("degeneracy of the folding code").

Regarding the problem of protein folding mechanisms, there are currently two basic hypotheses discussed in the literature: One is based on the assumption that folding is initiated by the formation of fluctuating elements of local secondary structure along the unfolded polypeptide chain *(15,16)*; subsequent interactions between these ordered regions (accompanied by simultaneous rearrangements) are then believed to lead to the formation of a compact globular structure. Alternatively, it is postulated that the unfolded polypeptide chain first collapses into a fairly compact "molten globule" state followed by the formation of specific secondary structures *(17)*. It has been pointed out elsewhere that these hypotheses are not necessarily mutually exclusive and could be reconciled to a unified theory on protein folding *(18)*. Taking these aspects into consideration, the question arises whether or not it is possible to derive a general strategy for the *"de novo"* design of new proteins from first principles that exhibit all characteristic properties of natural proteins, most importantly the ability to fold into a well-defined three-dimensional structure.

2. The TASP Approach to Protein *De Novo* Design

Even though the complexity of the molecular mechanisms of protein folding does not yet allow the prediction of tertiary structure from primary sequence, examples for the (attempted) *de novo* design of proteins are fairly abundant in the literature. There are two basic strategies that can be followed for the design of nonhomologous amino acid sequences with the putative potential to fold into a protein-like globular structure. One of them, which we shall call the "linear-chain" approach *(11)*, is closely guided by the structural principles realized in natural proteins. Thus, amphiphilic secondary structure elements are connected by turns and loops or less-ordered peptide segments to form a linear arrangement of amino acid residues. It is anticipated that, driven by hydrophobic forces, these amphiphilic segments will collapse into a supersecondary structure with a clearly defined arrangement of secondary structure blocks. It is not the objective of this chapter to discuss the various aspects of the design strategy in any detail; however, it should be pointed out that this approach does not include any conceptual features that would try to overcome the two most basic problems in protein *de novo* design, i.e., the competition between intramolecular folding and intermolecular aggregation, and the high loss in chain entropy associated with the folding process.

In order to address those inherent problems of polypeptide folding, we have developed a novel design concept that tries to transcend the basic principles of protein structure and folding into a different, i.e., nonlinear, covalent arrangement of amino acid residues *(19–22)*. To this end, amphiphilic peptide blocks are attached to a multifunctional template molecule that is designed to direct and reinforce intramolecular folding (vs intermolecular aggregation). The resulting branched macromolecules have been termed "Template-Assembled Synthetic Proteins" (or TASP) (Fig. 1), and may be considered as hybrids between synthetic polymers (branched chains, grafted polymers) and natural proteins.

It is important to realize that the amphiphilicity of the peptide blocks still represents a fundamental prerequisite for intramolecular folding that is driven by hydrophobic interactions. However, attachment of these segments to a conformationally restricted template will clearly result in an enhanced tendency for intramolecular interactions; furthermore, the final packing topology of a TASP molecule can be controlled in a straightforward manner via the number, type, and spatial arrangement of the attach-

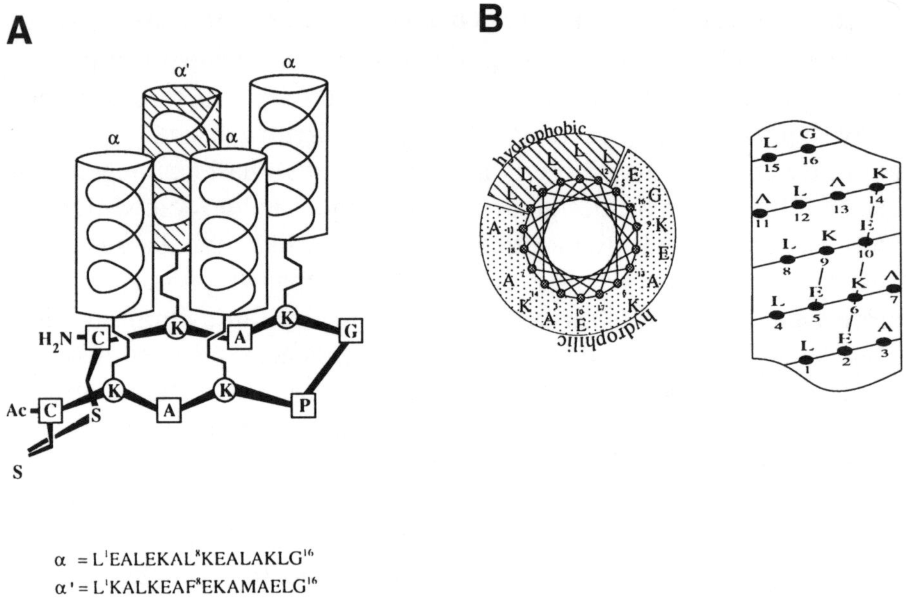

Fig. 1. **(A)** The TASP concept as exemplified by the schematic representation of a four-helix-bundle TASP T_4-(3α,α') incorporating the cyclic decapeptide template Ac-Cys-Lys-Ala-Lys-Pro-Gly-Lys-Ala-Lys-Cys-NH_2 and two different amino acid sequences α_{16} and α'_{16}. **(B)** Helical wheel *(23)* and helical net *(24)* representations of α_{16} illustrating amphiphilicity and favorable electrostatic interactions in this designed model helix. Ac = acetyl, K = lysine, A = alanine, P = proline, G = glycine, C = cysteine, E = glutamic acid, L = leucine, F = phenylalanine, M = methionine.

ment sites on the template. The thermodynamic aspects of this design strategy have been discussed elsewhere *(21)*. We simply want to mention that the branched-chain architecture of TASP molecules results in a reduced conformational entropy as compared to linear polypeptides of similar size, which is equivalent to a destabilization of the unfolded state. Because of its high volume density, the random-coil state of TASP molecules may resemble the "molten globule" state of natural proteins.

2.1. Templates for TASP Design

From the above discussion, it is obvious that the template (or carrier molecule) represents the key element of the TASP approach, since it serves to reinforce and direct the intramolecular association of the covalently attached peptide blocks. In principle, any multifunctional molecule (e.g., cyclic peptides, peptides with conformational constraints, saccharides, polycyclic aromatic, or aliphatic systems) with a proper spatial arrangement of the attachment sites may serve as template molecules. In our exploratory studies, we used linear oligopeptide templates whose sequences were designed based on the structural features of the cyclic peptide antibiotic Gramicidin S (GS). GS may be considered as the prototype of a conformationally constrained cyclic molecule; its crystal structure *(25)* shows two antiparallel β-sheet segments (Val-Orn-Leu) that are connected by two β-turn elements (D-Phe-Pro), which are the exact structural features required for a TASP template. Molecular modeling studies on the linear peptide KAKPGKAK using GS as a (modeling) template (i.e., substituting a Pro-Gly turn for the D-Phe-Pro turn in GS and Lys-Ala-Lys for Val-Orn-Leu) demonstrated that a low-energy conformation exists with all the Lys side chains pointing in the same direction and exhibiting the proper spacing for the construction, e.g., of a four-helix-bundle TASP *(21)*.

Some more recent designs of templates that were used in the construction of four-helix-bundle-type TASP molecules are depicted in Fig. 2. Among others, we have synthesized cyclic decapeptides of the general formula *cyclo*$(PGXAX)_2$ (X = Lys[Boc], Lys[Aloc], Lys[2-Cl-Z], Lys[Fmoc], Cys[SBu^t], or combinations thereof) (T_4, **IV, VI** in Fig. 2) as well as cyclic templates incorporating two β-turn mimetics (T_4', **II, III**) *(26,27)*. Furthermore, we have used the hexafunctional dodecapeptide *cyclo*$(Gly\text{-}Lys\text{-})_6$ (T_6) as the template for the construction of a six-helix-bundle TASP *(28)*. The use of α-cyclo dextrin derivatives (**V**) as well as templates exhibiting orthogonally protected attachment sites for the peptide blocks (**VI**) is presently elaborated in our laboratory.

2.2. Properties of TASP Molecules

All TASP molecules described to date are soluble in aqueous buffer solutions and occur as monomeric species, as could be shown by size-exclusion chromatography. Conformational studies by means of CD spectroscopy have generally established a strong secondary structure stabilizing effect of the template, especially for cyclic template molecules. As an example, Fig. 3 shows the CD spectrum of a fully symmetrical four-helix-bundle TASP (all helical building blocks are identical) in comparison to the spectrum of the isolated helical building block attached to

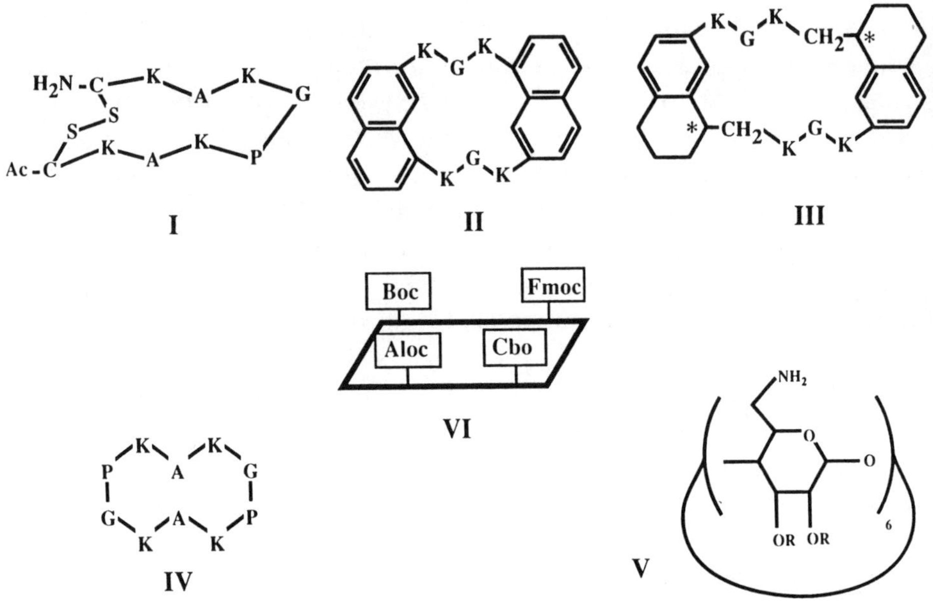

Fig. 2. Templates for TASP design. **I–IV** represent templates for the construction of α-helical bundle TASP molecules. The helical peptide blocks are covalently attached to the ε-amino side chains of the lysine residues (K). In **V**, the C^6-OH groups in α-cyclo-dextrine were transformed to NH_2-groups and can be used as attachment sites for the construction of a 6-α-helical bundle TASP. **VI** represents an orthogonally protected template in which the lysine side chains of the template (type **I–IV**) can be selectively deprotected. By this means, a four-helical bundle TASP carrying four different helical blocks can be constructed by fragment condensation *(see text)*.

the ε-amino group of Ac-Lys-NH_2 in aqueous solution at neutral pH. The dramatic difference in helicity between the TASP and the isolated segment convincingly demonstrates the profound helix-stabilizing effect of the cyclic template. These findings have been corroborated by denaturation experiments and 2D-NMR experiments, which shall not be discussed in this chapter *(22)*.

3. Materials

We generally obtain all the commonly protected amino acid derivatives, i.e., N^{α}-Z-, Boc-, or Fmoc-protected derivatives bearing standard side-chain-protecting groups, from commercial sources in Switzerland

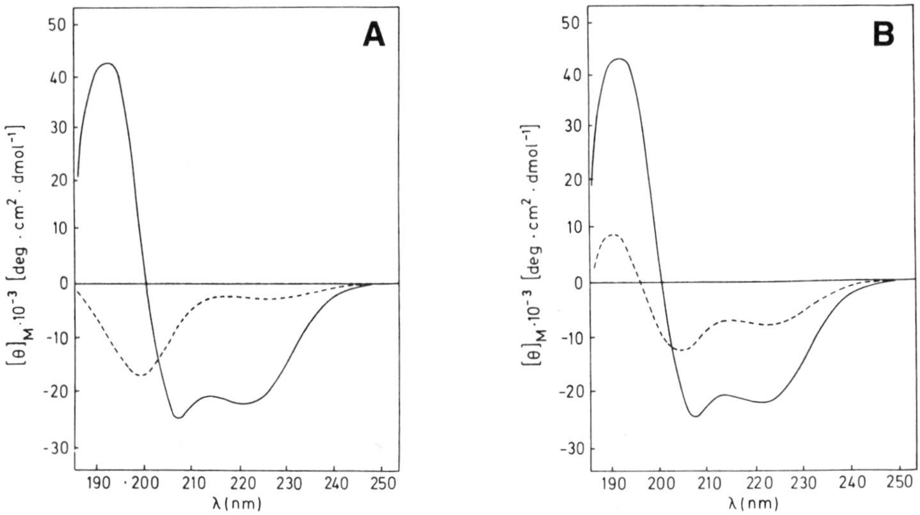

Fig. 3. (**A**) Chain-length dependence of α-helix formation as found by circular dichroism (CD); CD spectra of four-helix-bundle TASPs T_4-$(4\alpha_{15})$ (—) and T_4-$(4\alpha_{11})$ (---), c = $10^{-4}M$ in aqueous phosphate buffer, pH 7, at 25°C. (**B**) CD spectra of a four-helix-bundle TASP T_4-$(4\alpha_{15})$, c = $10^{-4}M$ (——) and the corresponding helical α_{15} attached to the ε-amino group of Ac-Lys-NH_2 (∫ Ac-Lys[α_{15}]-NH_2), c = $4 \times 10^{-4}M$ (---) in aqueous phosphate buffer, pH 7, at 25°C. For the template (T_4), cf. Fig. 1, α_{11}∫ H-KALKEALAKLG, α_{15}∫ H-EALEKALKEALAKLG (from ref. *22*).

(Bachem, Läufelfingen; Novabiochem, Bubendorf) or in the US (Bachem, Torrance, CA), and use them without further purification. Special derivatives that are not commercially available (e.g., allyloxycarbonyl [Aloc]-protected derivatives) are synthesized in our laboratory according to published procedures.

The same applies to the functionalized resins used for solid-phase peptide synthesis (SPPS). SPPS itself may be performed either manually according to established standard procedures *(29–33)* (as we tend to do in many instances) or on commercial peptide synthesizers.

Analysis and purification of TASP molecules, peptide blocks, and templates are routinely carried out by reverse-phase high-performance liquid chromatography (RP-HPLC) (*see* Chapters 3 and 4) and, if necessary, high-performance ion-exchange chromatography (IEC) (*see* Chapter 5). Any commercially available HPLC equipment may be used for this purpose (*see* Chapter 3). In our laboratory, we rely almost exclusively on HPLC equipment from Waters Inc. (for analytical as well as preparative purposes) in combination with columns filled with material manufactured by Vydac (Uppsala, Sweden). We have had satisfactory experiences with FPLC equipment and ion-exchange columns obtained from Pharmacia-LKB for analysis and purification of TASPs by IEC (*see* Chapters 2 and 5).

4. Methods

A large number of potential synthetic approaches is conceivable for the preparation of TASP molecules. However, not considering the more subtle aspects of the choice of anchoring groups in SPPS or of protecting-group strategy in general, two basic avenues can be followed for the synthesis of these branched macromolecules. The entire TASP molecule may be assembled by SPPS, or the TASP molecule may be assembled by fragment condensation techniques (*34; see also* Chapters 14 and 15, *PSP*), i.e., by the coupling of purified peptide blocks to the template in solution. In principle, a further alternative may consist of the coupling of purified fragments to the resin-bound template; however, we feel that this approach is certainly less advantageous than fragment condensation in solution. The preferred choice between those two strategies clearly depends on the objective of the research the TASP molecules are designed and needed for. Synthesis of TASPs by SPPS offers the advantage of greater speed since the molecules can be assembled in a matter of days. On the other hand, extensive purification of the crude products will be required, which may be very time-consuming; in addition, the quantities of really pure material obtained may be rather low. However, if there is no need for material of extremely high purity and/or small quantities of the TASP suffice for the purpose at hand (which is often the case for biological or immunological experiments), SPPS should be the method of choice.

Synthesis of TASPs by the fragment condensation approach in solution, on the other hand, offers the advantage of much better control over the purity of the final products, because the fragments as well as the template used in the coupling reaction can be purified and characterized before the final assemblage. If larger quantities of reliably pure material are required (e.g., for structural studies by NMR or X-ray crystallography), the fragment condensation approach may be more advantageous.

However, even at this point, it should already be pointed out that one of the greatest pitfalls of this latter approach may be the low solubility of protected peptide fragments, thus rendering the purification of intermediates very difficult (or even making it impossible) and also causing problems in the final coupling step.

In the following sections, we shall discuss these two basic approaches in greater depth, and a detailed example will be given for the successful synthesis of a TASP molecule by either of these strategies. In addition, we have included the synthetic protocol for the synthesis of a cyclic template molecule.

4.1. Solid-Phase Synthesis of TASPs (Scheme 1)

The synthesis of TASPs by SPPS poses a number of problems that are not normally encountered in the stepwise solid-phase synthesis of linear peptides, e.g., the requirement of N^{ε}-protecting groups of template lysines exhibiting orthogonal stability with respect to the N^{α}-protecting groups and the anchoring group between peptide and resin as well as the parallel synthesis of several peptide chains attached to the side chains of residues that are part of the same template peptide. Based on the current status of solid-phase methodology, the most reasonable option for the synthesis of TASPs seems to consist of a synthetic scheme employing HF-cleavable linker group (methylbenzhydrylamine, MBHA), a TFA-labile N^{α}-protecting group (*tert*-butyloxycarbonyl, Boc), HF-cleavable protecting groups for permanent side-chain protection, and a base-labile, but completely trifluoroacetic acid (TFA) stable protecting group (9-fluorenylmethoxycarbonyl, Fmoc) for temporary purposes (either on the ε- or the α-amino group of template Lys residues) (cf below).

It has been demonstrated very recently that this strategy allows the synthesis of symmetrical (all peptide blocks have identical sequence) as well as unsymmetrical (peptide blocks attached to different sites on the template have different sequences) TASP molecules. For the synthesis of symmetrical TASPs, the entire template (or its protected linear precursor; *see* Section 4.1.1.) can be assembled using N^{α}-Boc-Lys(Fmoc) for the incorporation of Lys residues (and the corresponding N^{α}-Boc derivatives of other template amino acids). After removal of the N^{ε}-Fmoc-protecting groups from lysine side chains, the peptide blocks attached to the ε-amino groups of these template lysines can be synthesized in parallel (N^{α}-protection; TFA-stable, but HF-labile side-chain protection).

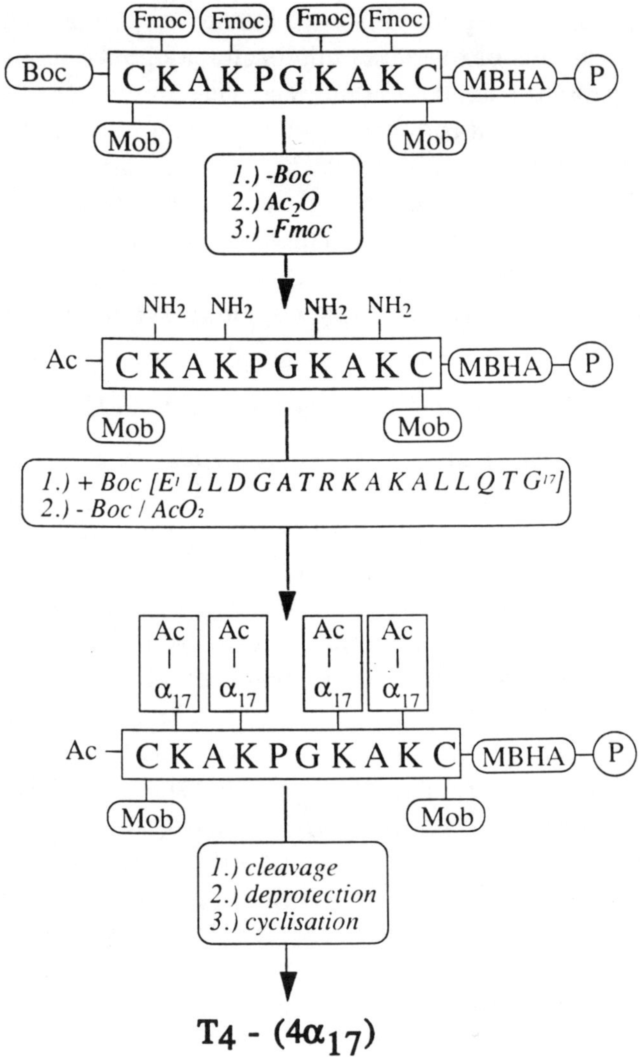

Scheme 1. Solid-phase peptide synthesis of a four-α-helix-bundle TASP T_4-($4\alpha_{17}$) *(22) (see text).*

If unsymmetrical TASPs are desired, the incorporation of at least the first Lys residue (counting from the C-terminus) has to be performed as the N^α-Fmoc-N^ε-Boc (or N^α-Boc-N_ε-Fmoc) derivative. If, in addition, the sequence of the peptide block attached to this first Lys side chain is different from the sequence that will be attached to the next Lys residue to be incorporated into the template, then the elongation of the template has to be halted at this point. Instead of cleavage of the N^{α}-Fmoc (N^{α}-Boc)-protecting group, the synthesis has to proceed with the removal of the N^{ε}-Boc (N^{ε}-Fmoc) group from the lysine side chain and subsequent assemblage of the peptide block using N^{α}-Boc (N^{ε}-Fmoc) protection. Once the synthesis of this block has been completed (by *N*-acetylation) the N^{α}-Fmoc (N^{α}-Boc) group of the template lysines can be removed, and the synthesis is continued with the extension of the template. If all the following template lysines bear peptide blocks of identical sequences (as is the case in the synthesis of T_4-[$3\alpha_{16}$,α'_{16}] described in the next section), then synthesis proceeds according to the strategy outlined for symmetrical TASPs, i.e., the remaining template amino acids are introduced as N^{α}-Boc-protected derivatives.

After the entire TASP sequence has been assembled, the peptide can be cleaved from the resin with anhydrous HF (Chapter 4) or trifluoromethanesulfonic acid (TFMSA; Chapter 5) and purified (*see* Section 4.1.1.).

If TASPs incorporating a cyclic template are desired, the template sequence has to include two Cys residues at the C- and N-termini, respectively (**I** in Fig. 2). This allows for oxidative cyclization via intramolecular disulfide bond formation; in principle, this step could be performed while the TASP is still attached to the resin (e.g., with iodine, if the acetamidomethyl [Acm] group is used for protection of cysteines) or even before assemblage of the secondary structure building blocks (in the case of symmetrical TASPs). However, these strategies have not yet been explored experimentally, and so far, we find it more convenient to cleave the TASP containing the linear precursor template from the resin and cyclize the crude product directly in solution (i.e., without any prior purification). TASPs based on templates cyclized via amide bonds are not accessible by this general strategy. This would require one more dimension of orthogonality since the side-chain-protecting groups would have to remain intact during cleavage of the peptide from the resin.

Thus, although conceivable in principle, an even more elaborate protection scheme would be necessary for this purpose involving protecting groups that are not routinely used in SPPS (and may thus be difficult to implement on commercially available peptide synthesizers). The development of such a strategy would constitute a research project in its own right.

4.1.1. Synthesis of T_4-($3\alpha_{16}$, α'_{16}) (Fig. 1)

This section describes in detail the solid-phase synthesis (including purification protocols) of an unsymmetrical four-helix-bundle TASP molecule of the formula Ac-Cys-Lys(α_1)-Ala-Lys(α_1)-Pro-Gly-Lys(α_1)-Ala-Lys(α')-Cys-NH_2 that incorporates a cyclic decapeptide template and helical building blocks of two different sequences, α_1 and α'; α_1∫ Ac-LEALEK ALKEALAKLG-; α'∫ Ac-LKALKEAFEKAMAELG *(22)*.

The synthesis was carried out manually starting from a methylbenzhydrylamine-1%-crosslinked polystyrene (MBHA) resin (0.45 mmol/NH_2/g) prepared by modified literature procedure (using *p*-toluoyl chloride instead of benzoyl chloride) *(35)*. The general synthesis protocol followed established SPPS methodology *(29,30)*:

1. Remove the Boc-protecting group (when present) with 80% TFA and 1% ethanedithiol (EDT) in CH_2Cl_2 (25 min).
2. Wash with 1% EDT in 2-propanol, MeOH, and CH_2Cl_2.
3. Neutralize with 10% ET_3N/CH_2Cl_2.
4. Carry out coupling with a twofold excess of activated amino acid derivative using either diisopropyl carbodiimide (DIC) (with or without 1-hydroxybenzotriazole as additive) or benzotriazole-1-*yl*-oxy-tris(dimethylamino)-phosphonium hexafluorophosphate (BOP) as activating agents (*see* Chapter 1). Depending on the solubility of the Boc amino acid, do couplings in CH_2Cl_2, CH_2Cl_2/DMF, or NMP/DMF mixtures.
5. Monitor each coupling step by the qualitative ninhydrin test (Chapter 8), and carry out recouplings as necessary.
6. Perform N-terminal acetylations with 10% Ac_2O in 1% pyridine/CH_2Cl_2. For the assemblage of the peptide building blocks, Lys side chains are protected by the 2-chlorobenzyloxycarbonyl (2-Cl-Z) group; template lysines are incorporated as the N^{α}-Boc-Lys(Fmoc) derivatives, except for the first lysine residue (Lys^9 of the template peptide), which is incorporated as the N^{α}-Fmoc-Lys(Boc) derivative. Cys side chains were protected by the *p*-methoxybenzyl (Mob) group and Glu side chains as cyclohexyl esters.
7. After coupling of Boc-Cys(Mob) and Fmoc-Lys(Boc), the N^{ε}-Boc-protecting group is removed, and α'_{16} is assembled according to the general protocol. Following removal of the N^{α}-Fmoc-protecting group (20% piperidine/CH_2Cl_2; 5 min and 10 min), complete the template using Boc-Lys(Fmoc) for the incorporation of the remaining lysines.
8. Cleave the peptide from the resin with anhydrous HF in the presence of 5% anisole and 5% ethyl methyl sulfide using a Kel-F line (0°C, 75 min). Evaporate HF, precipitate the peptide with anhydrous ether, and collect by filtration.

9. Remove the Fmoc groups, and synthesize the helices α_{16} in parallel using standard methods.
10. Redissolve the crude material obtained from 1.5 g of peptide resin in 60 mL 20% acetic acid.
11. Dilute this solution with 3 L of water, and then add dropwise to a vigorously stirred solution of $K_3(Fe[CN]_6)$ (200 mg) and NH_4OAc (10 g) in 1 L of water (total addition time ~ 4 h). During this period, keep the pH between 6.8 and 6.9 by simultaneous addition of 10% aq. NH_4OH. Stir the solution at 4°C overnight.
12. Adjust the pH to 5 by addition of AcOH, and add a slurry (~10 mL) of BioRad AG3-X4A anion-exchange resin (analytical grade, 100–200 mesh, Cl^- form). Stir the mixture for 10 min, filter, and apply the filtrate to an ion-exchange column (15-mL gel bed volume), filled with AG3-X4A.
13. Load the eluent on a cation-exchange column (30 mL gel bed volume), filled with BioRex 70 resin (analytical grade, 100–200 mesh, H^+ form). The peptide is initially retained. Elute with 50% AcOH (~400 mL).

Figure 4A shows an analytical RP-HPLC trace of crude T_4-($3\alpha_{16}$,α'_{16}) after cyclization and the subsequent concentration by conventional cation-exchange chromatography. It can be seen that the crude mixture of reaction products after cyclization clearly contains a readily identifiable major component rather than a number of principal products having formed in comparable amounts. This is in agreement with the fact that no special problems were encountered in the course of the parallel assemblage of helices α_{16} allowing the completion of the TASP syntheses within a matter of days using standard solid-phase methodology. The major peak was isolated by preparative RP-HPLC using triethylammonium phosphate (TEAP) buffer/CH_3CN system (gradient from 50%B to 90% B in 50 min at 105 mL/min; buffer A = 0.1% TEAP [pH 2.25], buffer B = 80% CH_3CN/20% H_2O) and then repurified on the same column, but applying a 0.1% TFA buffer/CH_3CN system (gradient from 50% B to 90% B in 20 min; buffer A = 0.1% TFA , buffer B = 0.1% TFA in 80% CH_3CN/20% H_2O). Yield: 34 mg of lyophilized material. The analytical RP-HPLC trace of this material is depicted in Fig. 4B and demonstrates the dramatic improvement in the purity of T_4-($3\alpha_{16}$,α'_{16}) that could be achieved in the RP-HPLC purification steps.

However, based on the appearance of the original chromatogram of the crude peptide (Fig. 4A), it seemed conceivable that up to 14% of the material still had to be accounted for by various coeluting impurities. This contention was confirmed by the reanalysis of RP-HPLC-purified TASPs

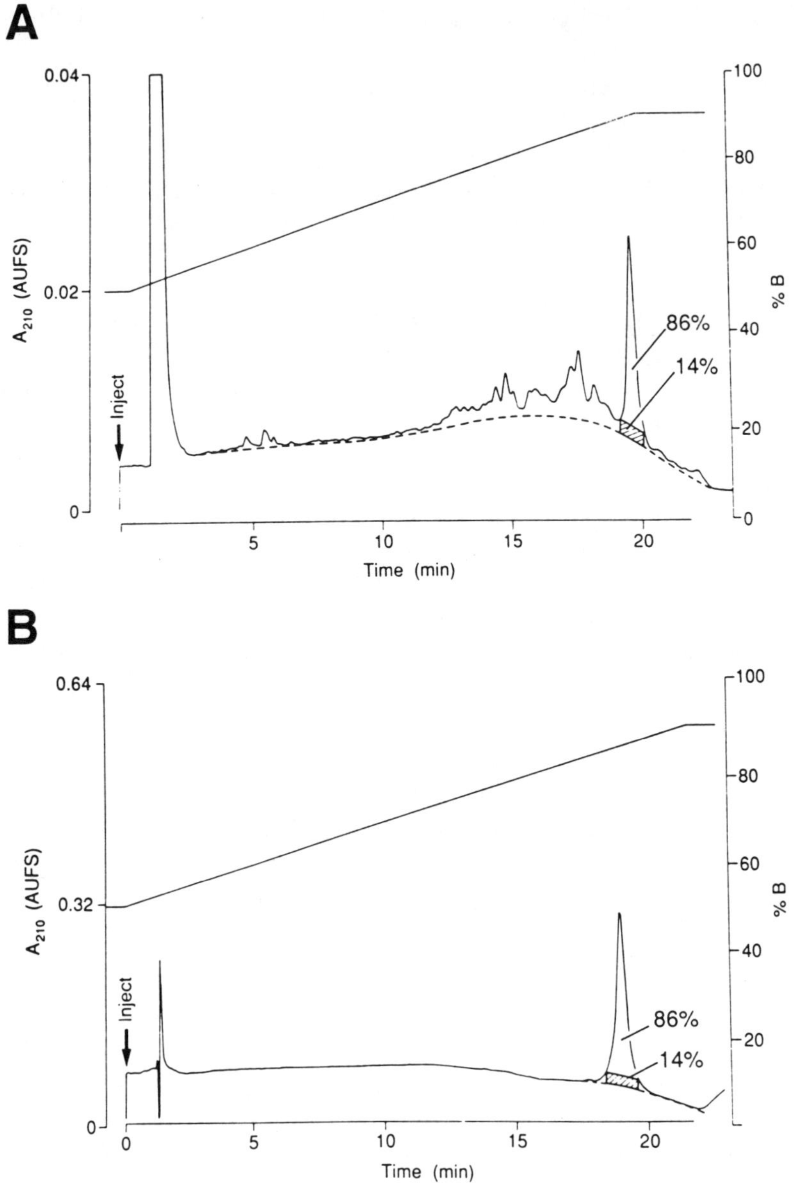

Fig. 4. (**A**) Analytical RP-HPLC of crude cyclized T_4-($3\alpha_{16,}\alpha'_{16}$); (**B**) analytical RP-HPLC after preparative RP-HPLC purification. Gradient from 50% B to 90% B over 40 min; buffer A = 0.1% TFA, buffer B = 0.1% TFA in 80% CH_3CN/20% H_2O (from ref. *22*).

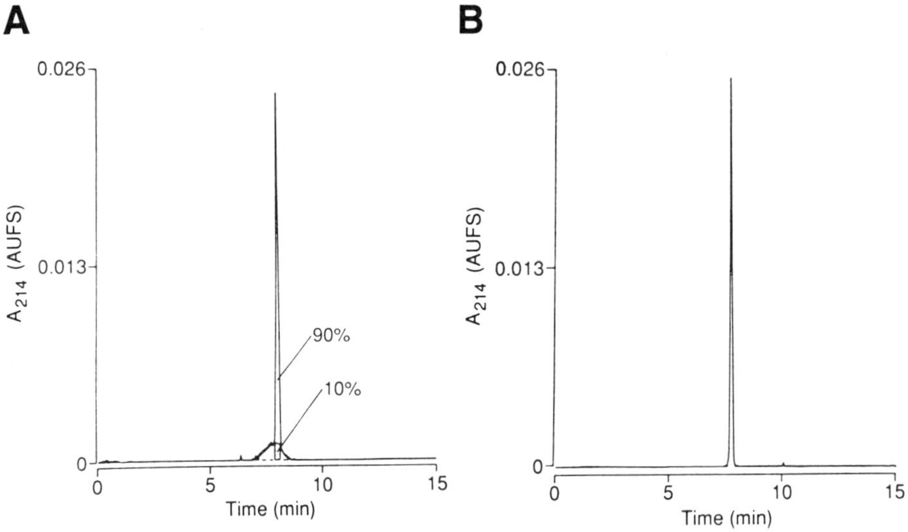

Fig. 5. Electropherogram of T_4-($3\alpha_{16},\alpha'_{16}$) after preparative RP-HPLC purification, and before (**A**) and after (**B**) further purification by high-performance IEC (from ref. *22*).

applying techniques that involve separation mechanisms distinctly different from those of RP-HPLC (high-performance IEC and capillary zone electrophoresis [CZE]) that indeed revealed a certain heterogeneity of these TASP preparations that had not been evident in the RP-HPLC analysis (Fig. 5A). Twenty-one milligrams of the RP-HPLC material were thus further purified by high-performance IEC on a Pharmacia Mono S HR 5/5 column (0.5 × 5.0 cm) (gradient from 0–80% B in 20 min at 1.0 mL/min; buffer A = 50 m*M* AcOH in 65% CH_3CN/35% H_2O [pH 5.0]; buffer B = 0.7*M* NaCl in A; UV detection at 214 nm) and then desalted by RP-HPLC (gradient from 40–95% B in 15 min at 1.0 mL/min; buffer A = 0.1% TFA, buffer B = 0.1% TFA in 80% CH_3CN/20% H_2O). Total yield from seven runs: 9.5 mg. The IEC step resulted in a markedly improved purity of T_4-($3\alpha_{16},\alpha'_{16}$) (Fig. 5B); we conservatively estimate that the purity of this preparation is at least 95%. No comparable levels of purity have been achieved in previous TASP syntheses by SPPS *(33)* that did not incorporate purification schemes based on two-dimensional chromatographic techniques. The purified TASP gave the correct molecular ion peak in LSI mass spectrometry, and amino acid analysis gave the expected ratios.

4.2. TASP Synthesis by Fragment Condensation

As indicated above, the major advantage of TASP synthesis derived from the fragment condensation approach consists of the potentially higher degree of control over the purity of the final product. In principle, the template and each of the secondary structure building blocks can be synthesized independently; they can then be purified to homogeneity and fully characterized before the final condensation step. The synthesis of the various components of the TASP can be performed either by classical solution techniques or by SPPS using anchoring groups that allow cleavage of fully protected peptide fragments from the polymeric support. Since several such linkers have become available over the last few years (even in combination with Fmoc/Bu^t strategy for chain elongation) *(32)*, we find it preferable to prepare the peptide blocks by SPPS, because the classical solution procedure would be much more tedious and time-consuming (however, *see* Section 4.2.1.1. for the synthesis of templates).

Two of the most serious potential problems inherent to the fragment condensation approach are the low solubility (or even insolubility) of protected peptide fragments as well as the racemization of the C-terminal amino residue. The latter problem can be easily taken care of by including a Gly-residue at the C-terminus of a peptide block, which can usually be accommodated in the design of the TASP without difficulties.

Regarding the solubility properties of protected peptide fragments, there are no general rules for their reliable prediction. We have had cases where fully (*tert*-butyl)-side-chain-protected 14-residue peptides were readily soluble and could thus be purified by semipreparative RP-HPLC using a CH_3CN-based solvent system *(27)*. On the other hand, in the example described in the following section, the poor solubility of the fully protected peptide fragment Ac-Lys(Boc)-Leu-Ala-Leu-Lys(Boc)-Leu-Ala-Leu-Lys(Boc)-Ala-Leu-Lys(Boc)-Leu-Ala-Leu-Lys(Boc)-Leu-Ala-OH effectively prevented its purification by HPLC methods; in fact, the solubility of this material was so low that we were not able to use it in the final fragment condensation reaction. The peptide had thus to be deprotected, and the ε-amino groups of the Lys residues were reprotected by the very hydrophilic β-methylsulfonyl ethoxycarbonyl (Msc) group. The reprotected peptide was then sufficiently soluble for the final condensation reaction. The use of highly polar, hydrophilic, or even solubilizing polymeric (polyethylene-glycol-based) side-chain-protecting groups (that are already used in the course of the solid-phase synthesis) may represent a general solution to overcoming the unsatisfactory solubility properties of protected peptide sequences. So far, this concept has not yet been evaluated in a systematic and comprehensive fashion, but studies along these lines are now in progress in our laboratory.

The solubility problem is less severe in the case of the template peptides, because it can usually be purified with all (if symmetrical TASPs are the target) or at least some of the Lys side-chains deprotected (making the peptide sufficiently water-soluble to carry out effective purification). If cyclic templates are desired, cyclizations via amide bonds may be performed after cleavage of the protected peptide from the resin with the crude material in highly polar, strongly solubilizing solvents, like DMF or even DMSO; no side-chain protection is required if cyclization occurs via disulfide bond formation. We have synthesized a variety of templates (Fig. 2) either by classical solution methods or by solid-phase techniques with equal success; however, SPPS offers the advantage of greater speed of synthesis, and in our hands, the purity of the material obtained has never been a problem for these relatively short sequences.

4.2.1. Synthesis of T_6-($6\alpha_{18}$) (Scheme 2)

This section describes in detail the synthesis of a six-helix-bundle TASP molecule incorporating the cyclic decapeptide template cyclo(GK GKGKGKGK)$_2$ with identical helical blocks (α_{18}) attached to the six Lys side chains; α_{18} = KLALKLALKALKLALKLA *(26)*. The molecule was obtained by a convergent fragment condensation approach involving the separate syntheses by solid-phase methods of the template molecule and the peptide block.

4.2.1.1. Synthesis of the Cyclic Template Cyclo(GK)$_6$

1. Assemble the protected linear precursor peptide on a polystyrene-1%-divinylbenzene support using the very acid-labile trialkoxybenzhydryl ester linker *(36)* in combination with an Fmoc/Bu^t protection scheme, i.e., using Fmoc-Gly and Fmoc-Lys(Boc) for the incorporation of Gly and Lys residues, respectively (starting from 6.5 g of Fmoc-Gly-anchor resin). Perform couplings via DCC/HOBT activation of protected amino acid derivatives (threefold excess) and an equimolar amount (with respect to activated amino acid derivative) of diisopropylethylamine (DIPEA) added to the reaction mixture in order to avoid premature cleavage of the peptide from the resin (by HOBT).

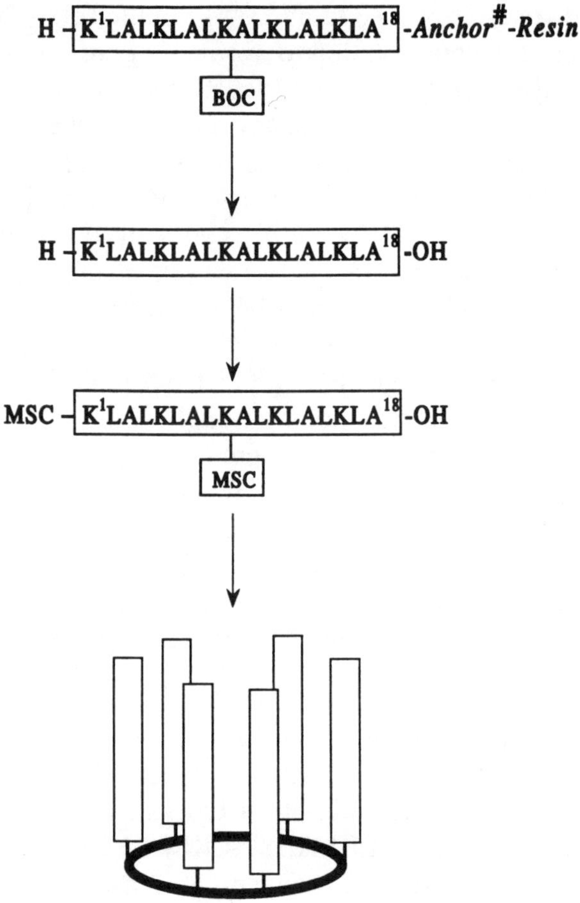

TASP T_6-($6\alpha_{18}$)

Scheme 2. Fragment condensation of a six-α-helix-bundle TASP T_6-($6\alpha_{18}$) *(28)* (*see text*; * = Trialkoxybenzhydryl ester).

2. Remove the Fmoc-protecting group from the N-terminal Lys(Boc)[1] by treatment with piperidine; cleave the protected peptide from the support by treatment with 250 mL 10% HOAc in CH_2Cl_2 for 90 min; remove the resin by filtration (the filtrate does not contain significant amounts of the desired protected peptide and can be discarded).

3. Wash the resin with 300 mL of CH_2ClCH_2Cl/CF_3CH_2OH (1/1) and subsequently with 170 mL of DMSO.
4. Evaporate the filtrate from the CH_2ClCH_2Cl washing to yield a highly viscous residue; redissolve in the filtrate of the DMSO washing step.
5. Add this solution dropwise to 1 L of ice-cold water with stirring (producing an oily precipitate that can neither be filtered nor spun down by centrifugation); concentrate the mixture on a rotary evaporator and lyophilize.
6. In order to remove traces of remaining HOAc, dissolve the product in 100 mL of DMSO and relyophilize to yield 4.23 g of crude $(Lys[Boc]Gly)_6$.
7. Dissolve this material in 200 mL of *N*-methyl pyrrolidone (NMP) together with 3.46 g of HOBT (10 Eq) (solution A); in parallel, dissolve 12.75 g (10 Eq) of the coupling agent PyBOP in 20 mL of NMP (solution B).
8. Start the cyclization reaction by adding 30 mL of solution A, 3 mL of solution B, and 6 mL of DIPEA to 100 mL of NMP. After 30 min, the Kaiser ninhydrin test should indicate the absence of any free amino groups. Repeat the addition of 30 mL of solution A, 3 mL of solution B, and 0.6 mL of DIPEA to the reaction mixture in 45-min intervals until all of solutions A and B are consumed. One hour after the last addition, pour the mixture into 1.5 L of water, and keep refrigerated over night.
9. Collect the precipitate by filtration, wash with 0.5 L of 0.1*M* $NaHCO_3$ solution and 0.5 L of water, and dry over P_2O_5. Suspend this material in 500 mL of water at 60°C for 1 h (stirring), filter the mixture, and dry the collected material. This procedure was repeated three times to yield 4.52 g of crude *cyclo*$(Lys[Boc]Gly)_6$.
10. Treat the protected cyclic peptide with 15 mL of 95% TFA for 2.25 h, after which time, pour the mixture into 150 mL of cold diisopropyl ether/petroleum ether (1/1), and refrigerate the suspension over night.
11. Collect the precipitated peptide by filtration, dissolve in 50 mL water, and lyophilize the solution. Yield: 5.03 g (>100%).
12. Purify crude cyclo $(KG)_6$ by preparative RP-HPLC on a 25 × 2 cm Nucleosil C_{18} column (7 µ, 300 Å) (gradient from 0–20% B in 40 min; buffer A = 0.1% TFA, buffer B = 0.1% TFA in MeOH). Inject 500-µL aliquots from a solution of the total material in 50 mL of water. Yield from 22 runs: 360 mg.
13. In order to convert the TFA salt of the peptide to its hexa-hydrochloride, dissolve 159 mg of pure cyclo$(KG)_6$ in 53 mL of 1*M* HCl, and lyophilize the solution. Relyophilize once from 1*M* HCl and once from H_2O.

4.2.1.2. Preparation of Lys(Msc)-Leu-Ala-Leu-Lys(Msc)-Leu-Ala-Leu-Lys(Msc)-Ala-Leu-Lys(Msc)-Leu-Ala-Leu-Lys(Msc)-Leu-Ala

1. Prepare Fmoc-Lys-Leu-Ala-Leu-Lys-Leu-Ala-Leu-Lys-Ala-Leu-Lys-Leu-Ala-Leu-Lys-Leu-Ala on an Fmoc-Ala-trialkoxybenzhydryl ester resin using standard Fmoc/Bu^t SPPS procedures (*see* Section 4.1.1.), and purified by preparative RP-HPLC in a typical experiment.
2. Reprotect the ε-amino groups of the lysine side chains of the 18-mer with the methylsulfonylethoxycarbonyl (Msc) group. Dissolve 550 mg of peptide (hexa-hydrotrifluoroacetate) in 15 mL of dimethylacetamide (DMA) followed by addition of a solution of 392 mg of MscONsu (7 Eq) in 1 mL of DMA and 436 mL (12 Eq) of DIPEA *(35)*. After 90 min at room temperature, add additional MscONsu (5.4 Eq). After 5 h, concentrate the reaction *in vacuo* to ~5 mL, then pour into 50 mL of ice-cold water, add 40 mL of 1*M* HCl, collect the oily precipitate by centrifugation, wash with three 40-mL portions of ether, and then redissolve in 70 mL of CF_3CH_2OH/DMSO 1/1. Remove the CF_3CH_2OH by evaporation, and lyophilize the remaining solution of the peptide in DMSO. Repeat the procedure once. Yield: 530 mg (89%).

4.2.1.3. Fragment Condensation

1. Dissolve Msc-protected 18-mer (200 mg, 6 Eq) and HOBT (23 mg, 12 Eq) in 2.44 mL DMA. Keep the viscous, turbid solution at 70°C for 10 min, and then sonicate for 3 min. This leads to a gel-like, slightly turbid solution.
2. Cool to 5°C, and then add DIC (76 mL, 42 Eq) (solution A).
3. In parallel dissolve *cyclo*$(GK)_6$ · 6HCl (15.6 mg, 1 Eq) in 0.1*M* NaOH (708 mL, 6 Eq), add 2.9 mL of DMA, and after cooling to 5°C, combine this solution with solution A and keep the mixture at 5°C. After 3 h, add a solution containing additional activated Msc-protected 18-mer; the latter is prepared by dissolving 100 mg (3 Eq) of 18-mer and 11.5 mg of HOBT in 1.44 mL of DMA *(see above)* followed by addition of 38 mL (21 Eq) of DIC to the precooled solution and subsequent preactivation for 2 h at 5°C. The kinetics of the coupling reaction is shown in Fig. 6.
4. After 3 d at 5°C, distribute the reaction mixture among six centrifugation tubes (11 mL/tube) and add 8.25 mL of Tesser solution (2*N* NaOH/dioxane/methanol [10/140/50]) *(37)* to each tube with vigorous mixing (Vortex mixer), in order to effect removal of the Msc-protecting group. After 1 min, briefly centrifuge the mixtures, and dissolve the residue in 16 mL of TFA (combined material). Finally, add 30 mL of MeOH and 30 mL of H_2O, and adjust the pH to ~5 with 2*M* NaOH. Keep this solution at 5°C.

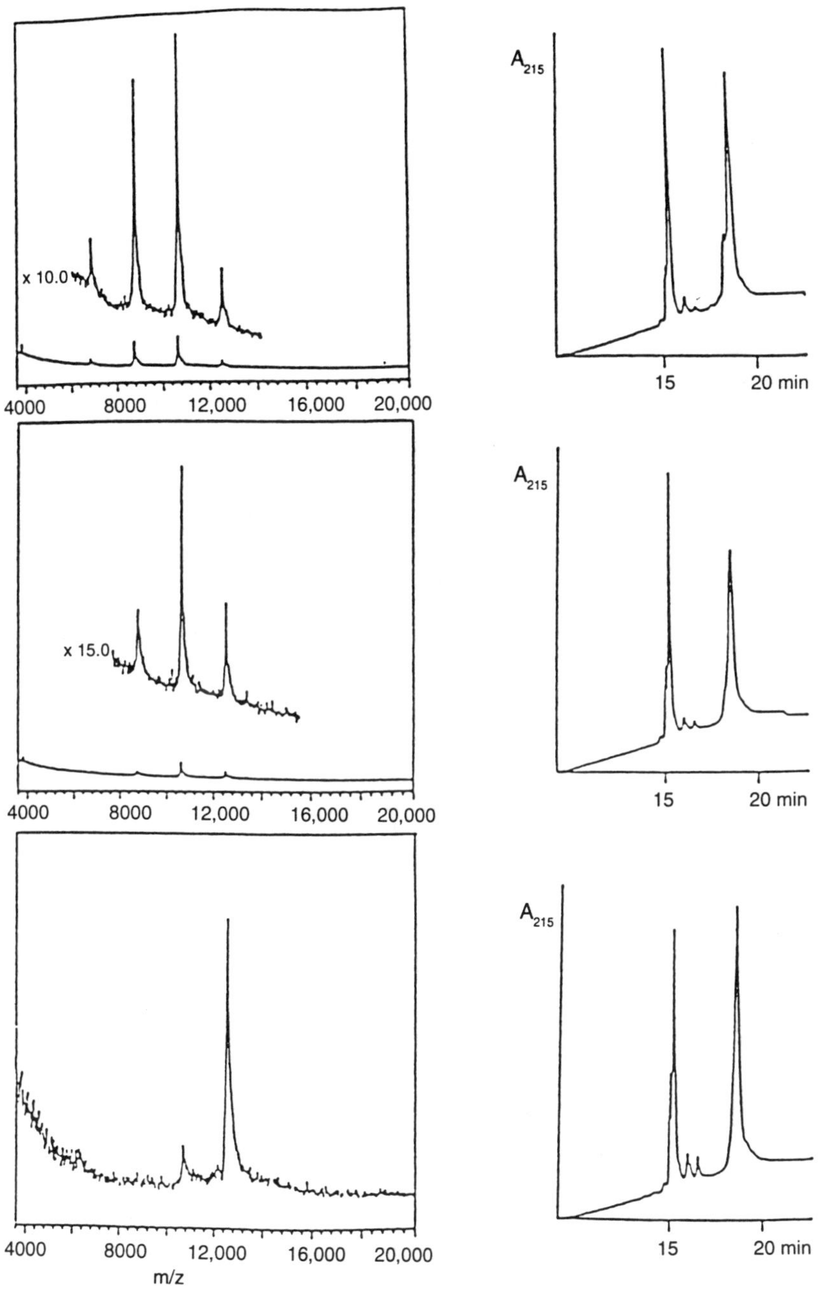

Fig. 6. Kinetics of the coupling reaction between Msc-protected α_{18} and *cyclo*$(GK)_6$ as followed by RP-HPLC and LDI-MS *(28)*.

5. Purify T_6-($6\alpha_{18}$) on a 25 × 1 cm Nucleosil C4 column (7 µ, 300 Å) at a flow rate of 5 mL/min. Inject 5-mL aliquots of the above solution in TFA/MeOH/H_2O, and employ a gradient from 40% B to 55% B over 15 min (A = 0.1% TFA, B = 0.1% TFA in CH_3CN). Yield: 24 mg of T_6-($6\alpha_{18}$). This material gave the correct molecular ion peak in LDI MS; amino acid analysis gave the correct ratios; enantiomer analysis after hydrolysis revealed that the C-terminal Ala residue had been racemized to ca. 3%.

5. Concluding Remarks

Because of the complexity of the matter, it should not be surprising if the foregoing discussion has not comprehensively covered all of the conceivable (and not even all of the realistic) approaches to the synthesis of TASP molecules. A variety of other synthetic strategies may be viable and are pursued in our laboratory, e.g., we are presently trying to assemble a TASP molecule representing a condensed version of the ROP protein (incorporating four different helical building blocks) by successive coupling of template fragments, each of which contains one of the template lysines with the appropriate (protected) peptide block already attached. Although this involves a tremendous synthetic effort for the synthesis of a specific target molecule, a more general perspective is offered by the application of enzymatic coupling methods or the prior thiol capture strategy to the synthesis of TASP molecules. Both of the latter techniques would allow fragment condensations in aqueous solution, thus obviating the need for protected fragments in the final condensation step(s). Both routes are already or will be under scrutiny in our laboratory in the near future.

All the above indicates that, although reasonably reliable and straightforward methods are now available for the synthesis of TASP molecules, the synthesis of TASPs of any desired sequence and three-dimensional arrangement is far from having become routine. A good deal of methods research and development are still required before such a general statement becomes accepted reality. In addition, as is most often the case in natural (and nonnatural) product synthesis, there may always be a better way to synthesize a specific target TASP. However, this shall not prevent us from simultaneously pursuing the ultimate goal of the TASP strategy for protein *de novo* design, i.e., the design and synthesis of branched polypeptide sequences that not only structurally, but also functionally resemble natural proteins.

Acknowledgment

This contribution was supported by the Swiss National Science Foundation.

References

1. Anfinsen, C. B. and Scheraga, H. A. (1975) Theoretical and experimental aspects of protein folding. *Adv. Prot. Chem.* **29,** 205–300; White, F. H., Jr. and Anfinsen, C. B. (1958) *Ann. NY Acad. Sci.* **81,** 515.
2. Fersht, A. (1985) *Enzyme, Structure and Mechanism*, 2nd ed., WH Freeman, San Francisco.
3. Knowles, J. (1987) Tinkering with enzymes: what are we learning? *Science* **236,** 1252–1258.
4. Pace, C. N. (1990) Measuring and increasing protein stability, *Trends Biotech.* **8,** 93–97.
5. Schimmel, P. (1989) Hazards of deducing enzyme structure–activity relationships on the basis of chemical applications of molecular biology. *Acc. Chem. Res.* **22,** 232,233.
6. Cf., e.g., Davies, J. E. and Gassen, H. G. (1983) Synthetische Genfragmente in der Gentechnik-die Renaissance der Chemie in der Molekularbiologie. *Angew. Chem.* **95,** 26; (1983) *Angew. Chem. Int. Ed. Engl.* **22,** 13–31.
7. Noren, C., Anthony-Cahill, S., Grifity, M., and Schultz, P. (1989) A general method for site-specific incorporation of unnatural amino acids into proteins. *Science* **244,** 182–188.
8. Wüthrich, K. (1989) Protein structure determination in solution by NMR spectroscopy. *Science* **243,** 45–50.
9. Wüthrich, K. (1989) The development of NMR spectroscopy as a technique for protein structure determination. *Acc. Chem. Res.* **22,** 36–46.
10. Ward, W. H. J. and Fersht, A. R. (1988) *Redesigning the Molecules of Life,* (Benner, S. A., ed.), Springer Verlag, Berlin.

11a. Regan, L. and DeGrado, W. F. (1988) Characterization of a helical protein designed from first principles. *Science* **241,** 946–948.

11b. Richardson, J. S. and Richardson, D. C. (1989) The de novo design of proteins. *Trends Biochem. Sci.* **14,** 304–309.

11c. Hecht, M. H., Richardson, J. S., Richardson, D. C., and Ogden, R. C. (1990) De novo design, expression, and characterization of Felix: a four-helix bundle protein of native-like sequence. *Science* **249,** 884–891.

12. Richardson, J. S. (1981) The anatomy and taxonomy of protein structure. *Adv. Prot. Chem.* **34,** 167–339.
13. Rossmann, M. G. and Argos, P. (1981) Protein folding. *Ann. Rev. Biochem.* **50,** 555–583.
14. Presnell, S. R. and Cohen, F. E. (1989) Topological distribution of 4-α-helix bundles. *Proc. Natl. Acad. Sci. USA* **86,** 6592–6596.
15. Anfinsen, C. B. and Scheraga H. A. (1975) Theoretical and experimental aspects of protein folding. *Adv. Prot. Chem.* **29,** 205–300.

16. Montelione, G. T. and Scheraga, H. A. (1989) Formation of local structures in protein folding. *Acc. Chem. Res.* **22,** 70–76.
17. Finkelstein, A. V. and Ptytsin, O. B. (1987) Why do globular proteins fit the limited set of folding patterns? *Prog. Biophys. Mol. Biol.* **50,** 171–190.
18. Baldwin, R. L. (1989) How does protein folding get started? *Trends Biochem. Sci.* **14,** 291–294.
19. Mutter, M. (1988) Nature's rules and chemist's tools: a way for creating novel proteins. *Trends Biochem. Sci.* **13,** 260–263.
20. Mutter, M., Altmann, K. H., Tuchscherer, G., and Vuilleumier, S. (1988) Strategies for the de novo design of proteins. *Tetrahedron* **44,** 771–785.
21. Mutter, M. and Vuilleumier, S. (1989) A chemical approach to protein design—Template Assembled Synthetic Proteins (TASP). *Angew. Chem. Int. Ed. Engl.* **28,** 535–554.
22. Mutter, M., Tuchscherer, G., Miller, C., Altmann, K. H., Carey, R., Wyss, D., Labhardth, A., and Rivier, J. R. (1992) Template-assembled-synthetic proteins (TASP) with four-helix bundle topology. Total chemical synthesis and conformational studies. *J. Am. Chem. Soc.,* **114,** 1463–1470.
23. Schiffer, M. and Edmundson, A. B. (1967) Use of helical wheel to represent the structures of proteins and to identify segments with helical potential. *Biophys. J.* **7,** 121–136.
24. Dunnil, P. (1968) The use of helical net-diagram to represent protein structures. *Biophys. J.* **8,** 865–875.
25. Hull, S. E., Karlsson, R., Main, P., Woolfson, M. M., and Dodson, E. J. (1978) The crystal structure of a hydrated Gramicidin S-Urea complex. *Nature* **273,** 443.
26. Ernest, I., Kalvoda, J., Rihs, G., and Mutter, M. (1990) Three novel mimics for the construction of sterically constrained protein turn models. *Tetrahedron Lett.* **31,** 4011–4014.
27. Ernest, I., Vuilleumier, S., Fritz, H., and Mutter, M. (1990) Synthesis of a 4-helix-bundle-like Template Assembled Synthetic Protein (TASP) by condensation of a protected peptide on a conformationally constrained cyclic carrier. *Tetrahedron Lett.* **31,** 4015–4018.
28a. Steiner, V. (1991) Synthesis and characterization of a 6-α-helix bundle TASP by fragment condensation, Ph.D Thesis, University of Basel.
28b. Carey, R. I., Dörner, B., Mutter, M., Labhardt, A. M., Steiner, V., and Rink, H. (1992) New routes to artificial proteins applying the TASP concept, in *Innovation and Perspectives in Solid Phase Peptide Synthesis* (Epton, R., ed.), SPPC, Birmingham, UK, pp. 125–131.
29. Stewart, J. M. and Young, J. D. (1984) *Solid Phase Peptide Synthesis*, 2nd ed., Pierce Chemical Co., Rockford, IL.
30. Barany, G. and Merrifield, R. B. (1980) Solid phase peptide synthesis, in *The Peptides—Analysis, Synthesis and Biology*, vol. 2 (Gross, E. and Meienhofer, J., eds.), Academic, New York, pp. 1–248.
31. Barany, G., Kneib-Cordonier, N., and Mullen, D. G. (1987) Solid phase peptide synthesis: a silver anniversary report (review). *Int. J. Pept. Protein Res.* **30,** 705–739.

32. Fields, G. B. and Noble, R. L. (1990) Solid phase peptide synthesis utilizing 9-fluorenylmethoxycarbonyl amino acids (review). *Int. J. Pept. Protein Res.* **35,** 161–214.
33. Mutter, M., Hersperger, R., Gubernator, K., and Müller, K. (1989) The construction of new proteins: V. A Template Assembled Synthetic Protein (TASP) containing both a α-helix-bundle and β-barrel-like structure. *Proteins* **5,** 13–21.
34a. Wünsch, E. and Müller, E. (eds.) (1976) Houben-Weyl, *Methoden der Organischen Chemie,* vol. XV/2, Thieme Verlag, Stuttgart.
34b. Kaiser, E. T., Mihara, H., Laforet, G. H., Kelly, J. W., Walters, L., Findeis, M. A., and Sasaki, T. (1989) Peptide and protein synthesis by segment—synthesis condensation. *Science* **243,** 187–192.
35. Penke, B. and Rivier, J. (1987) Solid phase synthesis of peptide amides on a polystyrene support using 9-fluorenylmethoxycarbonyl protecting groups. *J. Org. Chem.* **52,** 1197–1200.
36. Rink, H. (1987) Solid phase synthesis of protected peptide fragments using a trialkoxy-diphenyl-methylester resin. *Tetrahedron Lett.* **28,** 3787–3790.
37. Rajh, H. M., Mariman, E. C. M., Tesser, G. I., and Nirard, R. J. F. (1980) Tryptophan replacement in the C-terminal octapeptide of CCK-PZ. *Int. J. Pept. Protein Res.* **15,** 200–210.

CHAPTER 16

Chemical Synthesis of the Aspartic Proteinase from Human Immunodeficiency Virus (HIV)

Paul D. Hoeprich, Jr.

1. Introduction

An aspartic proteinase encoded by the human immunodeficiency virus (HIV) is expressed during the course of infection leading to acquired immunodeficiency syndrome (AIDS). This enzyme is essential for viral infectivity and proliferation; it selectively cleaves all viral protein components from a "polyprotein" precursor molecule *(1)*. It is essential for the production of fully competent virus and is unique to a class of retrovirally encoded aspartic proteinases with no apparent counterpart in the mammalian realm. As such, it is an ideal target for design of specific inhibitors as antiviral therapeutic agents for the treatment and perhaps cure of AIDS. Critical to this effort has been availability of the enzyme for structure–function analysis. The three-dimensional structure of the enzyme, derived from X-ray crystallographic efforts, was determined in 1989 by Wlodawer et al. *(2)*. Diffraction-quality crystals of the HIV-1 proteinase were obtained from chemically synthesized protein corresponding to the sequence of the SF-2 isolate from an AIDS-associated retrovirus (ARV-2) as reported by Sanchez-Pescador et al. *(3)*. This particular protein was synthesized using *t*-Boc/benzyl strategy starting with Phe-PAM resin following a highly optimized synthetic regimen described previously by Schneider and Kent. In their synthesis, cysteine occurring at positions 67 and 95 in the native sequence was replaced

From: *Methods in Molecular Biology, Vol. 36: Peptide Analysis Protocols*
Edited by: B. M. Dunn and M. W. Pennington

with α-aminobutyric acid (Abu), an isosteric analog of cysteine, i.e., a methyl group replaces the thiol moiety *(4)*.

Others have succeeded in synthesizing active enzyme, but not obtaining crystalline protein. In these attempts, minimally, each residue was double coupled with several amino acids being coupled three times, and a few added four and five times *(5,6)*. The results of an effort to synthesize the identical aspartic proteinase sequence as reported by Schneider and Kent, i.e., aminobutyric acid at positions 67 and 95, using Fmoc/*t*-butyl strategy, are described in Fig. 1.

2. Materials

The solid support/matrix, i.e., standard polystyrene-divinylbenze copolymer with 1% crosslinking containing Fmoc-L-phenylalanine, Fmoc-α-aminobutyric acid, and Fmoc-4-nitrophenylalanine, was purchased from Bachem Biosciences, Inc., Philadelphia, PA. The first amino acid in the synthesis, Fmoc-L-Phe residue 99 in the protein, was preloaded on the solid phase via *p*-alkoxybenzyl alcohol-type linker (Wang, *[7]*), with a substitution level of 0.6 mmol/g. All reagents for assembly of the protected polypeptide, i.e., Fmoc-amino acids, solvents, HBTU, and so forth, were from Applied Biosystems, Inc., Foster City, CA. Thioanisole, ethanedithiol, and phenol scavengers were obtained from Aldrich Chemicals, Madison, WI. Pepstatin agarose was purchased from Sigma Chemical Company, St. Louis, MO.

3. Methods

3.1. Solid Support

Choice of the "proper" solid support is critical to the success of long-chain assembly. As can be seen from Table 1, three different preloaded polystyrene resins were used in the course of synthetic studies on the proteinase. A critical parameter appears to be the "swellability" of this support, i.e., the degree to which a known amount of polystyrene resin particles will swell in the presence of a given solvent or mixture of solvents *(8)*. In the studies with the HIV-1 proteinase synthesis, all supports investigated were purported to be copolymers of polystyrene and divinylbenzene 1% crosslinking and comparable substitution level of Fmoc-L-Phe. As can be seen, the resin that showed maximal swelling in *N*-methylpyrrolidine (NMP) was the support that gave the best yield of protected polypeptide resin. At the outset of long-chain synthetic effort, one can easily determine the swelling characteristics of a given resin by

P Q I T L W Q R P L V T I R I G G Q L K E A L L D T G A D D

2 2 2 2 2 2 2 2 2 2 2 2 2 2 2 2 2 2 2 2 2 2 2 2 2 2 2 2 2 2

T V L E E M N L P G K W K P K M I G G I G G F I K V R Q Y D

2 2 2 2 2 2 1 1 1 1 1 1 1 1 1 1 1 1 1 1 1 1 1 1 1 1 2 1 1 1

Q I P V E I **Ab** G H K A I G T V L V G P T P V N I I G R N L L T Q I G **Ab** T L N F

1 1 1 1 1 1 1 1 2 1 1 1 1 1 2 2 2 1 1 1 1 1 1 2 2 2 3 3 2 2 1 1 1 1 1 1 1 1

Side chain protecting groups:

Q & N -- trityl	R -- Pmc	
D & E -- t-butyl	K -- t-Boc	**Ab** -- α-aminobutyric acid
S, T & Y -- t-butyl	H -- Bum	

Fig. 1. Coupling strategy and side-chain protecting groups used in the protease synthesis. The numbers under each residue refer to the number of couplings needed at that position.

Table 1
Resin Swelling

	Swelling, NMP	Yield, HIV Prt	Sample
Resin-HMP	7.8 mL/g	—	
Resin-HMP-Phe-Fmoc	8.0 mL/g	1.5 g	1
0.6 mmol/g	6.0 mL/g	1.0 g	2
	6.2 mL/g	0.8 g	3

using a graduated "Merrifield-type" peptide synthesis vessel, a known amount of resin, and a series of "standard" solvents, for example, in the study summarized in Table 1.

1. Weigh out 0.25-g samples of three Fmoc-Phe-HMP-resins.
2. Equilibrate this material in each case with NMP by three 30-min washes with 5-mL portions of the solvent. After the final wash, drain the vessel (via aspiration), and determine the volume occupied by the swollen resin from graduated markings on the vessel. Correct this value by the volume obtained in a nonresin swelling solvent, e.g., methanol. Normalize the resulting net swelling values to mL/g. The proteinase, or the peptide of interest, is then synthesized once on each of the supports. Clearly, the resin that gave the best swelling behavior with just the first amino acid attached was the support that gave the highest yield of protected peptide on the resin at the end of the synthesis.

3.2. Fmoc Removal

1. Remove the N^{α}-fluorenylmethoxycarbonyl (Fmoc) group by two or three sequential treatments with 20% piperidine/NMP each lasting 3–7 min with a regular increase in total time of deprotection (add-time feature of Applied Biosystems Peptide synthesizer 431A) every ten steps.
2. Monitor the extent of this reaction spectroscopically at 305 nm by periodically detecting the presence of Fmoc byproduct, the dibenzofulvene-piperidine adduct.

Details of the Fmoc removal protocol and the spectroscopic monitoring system as adapted to a 431A ABI Peptide Synthesizer have been described in detail *(9)*. Briefly, a synthesis cycle commences with two 3-min treatments of the resin with a piperidine/NMP solution (approx 20% piperidine v/v). At the end of each 3-min period, an aliquot (10–15 mL) of the piperidine/NMP solution is removed and sent automatically via the auxillary waste line to an external HPLC-type UV detector; simultaneously, a chart recorder is turned on using one of the two relay switches on the backside of the 431A instrument, and a peak is recorded. This monitoring of Fmoc removal can be repeated as many times as one wants within a given cycle simply by defining the number of "loops" in step 1 of the B module or Fmoc removal module (*see* ref. *9*). The result is a "picture" of the entire synthesis as shown in Fig. 2.

3.3. Solid-Phase Peptide Synthesis

Synthesis of the protein was carried out using an ABI 431A peptide synthesizer. *FastMoc*™ chemistry was used throughout *(10)*. Peptide bond formation was accomplished by *in situ* activation using 2-(1H-benzotriazol-1-*yl*)-1,1,3,3-*tetra*-methyluronium hexafluorophosphate (HBTU) as described previously *(11,12)*. The modification suggested by Gausepohl et al. where the ratio of Fmoc amino acid:HBTU is 1.0:0.8 was used, thus avoiding "capping" of the free α-amino group by minimizing the potential reaction with the uronium reagent itself *(13)*. The assembly process was optimized by taking into consideration previously published information, that is, coupling difficulty in the vicinity of Arg 87, as well as a more systematic periodic sampling tactic. The latter was simply to remove a 25–50 mg portion of protected peptide resin after 10 cycles. Each sample was cleaved from the solid support, HPLC was run, and "peaks" collected. Amino acid composition and mass analysis of the main peaks ensued. These data combined with resin sampling and quantitative ninhydrin analysis information yielded a reasonably clear picture of synthetic progress, i.e., incomplete coupling reactions, deletions, and so on. In actual practice, this tactic proved to be useful only through the first 30 residues; the cleaved peptide beyond the 30-mer stage was quite insoluble, thus limiting analytic workup. In the latter third of the synthesis, each residue was double-coupled.

Insight into the mode of action of this uronium salt was noted during the course of this work in model reaction using doubly labeled O^{18} glycine. The Fmoc derivative of the latter was condensed with an equivalent of phenylalanine ethyl ester in the presence of 0.8 Eq of HBTU and excess HOBT and DIEA (1.5 Eq). The tetramethylurea formed during the reaction (possible leaving group?) contained O^{18} in the urea oxygen. This result is consistent with a possible mechanism proposed by Knorr et al. *(14)*.

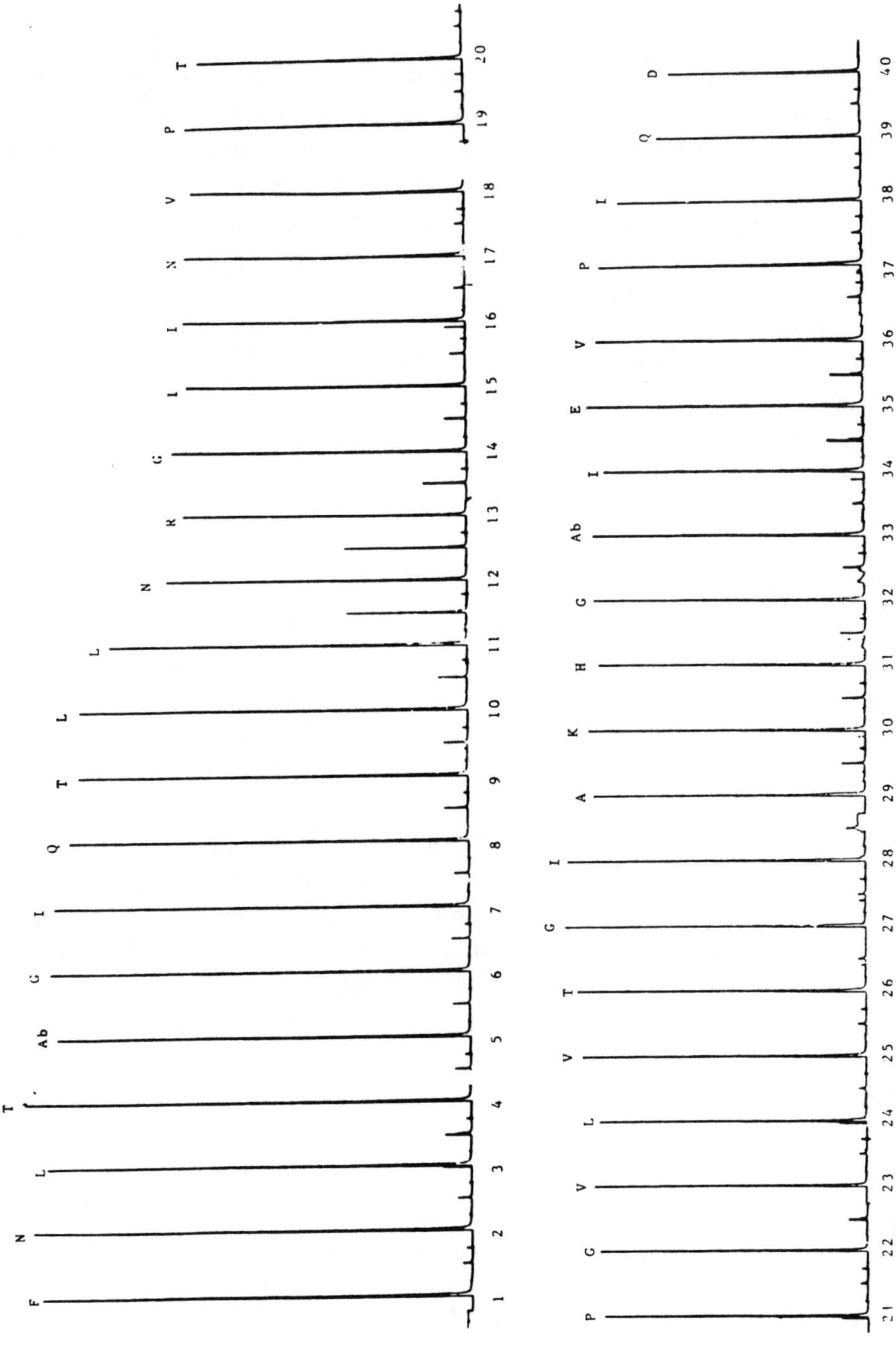

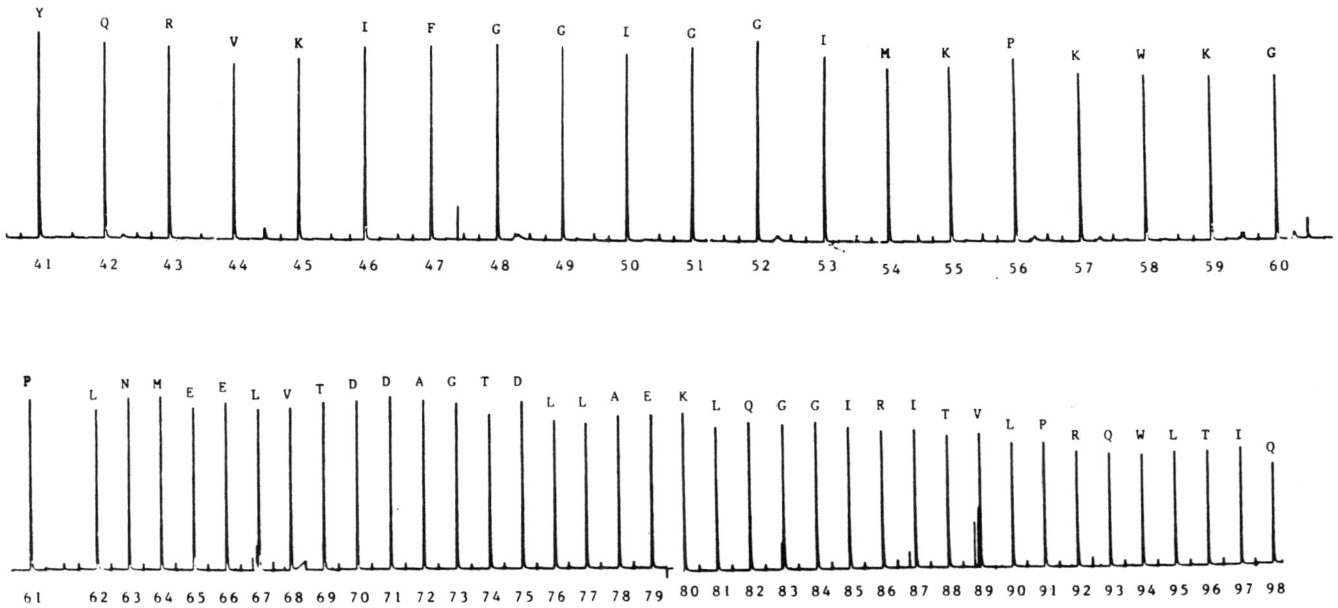

Fig. 2. UV monitoring profile HIV-1 Protease (antibodies 67, 95) 1–99. The chart recorder speed was decreased during cycles 62–98.

Table 2
Capping Cycle, 0.25 mmol, for Synthesis on the ABI Model 431A

Step	Fxn #	Function name	Time	Add
1	14	#10 B VB	1	0
2	56	#10 B RV	15	1
3	2	Vortex On	30	0
4	3	Vortex Off	1	0
5	42	Drain RV	15	0
6	50	#10 RV-DRN	5	0
7	42	Drain RV	10	0
8	77	PRS #4	10	0
9	17	#4 B VB	1	0
10	52	#4 B RV	15	0
11	56	#10 B RV	2	0
12	40	Mix RV	5	0
13	93	GAS To RS	3	0
14	41	Vent RV	2	0
15	98	Begin Loop	20	0
16	2	Vortex On	20	0
17	3	Vortex Off	10	0
18	99	End Loop	1	0
19	41	Vent RV	2	0
20	9	Gas T VB	2	0
21	42	Drain RV	20	0
22	98	Begin Loop	4	1
23	56	#10 B RV	15	1
24	2	Vortex On	20	0
25	3	Vortex Off	10	0
26	42	Drain RV	10	0
27	50	#10 RV-Drn	5	0
28	42	Drain RV	10	0
29	99	End Loop	1	0

Follow each coupling by a 10-min capping step using a solution of 0.5*M* Ac_2O, 0.125*M* DIEA, and 0.2% HOBT (w/v) in NMP. For convenience, prepare a "stock" solution of this capping reagent, and place in bottle position 4. A special cycle for capping may be used that adds an appropriate volume of the mixture directly to the reaction vessel (Table 2). Weigh washed/dried peptide resins to determine overall weight gain, which should be on the order of 1.6–1.8 g. These numbers suggest an **average** step yield or % coupled/cycle just >99% over 98 residues.

3.4. Deprotection and Cleavage

1. Cleave the protein from the solid support with simultaneous removal of side-chain-protecting groups using a mixture of TFA (85%), H_2O (5%), thioanisole (5%), EDT (5%), and phenol (5% w/v), after King et al. *(15)*.
2. Treat 0.8 g protected peptide/protein-resin with 10 mL of the cleavage mixture, and allow to react with agitation for 2.5–2.75 h (Merrifield-type vessel and reciprocating shaker) at room temperature.
3. Remove the mixture containing the protein and cleavage byproducts from the resin solid support by filtration, and precipitate the peptide/protein product by the addition of cold (4°C) methyl *t*-butyl ether.
4. Vortex the ether precipitate quickly, and then centrifuge to pellet the crude protein.
5. Decant the ether, containing various scavengers and cleavage-derived byproducts.
6. Redissolve the precipitate in 1 mL TFA, and precipitate the protein by ether addition followed by centrifugation and decantation of the ether phase an additional three times. After the final precipitation, take up/dissolve the pelleted material in 50% acetic acid/water, dilute out to 10% acetic acid with the addition of water, freeze in a dry ice/acetone bath, and lyophilize. The overall yield at this point is 325 mg or approx 58% of theoretical.

3.5. Purification and Characterization

1. Dissolve the lyophilized crude material (200 mg) in 50% acetic acid (3–5 mL), and pass over a column of Sephadex G-50 column (2.5 × 100 cm) equilibrated in the same solvent at a flow rate of 60 mL/h; collect 3-mL fractions after passage of 110 mL void volume.
2. Pool and lyophilize high-molecular-weight material, i.e., early eluting fractions as monitored by A_{280} and SDS-PAGE using a Tris-Tricine buffer system *(16)*. A typical gel is shown in Fig. 3.
3. Dissolve the crude material (approx 118 mg of desalted crude polypeptide recovered from an initial chromatography of 200 mg crude material) in 50% acetic acid and rechromatograph over the same column, pool as described above, and lyophilize to give "semi-pure" polypeptide (80 mg), i.e., essentially one band on a 12.5% SDS-PAGE gel run in Tris-Tricine buffer system.
4. Refold this G-50 "semipure" protein following the procedure described by Tomasselli et al. *(17)* in the presence of BSA (0.5% w/v). Many attempts to effect final purification ensued; the combination of pepstatin agarose *(18)* followed by reverse-phase HPLC appears to be most convenient (Fig. 4).

The enzyme was characterized by sequence analysis, i.e., NH_2-terminal, and following both tryptic and CNBr fragmentation. Additionally,

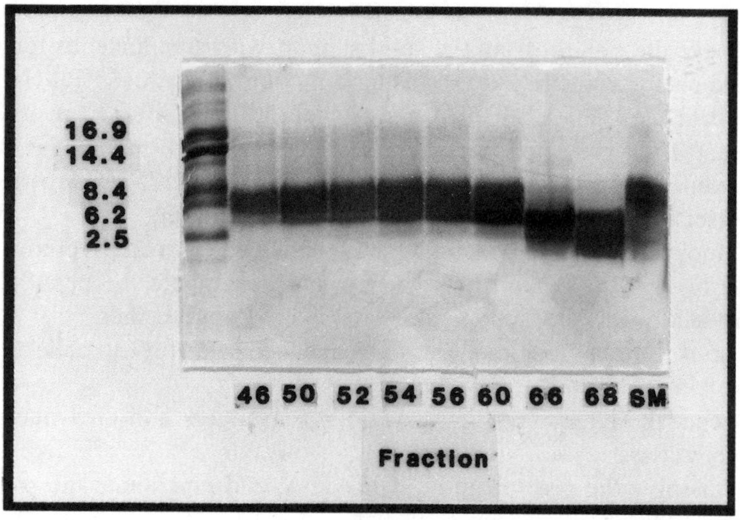

Fig. 3. SDS-PAGE analysis of G-50 Sephadex gel filtration derived fractions. Electrophoresis conditions: 12.5% acrylamide, Tris-Tricine buffer system, Bio-Rad mini-gel apparatus, 7 mA constant current (50–90 V) for approx 1 h. The upper band is the protease; early fractions from G-50 contain predominantly this material.

the fragment mixtures resulting from the trypsin and CNBr treatments were characterized by mass analysis. The data from these studies were consistent with the structure shown in Fig. 1. Additionally, results from amino acid analysis, isoelectric focusing electrophoresis (pI = 9.5), and capillary electrophoresis again were consistent with a homogeneous protein with a primary sequence corresponding to that of (antibodies 67, 95) HIV-1 aspartic proteinase. Matrix-assisted laser desorption mass spectroscopy (Fig. 5) clearly showed a molecular species that has the expected molecular mass. Additionally, a portion of the enzyme was sent to Kent laboratory for analysis by electrospray mass spectroscopy, where it was confirmed that indeed the correct molecule had been synthesized.

Proteolytic activity of the enzyme was measured using synthetic peptide substrate molecules representing known polyprotein "processing" sites. Initial measure of enzymatic activity was made using a discontinuous assay, i.e., periodic HPLC analysis of a mixture of enzyme and peptide substrate (ATLNF-PISPW, in this case, with the scisscle bond being between the Phe-Pro). Results from this assay defined the following

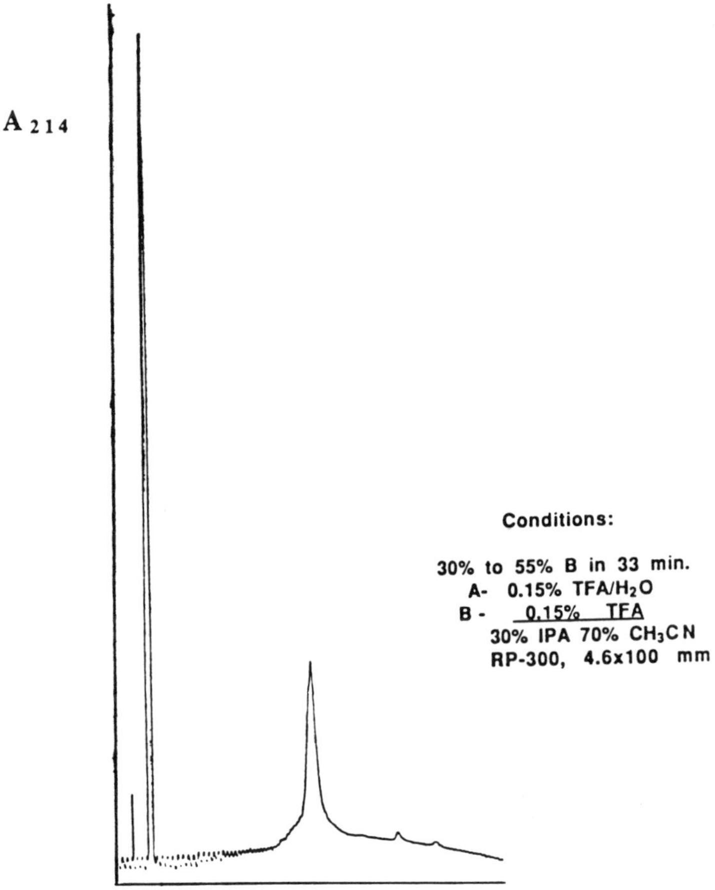

Fig. 4. HPLC tracing of purified protease. The column was a C_8-reverse phase (RP-300, ABI) packing; the gradient conditions were those of Tomaselli et al. *(17)*.

kinetic parameters: $K_m = 2.20$ m*M* and $k_{cat} = 200$ min^{-1}; for this series of experiments, the actual enzyme concentration was determined by active site titration. These data are consistent with published kinetic parameters for this enzyme and similar synthetic peptide substrates *(5)*. Two additional assays were used to measure activity of synthetic enzyme. One, a UV/Vis chromophoric peptide substrate Ac-KASQF(NO_2)-PVV-amide (after the work of Nashed et al., *19*), was prepared. 4-Nitrophenylalanine was substituted for tyrosine at the scisscle bond, —Tyr-Pro—. As

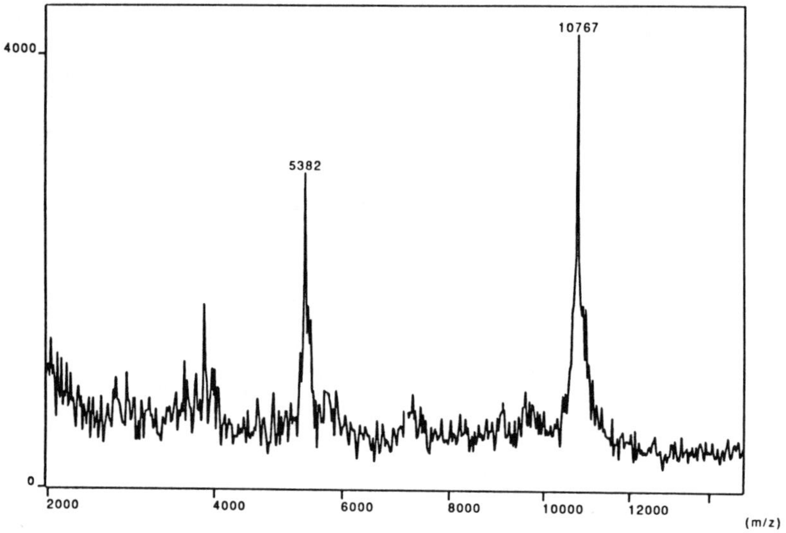

Fig. 5. Matrix-assisted laser desorption mass spectrometry of synthetic protease. MH^+ and $MH2^+$ peaks are shown.

enzymatic hydrolysis proceeded, an increase in absorbance at 316 nm was measured in a continuous manner. Companion HPLC tracings of the intact peptide substrate and hydrolysis products as a function of time confirmed the expected bond cleavage (Fig. 6). This assay was used on a routine basis as well. Briefly, to 400 mL of a 450-m*M* solution of substrate dissolved in 100 m*M* Na acetate, pH 5.5, 1*M* NaCl was added a 50-µL sample aliquot containing enzyme. This sample was always dissolved in "refolding" buffer, i.e., 50 m*M* Na acetate, pH 5.5, 10% glycerol, 5% ethylene glycol, 0.1% Triton X-100 (reduced), 1 m*M* EDTA, and 0.8*M* urea. The dilution took place in a cuvet. The latter was inverted twice and placed in a diode array detector (Hewlett-Packard 8451A) and the increase in A_{316} was recorded (*see also* Chapter 13 for a fuller description of assays employing chromogenic substrates).

The other assay, also continuous in nature, was based on fluorescence. A synthetic peptide (NMA-SNQY-PIVQK[DNP]-amide), derived from the p17/p24 cleavage site in the polyprotein, is "capped" with *N*-methyl anthranilic acid (NMA) on the NH_2-terminal serine. This substrate also contained an additional residue, lysine ε-*N*-dinitrophenyl (DNP), as a COOH terminal amide. This modified peptide can be hydrolyzed by the proteinase between

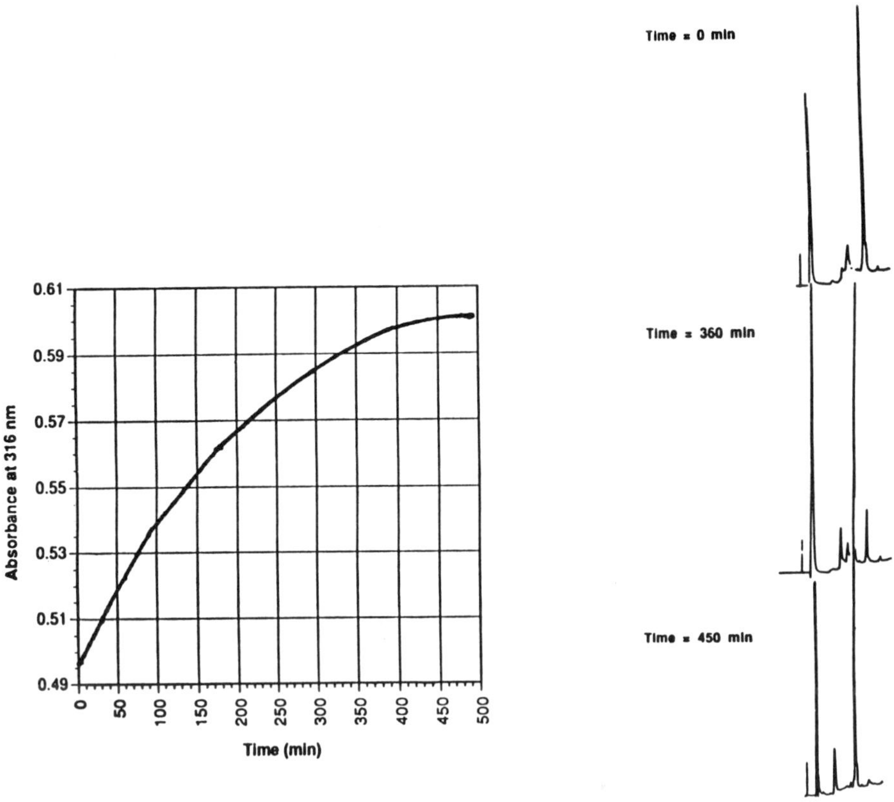

Fig. 6. The trace on the left shows a typical curve for the enzyme assay described in the text, i.e., increase in A_{316} over time. The three HPLC profiles on the right confirm that the substrate, shown as a single peak at $t = 0$ is completely hydrolyzed, i.e., two earlier eluting peptides, over the time course shown in the tracing.

—Tyr— and —Pro—. In the uncleaved peptide, there was no measureable fluorescence. The DNP group quenched the fluorescence from NMA. When enzymatic hydrolysis occurred, the NMA fluorophore, no longer in proximity to the DNP group such that its fluorescence was no longer absorbed by the DNP moiety or "quenched," could be measured directly and continuously. The technique and buffers used for this fluorogenic assay are essentially the same as described for the UV/Vis continuous assay.

The automated stepwise assembly of 99 Fmoc-protected amino acids into a functional enzyme using HBTU-mediated, i.e. *FastMoc* coupling chemistry has been accomplished. An efficient peptide bond-forming reaction matched with the relatively mild deprotection and cleavage steps unique to Fmoc/*t*-butyl-based synthetic strategy can be used to prepare large polypeptides and proteins. This combination represents another approach toward fulfilling the vision of Merrifield as to the ultimate application of the solid-phase peptide synthesis principle—the reliable preparation of proteins and novel protein analogs, e.g., containing unnatural amino acids, to explore structure–function relationships *(20)*.

4. Discussion

The HIV-1 proteinase has been prepared in a stepwise manner using an ABI 431A automated peptide synthesizer and *FastMoc* chemistry. Starting with 0.25 mmol of commercially available Fmoc-Phe-HMP-resin (Bachem Biosciences, Philadelphia, PA), the total synthesis was completed over the course of approx 10 d. The number of couplings/residue and side-chain-protecting groups is shown in Fig. 1. Multiple couplings were done in regions of the protein that were known to be problematic from previous work *(5)* and "discovered" during optimization in conjunction with UV monitoring. Monitoring the synthesis by detecting the presence of dibenzofulvene-piperidine adduct generated during removal of the Fmoc group proved to work well and provided insight as to where difficult coupling events might occur given that with Fmoc synthetic chemistry, difficult coupling almost invariably follows a difficult deprotection *(21)*. During the course of the synthesis, "difficult" sequences/stretches were encountered twice that impacted the coupling yield. These regions spanned cycles 11–13 (residues 85–87 from the protein sequence) and cycles 38 and 39 (residues 61 and 62). Figure 7 shows a detailed portion of the monitoring trace in the vicinity of cycles 11–13. As can be seen, difficulty in removing the Fmoc group was encountered. Ninhydrin assays showed significantly diminished acylation (data not shown) in the same region. This effect was apparently the result of a physical change in the nature of the solid support. Microscopy of the polystyrene beads after cycle 11 revealed a general spherical morphology with apparent diameter slightly less than the starting resin particle size. Remarkably, with the addition of two additional amino acids, the beads showed regions of invagination and had a more compact structure (Fig. 8). This dramatic change in morphology might be the result of β-sheet-induced aggregation of the growing peptide chains causing a collapsed gel structure and/or "desolvation" of the protected peptide chain resulting in reduced coupling efficiencies.

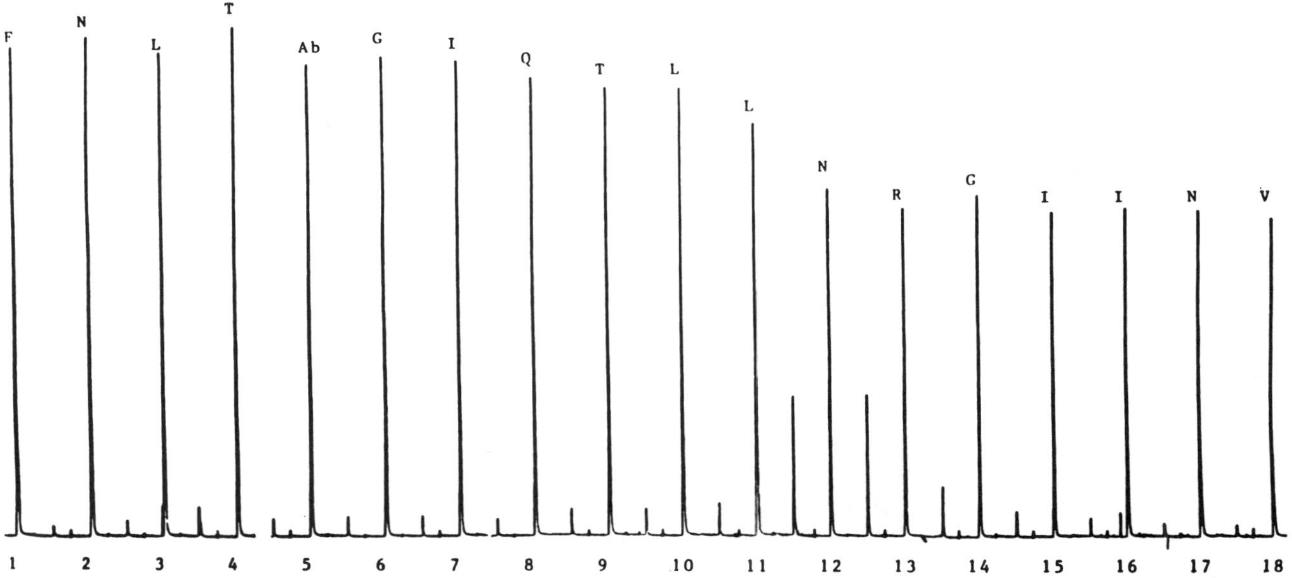

Fig. 7. UV monitoring of Fmoc removal reaction; detail of the first 18 cycles (residues 82–99 HIV-1 Protease). Beginning with cycle 11, appreciable absorbance in the second piperdine treatment is seen, i.e., large secondary peak. This trend continues into the next two cycles, then the synthesis plateaus. This effect is real in that "% coupled" values according to quantitative ninhydrin drop from 99+ to around 60% at cycles 1 (Arg 87).

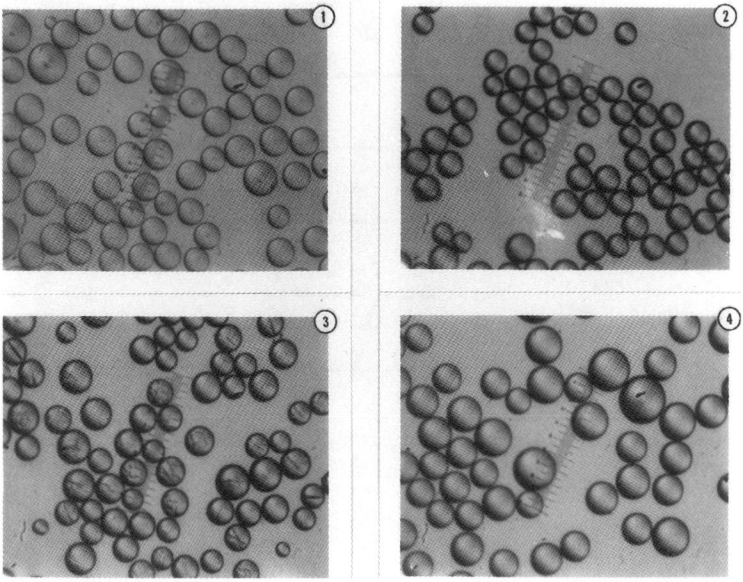

Fig. 8. Photomicrographs of resin samples: (1) pre-loaded Fmos-Phe-resin, (2) after cycle 11, (3) after cycle 13, and (4) after cycle 98.

Following each cycle, the resin was capped by acetylation as previous described (*see* Section 3.3.). The cycle for "capping" at the 0.25-mmol scale is given in Table 2.

During the use of these cycles and the capping reagent mixture, a brownish color was imparted to the reaction vessel, reaction vessel filters, and the lines from bottle position 4. This coloration was/is most likely the result of oxidation of the triazole in the HOBT reagent and has not had a detrimental effect on the synthesis itself.

Acknowledgments

I would like to thank Professor Charles Craik of the Department of Pharmaceutical Chemistry, U.C. San Francisco, and his two postdoctoral fellows Diane DeCamp and Rafael Salto for doing the initial characterization of enzyme activity and isoelectric focusing studies. I would like to thank L. Zieski, K. Hsi, L. Cheng, and D. Hawke of Applied Biosystems. Inc. for their contributions to the analytical part of this work.

References

1. Kohl, N. E., Emini, E. A., Schleif, W. A., Davis, L. J., Heimbach, J. C., Dixon, R. A. F., Scolnick, E. M., and Sigal, I. S. (1988) Active human immunodeficiency virus protease is required for viral infectivity. *Proc. Natl. Acad. Sci. USA* **85,** 4686–4690.
2. Wlodawer, A., Miller, M., Jaskolski, M., Sathyanarayana, B. K., Baldwin, E., Weber, I. T., Selk, L. M., Clawson, L., Schneider, J., and Kent, S. B. H. (1989) Conserved folding in retroviral proteases: crystal strucuture of a synthetic HIV-1 proteinase. *Science* **245,** 616–621.
3. Sanchez-Pescador, R., Power, M. D., Barr, P. J., Steimer, K. S., Stempien, M. M., Brown-Shimer, S. L., Gee, W. W., Renard, A., Levy, J. A., Dina, D., and Luciw, P. A. (1985) Nucleotide sequence and expression of an AIDS-associated retrovirus (ARV-2). *Science* **227,** 484–492.
4. Schneider, J. and Kent, S. B. H. (1988) Enzymatic activity of a synthetic 99 residue protein corresponding to the putative HIV-1 protease. *Cell* **54,** 363–367.
5. Nutt, R. F., Brady, S. F., Darke, P. L., Ciccarone, T. M., Colton, C. D., Nutt, E. M., Rodkey, J. A., Bennerr, C. D., Waxman, L. H., Sigal, I. S., Anderson, P. S., and Veber, D. F. (1988) Chemical synthesis and enzymatic activity of a 99-residue peptide with a sequence proposed for the human deficiency virus protease. *Proc. Natl. Acad. Sci. USA* **85,** 7129–7133.
6. Copeland, T. D. and Oroszlan, S. (1988) Genetic locus, primary structure, and chemical synthesis of Human Immunodeficiency Virus protease. *Gene Anal. Techn.* **5,** 109–115.
7. Wang, S. S. (1973) *p*-Alkoxybenzyl alcohol resin and *p*-Alkoxybenzyloxy-carbonylhydrazide resin for solid phase synthesis of protected peptide fragments. *J. Am. Chem. Soc.* **95,** 1328–1333.
8. Fields, G. B. and Fields, C. G. (1991) Solvation effects in solid phase peptide synthesis. *J. Am. Chem. Soc.* **113,** 4202–4207.
9. Otteson, K. M., MacDonald, R. L., Noble, R. L., and Hoeprich, P. D. (1991) U.V. deprotection monitoring with *FastMoc*™-SPPS on the Model 431A peptide synthesizer. Applied Biosystems, Inc., Research News, Peptide Synthesis, December 1991.
10. *FastMoc*™ chemistry: HBTU activation in peptide synthesis on the Model 431A, User Bulletin No. 33, Applied Biosystems, Inc., November 1990.
11. Dourtglou, V., Ziegler, J. C., and Gross, B. (1978) L'Hexafluoro-phosphate de *O*-Benzotriazolyl-*N,N*-tetramethyluronium: Un Reactif de Couplage Peptidique Nouveau et Efficace. *Tetrahedron Lett.* **15,** 1269–1272.
12. Fields, C. G., Lloyd, D. H., Macdonald, R. L., Otteson, K. M., and Noble, R. L. (1991) HBTU activation for automated Fmoc solid-phase peptide synthesis. *Peptide Res.* **4,** 95–101.
13. Gausepohl, H., Pieles, U., and Frank, R. W. (1992) Schiff base analog formation during *in situ* activation by HBTU and TBTU, in *Peptides: Chemistry and Biology. Proceedings of the 12th American Peptide Symposium* (Smith, J. A. and Rivier, J. E., eds.), Escom, Leiden, pp. 523,524.

14. Knorr, R., Trzeciak, A., Bannwarth, W., and Gillessen, D. (1989) New coupling reagents in peptide chemistry. *Peptides 1988: Proceedings of the 20th European Peptide Symposium* (Jung, G. and Bayer, E., eds.), W. de Gruyter, Berlin, pp. 37–39.
15. King, D. S., Fields, C. G., and Fields, G. B. (1990) A cleavage method which minimizes side reactions following Fmoc solid-phase peptide synthesis. *Int. J. Peptide Protein Res.* **36,** 255–266.
16. Application in SDS-PAGE, Electroblotting and protein sequencing, User Bulletin No. 42 Applied Biosystems, April 1991.
17. Tomasselli, A. G., Olsen, M. K., Hui, J. O., Staples, D. J., Sawyer, T. K., Heinrikson, R. L., and Tomich, C. C. (1990) Substrate analogue inhibition and active site titration of purified recombinant HIV-1 protease. *Biochemistry* **29,** 264–269.
18. Rittenhouse, J., Turon, M. C., Helfrich, R. J., Albrecht, K. S., Weigl, D., Simmer, R. L., Mordini, F., Erickson, J., and Kohlbrenner, W. E. (1990) Affinity purification of HIV-1 and HIV-2 proteases from recombinant *E. coli* strains using pepstatin-agarose. *Biochem. Biophys. Res. Commun.* **171,** 60–66.
19. Nashed, N. T., Louis, J. M., Sayer, J. M., Wondrak, E. M., Mora, P. T., Oroszlan, S., and Jerina, D. M. (1989) Continuous spectro-photometeric assay for retroviral proteases of HIV-1 and AMV. *Bichem. Biophys. Res. Comm.* **163,** 1079–1085.
20. Merrifield, B. (1986) Solid phase synthesis. *Science* **232,** 341–347.
21. Atherton, E. and Sheppard, R. D. (1989) *Solid Phase Synthesis—A Practical Approach.* IRL Oxford University Press, Oxford.

Chapter 17

Multiple and Combinatorial Peptide Synthesis

Chemical Development and Biological Applications

Philip C. Andrews, Daniele M. Leonard, Wayne L. Cody, and Tomi K. Sawyer

1. Introduction

The rapid synthesis of peptides to support both exploratory peptide lead discovery and analog structure–activity studies has been the subject of intense research and technology development over the past few years *(1–4)*. Two general approaches have been advanced that may be classified as follows: (1) multiple peptide synthesis (MPS) and (2) combinatorial peptide synthesis (CPS). The peptide chemistry aspect of such multiple or combinatorial approaches typically integrates experimental methodologies that have been well established to date for the preparation of single peptides by solid-phase techniques. However, there exist novel and, in some cases, proprietary chemistry methods and/or materials that are essential to each of these approaches *(vide infra)*. A common objective of the multiple or combinatorial approaches is to expedite the process and quantitative scope of synthetic peptide production in a manner that interfaces well with biological testing.

In simple terms, MPS may be considered the generation of a finite number of peptides using manual methods and/or automated instrumentation (Table 1) that yields peptides as individual products. A variety of strategies may be envisaged to accomplish this objective. For example, a series of peptide analogs may be synthesized using compartmentalized peptide-

From: *Methods in Molecular Biology, Vol. 36: Peptide Analysis Protocols*
Edited by: B. M. Dunn and M. W. Pennington Copyright ©1994 Humana Press Inc., Totowa, NJ

resin intermediates wherein the primary structural identity of the series can be achieved by simultaneous coupling with the designated amino acids in a stepwise manner. Such an MPS strategy would be particularly useful to yield a series of individual peptides having a single-site modification. The generation of a peptide mixture wherein only a single amino acid residue is varied among the series of peptide analogs may also be easily achieved by MPS chemistry. In fact, the design of synthetic peptide mixtures has become a sophisticated technology and "art." Specifically, the simultaneous coupling of a single amino acid to a mixture of synthetic peptide-resin intermediates and simultaneous coupling of multiple amino acids to a single synthetic peptide-resin intermediate provide two possible strategies to create so-called peptide libraries. By definition, such peptide libraries are the products of combinatorial peptide synthesis (CPS), which typically embodies the generation of a large number (e.g., 10^3–10^7) of peptides wherein two or more amino acid residues are varied within the primary structure of each of the individual component peptides. Both manual methods and automated instrumentation may be used to prepare peptide libraries. Overall, the transition from single to multiple and multiple to combinatorial peptide synthesis has rapidly developed, and MPS/CPS chemistry itself has catalyzed the development of peptidomimetic libraries *(5)*, which may also be of significant impact to drug discovery strategies.

An overview of MPS/CPS methods as well as chemical scope and biological applications of such approaches in the field of peptide research, including exploratory lead finding and analog structure–activity development studies, is described below. Representative examples of the peptide chemistry strategy used for some synthetic methods are given, and relevant literature (including patents) is included.

2. Multiple and Combinatorial Peptide Chemistry Methods

2.1. Synthetic Methods

2.1.1. Proprietary Nature of Some Developed MPS / CPS Methods

It is quite evident that the chemical development and use of MPS/CPS approaches have emerged as a powerful technology for both basic research and drug discovery. To the extent that some of the MPS/CPS chemistry strategies described in this chapter may be proprietary it was deemed appropriate that known patents (or patent applications) be referenced.

2.1.2. MPS/CPS by "Multipin" or "Tea-Bag" Methods

The Geysen "multipin" method employs polyacrylate-grafted polyethylene rods arranged in a standard 96-well plate that can be used for either the synthesis of 96 individual peptides or mixtures. Simultaneous peptide synthesis using conventional solid-phase synthesis may be performed using readily available materials and instrumentation. The peptide-pin intermediates may be directly used for biological screening *(6)*. Cleavage of the peptides from the pin may be optionally used with appropriate linkers to generate free peptide products *(7)*. Several patents *(8)* related to the chemistry and/or biological testing of such peptides prepared by the Geysen pin method exist. Furthermore, the pin technology has been commercialized by Zeneca Cambridge Research Biochemicals, Inc. (Wilmington, DE).

The Houghten "tea-bag" method *(9)* employs the compartmentalization of peptide-resin intermediates in porous polypropylene bags that can be subjected to simultaneous peptide synthesis using conventional solid-phase synthesis. The peptide-resin intermediates may, therefore, be exposed to common steps, such as deprotection and washing. Similarly, if an identical amino acid is being coupled to different peptide resins, the tea-bag method permits simultaneous coupling by simply mixing the appropriate peptide-resin bags with the desired reagents. In contrast, dissimilar amino acid coupling is accomplished by separation of the peptide-resin bags. Cleavage of up to 24 peptides may be achieved by using a commercially available, multivessel HF apparatus *(10)*, or alternatively, an Fmoc synthesis/TFA cleavage strategy may be used. The Houghten tea-bag method provides the opportunity for the preparation of relatively large amounts of peptide-resin intermediates or free peptide products. A semiautomated method has been described *(11)* that provides the simultaneous synthesis of up to 120 different peptides. Furthermore, the tea-bag method has been extended to CPS chemistry in terms of rapid synthesis of peptide libraries (*12*; also refer to Section 2.1.3.). Patents related to the chemistry and/or biological testing of such peptides prepared by the Houghten tea-bag method exist *(13)*.

2.1.3. MPS/CPS by "Split Synthesis" Methods

The development of peptide libraries containing a large number of component peptides by modification of conventional, solid-phase synthesis of single peptides has been advanced by several academic *(14–16)* as well as pharmaceutical/industrial *(12,17–19)* research groups. It is noted that the term "split synthesis" also refers to other descriptors, such as "portioning-mixing" *(14)* and "divide, couple, and recombine" *(12)*. The underlying concept of the split synthesis method is that a single peptide (i.e., single sequence of identical peptides) is prepared on a single solid-support matrix component (e.g., resin bead). In this approach, a solid support is divided into equal portions, and individual portions are then subjected to coupling with a desired amino acid that is variable at a particular site in the primary structure of the peptide library being prepared. This step affords the possibility of uniform and quantitative coupling prior to the recombining of the peptide-resin portions. This yields a statistical distribution of selected amino acids that are varied at any particular site in the target sequence. Similar to the above MPS/CPS methods, the split-synthesis approach simplifies the overall chemistry effort by reducing the number of deprotection and washing steps, since these may be accomplished in a single reaction vessel. With respect to the solid-support matrix, which may be used to perform such split synthesis, it is noted that methods exist that include the application of resin beads that are compatible with aqueous solvent systems and, therefore, permit the facile biological screening of the peptide-bound beads *(17)*. Several patents *(13b,20)* related to the chemistry and/or biological testing of such peptides prepared by the aforementioned split-synthesis methods exist.

2.1.3.1. Example: Split-Synthesis Preparation of a Tetrapeptide Library, AA_3-AA_2-AA_1-Ala (Where AA_1–AA_3 Are Randomized by Glu, Phe, and Lys)

An example of a split synthesis is taken from a recent report by Furka and coworkers *(14)* in which a tetrapeptide library was prepared, having a common C-terminal Ala residue, and varied at the first three amino acid sites by Glu, Phe, and Lys. The design of this tetrapeptide library is, therefore, a $3 \times 3 \times 3 \times 1$ or 27-peptide component library that was divided into three individual mixtures (nine peptides/each) at the very last step. Scheme 1 illustrates the process of synthesis, splitting, and combining for this particular tetrapeptide library in terms of the stepwise development of peptide-resin intermediates and the predetermined amino acid sequences in each mixture (after each of the three splitting steps). This split synthesis was carried out manually using Boc-Glu(Bn), Boc-Phe, and Boc-Lys(Z), and a Boc-Ala-resin that was prepared from BioBeads S-X1 chloro-

Scheme 1
Method of Preparation of a Tetrapeptide Library by Split Synthesis[a]

Materials and/or steps	Specific case for AA_3-AA_2-AA_1-Ala Library[b]
1. Boc-AA[c]	Boc-Ala, Boc-Glu(Bn), Boc-Phe, Boc-Lys(Z), and chloromethylpolystyrene-divinylbenzene resin
2. Boc-AA-R	Boc-Ala-R
3. Split Boc-AA-R	The Boc-Ala-R is split into three portions
4. Boc-AA_1-AA-R	Mixture 1: Boc-Glu(Bn)-Ala-R
	Mixture 2: Boc-Phe-Ala-R
	Mixture 3: Boc-Lys(Z)-Ala-R
5. Combine Boc-AA_1-AA-R	The three above dipeptide resins are mixed
6. Split Boc-AA_1-AA-R	The combined dipeptide-resin mixture is split into three portions
7. Boc-AA_2-AA_1-AA-R	Mixture 1: Boc-Glu(Bn)-AA_1-Ala-R
	AA_1 = Glu(Bn), Phe, and Lys(Z)
	Mixture 2: Boc-Phe-AA_1-Ala-R
	AA_1 = Glu(Bn), Phe, and Lys(Z)
	Mixture 3: Boc-Lys(Z)-AA_1-Ala-R
	AA_1 = Glu(Bn), Phe, and Lys(Z)
8. Combine Boc-AA_1-AA-R	The three above tripeptide resins are mixed
9. Split Boc-AA_1-AA-R	The combined tripeptide-resin mixture is split into three portions
10. Boc-AA_3-AA_2-AA_1-AA-R	Mixture 1: Boc-Glu(Bn)-AA_2-AA_1-Ala-R
	AA_1 = Glu(Bn), Phe, and Lys(Z)
	AA_2 = Glu(Bn), Phe, and Lys(Z)
	Mixture 2: Boc-Phe-AA_2-AA_1-Ala-R
	AA_1 = Glu(Bn), Phe, and Lys(Z)
	AA_2 = Glu(Bn), Phe, and Lys(Z)
	Mixture 3: Boc-Lys(Z)-AA_2-AA_1-Ala-R
	AA_1 = Glu(Bn), Phe, and Lys(Z)
	AA_2 = Glu(Bn), Phe, and Lys(Z)

[a]This tetrapeptide library consists of three mixtures (nine individual peptide components/mixture).

[b]Adapted from Furka and coworkers *(14)*.

[c]AA = amino acid; R = resin.

methyl resin (Bio-Rad Laboratories, Richmond, CA). The scale of the synthesis was 0.27 mmol (360 mg Boc-Ala resin substituted at 0.74 Ala/g). Boc-amino acid couplings were performed by a standard method *(21a)*, with modifications *(21b,c)*, using diisopropylcarbodiimide and 1-hydroxybenzotriazole, and a twofold excess of preactivated Boc-amino acid derivatives dissolved in CH_2Cl_2-DMF (3:1 v/v) was used to perform each of the coupling reactions (refer to steps 4, 7, and 9 of Scheme 1). The coupling reaction was monitored by the Kaiser ninhydrin test *(22)*. Beginning with the Boc-Ala resin, three 120-mg portions were taken and added to three individual reaction vessels (hereinafter referred to as Mixtures 1, 2, and 3, respectively), and the TFA cleavage of the Boc group was then performed. Three Boc-protected dipeptide-resin intermediates were prepared by coupling Boc-Glu(Bn), Boc-Phe, or Boc-Lys(Z) to the Ala-resin intermediates of Mixtures 1, 2, and 3, respectively. Following completion of the synthesis of each of the three Boc-dipeptide-resin intermediates, Mixtures 1, 2, and 3 (each suspended in 10 mL of DMF) were combined into a common vessel, and the Boc-dipeptide-resin mixture was shaken for 10 min to assure good mixing. The Boc-dipeptide-resin/DMF suspension was then quickly divided, in order to avoid sedimentation, into three 120-mL portions (refer to step 6 of Scheme 1) that would have equimolar amounts of each of the component Boc-AA_1-Ala-resin intermediates wherein AA_1 was Glu(Bn), Phe, or Lys(Z). Following TFA cleavage of the Boc group from each dipeptide resin of Mixtures 1–3, an identical coupling reaction protocol was performed as described above to yield the Boc-tripeptide-resin intermediates (refer to step 7 of Scheme 1). The Boc-tripeptide-resin intermediates of Mixtures 1, 2, and 3 (each suspended in 10 mL of DMF) were then combined into a common vessel, and the Boc-tripeptide-resin mixture was shaken for 10 min to assure good mixing. Similar to the previous splitting step, the Boc-tripeptide-resin/DMF suspension was then quickly divided into three 120-mL portions (refer to step 9 of Scheme 1). Following TFA cleavage of the Boc group from each tripeptide resin of Mixtures 1–3, a final coupling reaction protocol was performed as described previously to yield the Boc-tetrapeptide-resin intermediates (refer to step 10 of Scheme 1). Each of these mixtures contained nine equimolar tetrapeptide components, which were varied at AA_1 and AA_2 by Glu(Bn), Phe, and Lys(Z), respectively. Following TFA cleavage of the Boc group from each tetrapeptide resin of Mixtures 1–3, cleavage of the side-chain protecting groups and the peptides from the resin was carried out using trifluoromethanesulfonic acid. The three separate solutions of crude tetrapeptides were each filtered from the resin, and each solution was washed with TFA into dry ether (standing overnight at –20°C) followed by filtration, washing (twice with ether), and the collected precipitate dried over KOH and then over P_2O_5.

2.1.4. MPS/CPS by "Multiple Coupling" Method

Multiple amino acid coupling to a given peptide resin (or subsequent peptide-resin mixture) within a single reaction vessel may also provide a strategy to produce peptide libraries. However, this has not been a major MPS/CPS chemistry method as compared to the aforementioned technologies. A noteworthy advantage of the multiple coupling method is the fewer number of overall required steps to produce a peptide-resin mixture relative to multipin, tea-bag, or split-synthesis methods. The major disadvantage is the potential for disproportionate coupling of different amino acids that would be competing with each other for reactive sites and, subsequently, the expected lack of ability to produce peptide libraries uniformly consisting of stoichiometric equivalents (mole-wise) of component peptides. Recently, an approach using substoichiometric amounts of each activated amino acid to a peptide-resin intermediate with extended coupling time followed by a repeat coupling cycle with identical reagents has been described *(23,24)* for the preparation of peptide libraries. This multiple substoichiometric addition-multiple coupling (or MSA-multiple coupling) strategy is expected to be similarly useful to peptide lead discovery or analog structure–activity development as those methods described above. In contrast to the "single peptide-single resin bead" property of the aforementioned methods, however, the MSA-multiple coupling method creates heterogeneity of peptides on a single resin bead or other solid-support matrix. Nevertheless, cleavage of the peptides from the solid-support gives rise to the same heterogeneity of peptides as would be observed for other MPS/CPS chemistry methods (e.g., tea-bag) to the extent of amino acid variation within the target series of peptides. In this regard, it is noted that multiple amino acid coupling to a single resin to prepare a series of single-site-modified peptides has been reported *(25)*, and this study utilized both HPLC and fast atom bombardment-mass spectrometry to isolate and characterize each individual analog.

2.1.4.1. Example: MSA-Multiple Coupling Preparation of a Heptapeptide Library, AA_3-AA_2-AA_1-Tyr-Ala-Ala-Lys (where AA_1-AA_3 Are Randomized by Ala, Asp, Glu, Phe, Gly, His, Ile, Lys, Leu, Met, Asn, Pro, Gln, Arg, Ser, Thr, Val, Trp, Tyr)

An example of an MSA-multiple coupling method is taken from a recent report by Andrews and coworkers *(24b)* in which a heptapeptide library was prepared having a common C-terminal Tyr-Ala-Ala-Lys sequence and varied at the first three amino acid sites by 19 of the common amino acids (Cys excluded to avoid disulfide formation). The design of this heptapeptide library is, therefore, a 19 × 19 × 19 × 1 × 1 × 1 × 1 or 6859-peptide component library that was prepared as a single mixture. The peptide library was made using an ABI Model 431 peptide synthesizer. One millimole of RINK resin was loaded in a 65-mL reaction vessel. Standard Fmoc cycles were used except that an extra 8-mL vol of NMP was delivered to the reaction vessel for the deprotection and coupling steps. Positions containing single amino acids (i.e., Tyr-Ala-Ala-Lys) were triple coupled at 1 mmol, and the coupling times were doubled. Sites at which multiple amino acids were to be incorporated were double coupled at 0.8 mmol, and the coupling times were increased threefold. The solvent used for coupling and deprotection was *N*-methyl-pyrrolidinone (NMP) containing 5*M* ethylene carbonate. Side-chain-protecting groups included the following: Ser(^{t}Bu), Thr(^{t}Bu), Tyr(^{t}Bu), Asp(^{t}Bu), Glu(^{t}Bu), Arg(Pmc), Lys(Boc), His(Trt), and Trp(Boc). Peptide-resin cleavage and side-chain deprotections were performed using 85% TFA 5% ethanedithiol-10% thioanisole at room temperature under nitrogen for 1 h. Samples were then dried under a nitrogen stream. No attempt was made to further remove byproducts of the synthesis and cleavage in order to prevent loss of hydrophobic peptides by adsorption or incomplete precipitation. The mass spectrum of this peptide library and the theoretical distribution of masses are shown below (Section 2.2.).

2.1.5. MPS/CPS by Other Synthetic Methods

A highly sophisticated MPS/CPS chemistry method has been described *(26)* that integrates solid-phase peptide synthesis with photolithography and miniaturization (*26*; a process designated as VLSIPS or very large-scale immobilized polymer synthesis). The VLSIPS approach utilizes photochemically labile protecting groups within the scope of light-directed, spatially addressable, and parallel chemical synthesis to prepare a large number (>10^3) of peptides at defined sites. Typically, micromolar levels of final peptide product are produced. A patent related to the chemistry and application of such immobilized peptides prepared by the Affymax VLSIPS method has been issued *(27)*. A number of other variations of the solid-phase matrix have been recently reported in conjunction with MPS/CPS chemistry strategies. For example, cellulose disks stacked in a reaction column have been used to produce individual

Table 1
Some Commercialized MPS/CPS Instruments

Instrument	Company	Comments
MPS-396™	Advanced ChemTech	96 Single peptides; Fmoc chemistry
MPS-357™	Advanced ChemTech	Combinatorial peptide libraries; Fmoc chemistry
AMS-422™	Gilson/Abimed	48 Single peptides; Fmoc chemistry
Symphony™	Rainin/Protein Technologies	12 Single peptides, Fmoc chemistry
PSSM-8™	Shimadzu	96 Single peptides, Fmoc chemistry
SMPS-350™	Zinser	48 Single peptides, Fmoc chemistry

peptides and peptide libraries *(28)*. The chemical nature of these disks permits multiple couplings to be conducted under continuous-flow conditions. Patents related to the chemistry of such cellulose disk-based MPS/CPS chemistry exist *(29)*. In a similar context, cellulose paper sheets *(30)* have been used to prepare both immobilized and free peptides by a "spot synthesis" method. Cotton fabric *(31)* has been shown to be a useful solid-phase matrix for MPS/CPS chemistry. It has been noted *(31b)* that some difficult amino acid sequences may be prepared using cotton fabric in high purity, although with lower chemical yields compared to conventional polystyrene-divinylbenzene copolymer. The application of polystyrene-grafted polyethylene film *(32)*, which can be used in separable sheets, to MPS/CPS chemistry methods has been illustrated, and a patent related to the chemistry and applications of such film-based MPS/CPS chemistry exists *(33)*.

2.1.6. MPS / CPS by Automated or Semiautomated Instrumentation and Related Technologies

There exists an increasing number of MPS/CPS instruments being marketed (Table 1), and the continued development and commercialization of such instruments exemplifies the high interest in MPS/CPS chemistry. Among the first commercialized technologies for performing multiple peptide synthesis was the RaMPS™ apparatus/kit marketed by E.I. duPont de Nemours and Company (Wilmington, DE), which is semiautomated in terms of the vacuum-based removal of solvents and reagents *(34)*. The RaMPS™ systems is designed to synthesize peptides concurrently. During recent years, several automated MPS instruments have been developed using robotic sample processing technologies and commercialized. Specifically, the Zinsser (Frankfurt, Germany) SMPS-350™ employs a modified Tecan™ robotic sample processor with two needles (one for delivery and the other for vacuum-based removal) to support delivery or removal operations (reagents, amino acids, and solvents) in an x-y-z directionality format over a set of 48 reaction vessels *(35)*. The Abimed AMS-422™ employs a Gilson (St. Louis, MO) M222 robotic sample processor with multiple needles to enable parallel distribution of reagents and solvents into 48 reaction vessels *(36)*. The Shimadzu (Columbia, MD) PSSM-8™ is an eight-channel automated MPS that utilizes robotically based, multineedle apparatus for pipeting solvents and reagents, and the instrument is computer interfaced *(37)*. This MPS instrument is designed to synthesize eight peptides in parallel fashion. The Advanced ChemTech (Louisville, KY) Model 396™ Multiple Peptide Synthesizer is an automated MPS instrument capable of preparing 96 peptides simultaneously in parallel fashion by use of a roboticized sample processor *(38a)*. A very recently reported *(38b)* second-generation instrument, the Advanced ChemTech Model-357 Librarian™, is the first marketed, automated MPS/CPS instrument. The Rainin (Woburn, MA) Symphony™ MPS instrument *(39)* is an automated MPS instrument capable of preparing 12 peptides simultaneously in parallel fashion. Some commercialized MPS technologies involving manual methods and special apparatus/kits exist, including the Zeneca/Cambridge Research Biochemicals Spots™ and Pins Technology™.

Other automated or semiautomated MPS/CPS instruments as well as manually operated MPS/CPS apparatus have been developed by various research groups. Descriptions of such instruments or apparatus have appeared in the literature *(40–44)*. Noteworthy is an automated MPS instrument (referred to as the Compas™) utilizing cotton fabric as an integrated reaction vessel and carrier that has been reported *(42)*. Also, a method to perform MPS using color-monitored, solid-phase peptide synthesis under low-pressure continuous-flow conditions has been developed *(43)* utilizing manual methods and simple instrumentation to synthesize approx 10 peptides in parallel fashion. Finally, the development of a fully automated multichannel peptide synthesizer (specifically, a prototype instrument that can produce four peptides simultaneously) that includes the integration of TFA cleavage capability has been very recently described *(44)*.

2.2 Characterization Methods

As expected, the characterization of a series of a peptide mixture generated by MPS/CPS chemistry methods becomes quite difficult, if not nearly impossible, as the complexity of the mixture increases. For MPS methods that produce peptides in separable fashion, the problem is simply a factor of performing more purifications and analyses to ensure the quality and integrity of the final products. Nevertheless, with the increased multiplicity of synthetic peptides being prepared (i.e., individual peptides or peptide mixtures), it has been quite evident that purification and/or characterization of the final products has not been given stringent attention. This is not surprising since the concept of accelerated peptide synthesis to address specific biological studies would be severely impaired by additional chemistry steps, particularly purification, that are typically time-consuming. Nevertheless, optimization of the synthesis is important to minimize the extent of undesired side reactions (e.g., incomplete coupling or deprotection of the series or mixture of peptides being prepared). Monitoring the coupling and deprotection efficiency at each step of synthesis can identify some, but not all, problems. Physical methods of analysis of the final peptide product may be useful, and such methods include amino acid analysis (AAA), Edman degradation, and mass spectrometry (MS). Amino acid analysis can be applied to the soluble and resin-bound peptide libraries, and the ratios of amino acids in the library can be determined with reasonable accuracy depending on the absolute size and amino acid complexity of the library. Edman degradation may similarly be applied to both soluble or resin-bound peptide libraries, but the yields of some of the amino acids may be too variable to provide useful information. Both AAA and Edman degradation are handicapped by not generally being methods used when unusual or chemically labile amino acids (e.g., phosphotyrosine or pTyr) are components in the synthetic peptide library. Furthermore, N-terminal acyl groups and side-chain-protecting groups also compromise the effectiveness of these analytical methods. Even D-amino acids can create problems unless chiral AAA methods are available, and stereochemical purity would be questionable unless all the component amino acids were uniform (i.e., no L- and D-isomers of the same amino acids). Particularly noteworthy to the structural characterization of peptide libraries is the use of mass spectrometric methods, and recent studies *(23,24)* have illustrated the application of electrospray MS (ES-MS) to the analysis of soluble peptide libraries. The ES-MS apparently is less susceptible to ion suppression as compared to fast

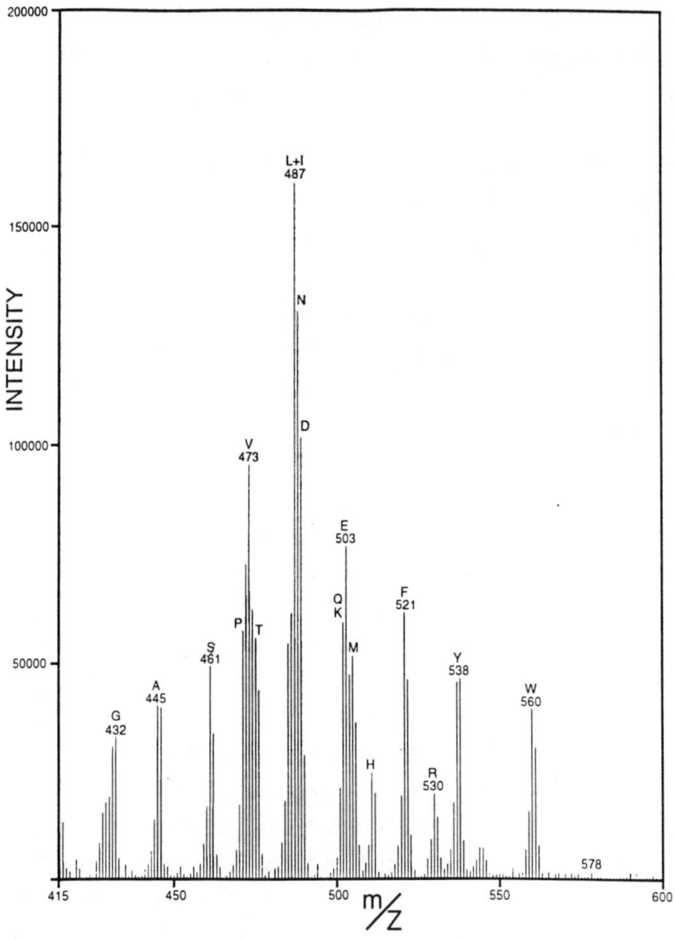

Fig. 1. ES-MS analysis of a multiple peptide mixture Xaa-Gly-Gly-Gly-Gly-Gly-Lys that is modified at the N-terminus (Xaa) by 19 amino acids, excluding Cys, showing that each of the major component peptides are present.

atom bombardment MS or matrix-assisted laser desorption MS. Other multiple analysis methods have been described (for recent review, *see* ref. *3*).

2.2.1. Example: ES-MS Analyses of Synthetic Peptide Libraries

An example of ES-MS analysis for a peptide library Xaa-Gly-Gly-Gly-Gly-Gly-Lys is shown in Fig. 1. All the major components of the library are apparent, indicating that ion suppression is not a serious problem for this particular mixture. Mass clustering for the individual amino

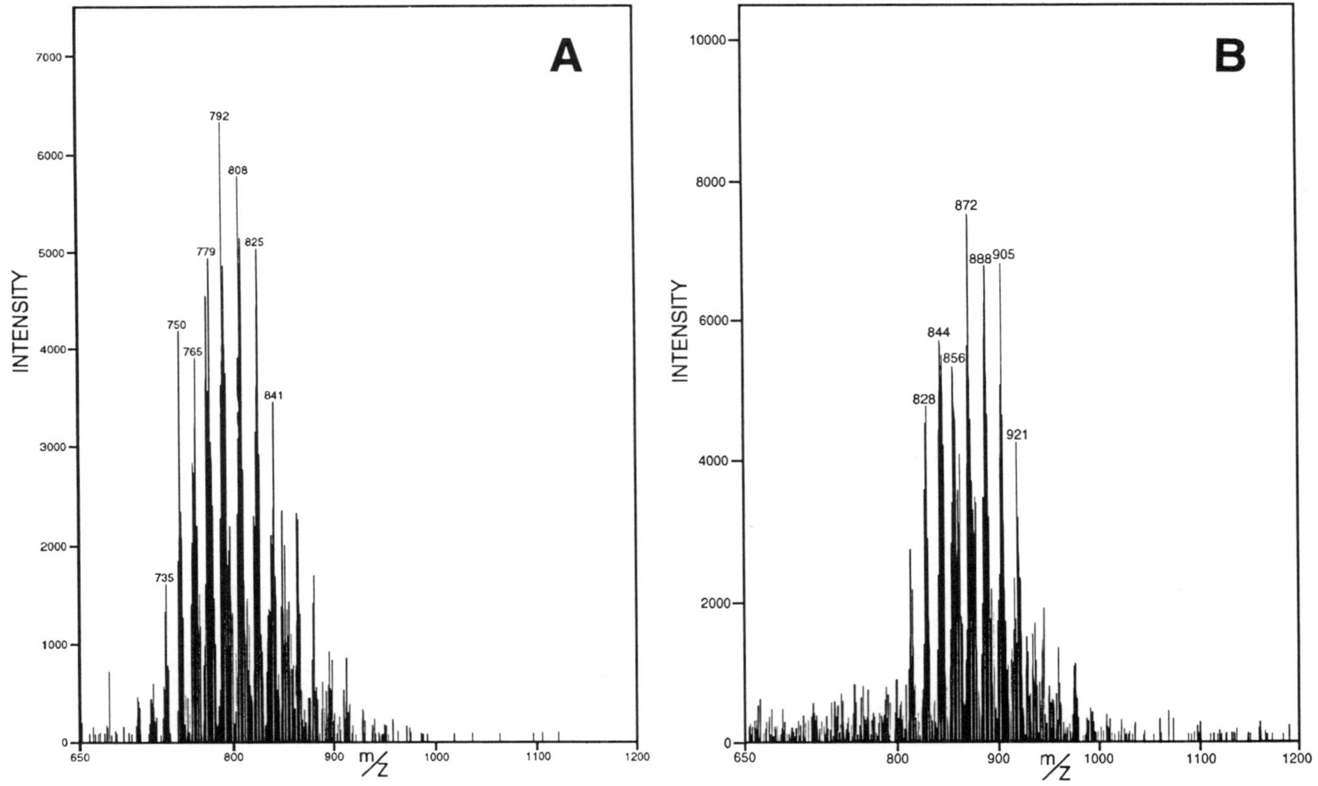

Fig. 2. ES-MS analyses of two combinatorial libraries each containing 6859 component peptides. **(A)** Mass spectrum of a heptapeptide library Xaa_3-Xaa_2-Xaa_1-Tyr-Ala-Ala-Lys-NH_2 and **(B)** Mass spectrum of a phosphotyrosine-modified heptapeptide library Xaa_3-Xaa_2-Xaa_1-pTyr-Ala-Ala-Lys-NH_2.

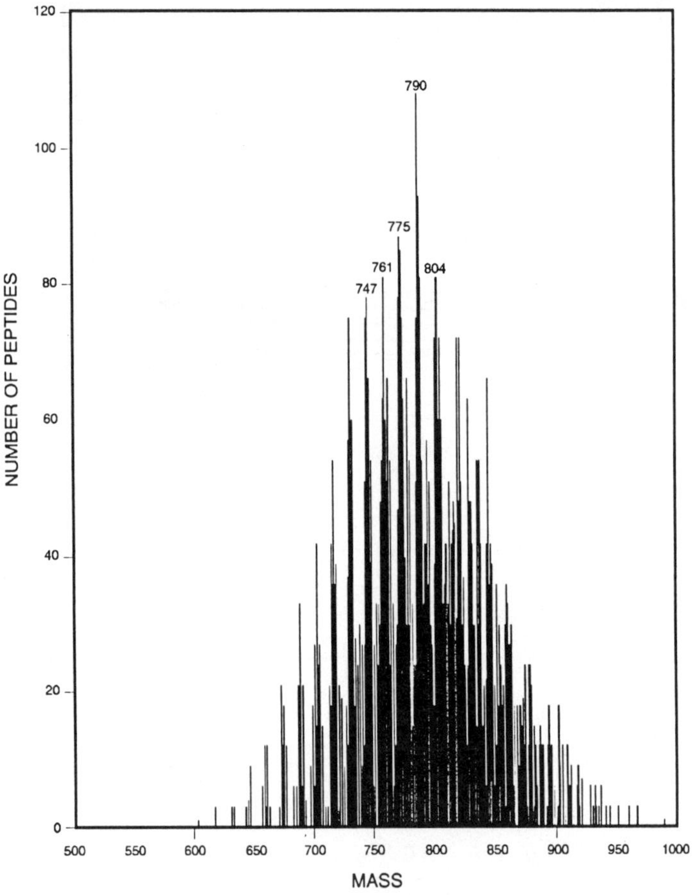

Fig. 3. Theoretical mass distribution of the singly protonated ions of the peptide library Xaa_3-Xaa_2-Xaa_1-Tyr-Ala-Ala-Lys-NH_2. Calculations were made using masslp, a module of PROCOMP 2.0.

acids results in mass clustering for the peptides in a library. In Fig. 2A, the spectrum of a complex peptide library Xaa_3-Xaa_2-Xaa_1-Tyr-Ala-Ala-Lys containing 6859 component peptides is shown. The theoretical distribution of masses for the same peptide library is shown in Fig. 3. Finally, the application of ES-MS analysis to characterize a peptide library Xaa_3-Xaa_2-Xaa_1-pTyr-Ala-Ala-Lys, which contains a chemically labile constituent, pTyr, is shown in Fig. 2B. A comparison of the ES-MS spectra of the pTyr- vs Tyr-containing peptide libraries (Fig. 2B vs 2A, respectively) unambiguously shows an 80-mass-unit shift as expected for the presence of a phosphate ester. Theoretical calculations indicate that ES-MS analysis can be an effective method for evaluation of crude peptide libraries even when they contain more than 1,000,000 component peptides *(23)*. Partial resolution of peptide library components can also be useful to evaluate the quality of the library. The use of HPLC coupled to ES-MS has been recently reported *(4)*.

3. Biological Applications and Scope of MPS/CPS Chemistry

A tremendous advantage of MPS/CPS chemistry is its integration with biological screening methods. In this regard, it is noted that two strategic applications for the aforementioned MPS/CPS approaches include: (1) peptide exploratory lead finding and (2) peptide analog structure–activity studies. Essentially all of the above MPS/CPS methods can be exploited to advance peptide (and related peptide-like) libraries for high-volume biological screening. However, it is the technology and/or process of identifying and characterizing (chemically) the lead compounds generated by these MPS/CPS approaches that is paramount to either of the two aforementioned strategic applications. Screening a variety of ligates (i.e., antibodies, receptors, and enzymes) has been performed using MPS/CPS-generated peptides *(vide infra)*, and two major processes of screening have emerged to test peptide libraries: serial *(5,17)* and parallel *(17)*. In the serial process, a peptide library consisting of different mixtures (or sublibraries) of peptides having one or more amino acid residues defined is tested, and the most active mixture(s) is advanced for an iterative process of synthesis and screening that ultimately results in the successive definition of each amino acid within the peptide ligand(s). A recently described variation of the serial process utilizes positional scanning libraries *(45)*. In the parallel approach *(17)*, which includes phage library display screening *(46)*, the simultaneous screening of a peptide library can be achieved using peptide-bound solid-support technology. Experimentally, this approach works on the principle that a single peptide sequence exists per individual resin bead and that the biologically active peptide-bead components can be physically separated from each other and structurally identified (e.g., microsequencing). It should be noted, however, that not all screens are amenable to peptide-bound solid-support matrices. Finally, the use of soluble (or free) peptide libraries provides a wider scope of chemical variation of the synthetic peptides from the standpoint that the C terminus is not involved in linkage to the matrix.

The chemical scope of these various MPS/CPS methods continues to become more sophisticated as the building blocks used extend beyond the natural amino acids and include novel amino acids, dipeptide isosteres, secondary structure mimetics, and so forth. Beyond the building blocks used, the chemical scope may be elaborated by other synthetic modifications, including varying N- and/or C-terminal functionalization, modified backbone/side-chain functionalization, and cyclization. A recent report *(47)* illustrates an intriguing study based on the transformation of oligo-Gly template to systematically transformed *N*-functionalized peptide analogs (i.e., "peptoids" as termed by the authors). Several reports *(48–50)* of encoded combinatorial (peptide) libraries illustrate the possibility that systematic identification and indirect structural determination of MPS/CPS-generated peptides or peptide-like compounds are possible, and such encoding techniques permit the chemical transition of peptide libraries toward peptidomimetic/nonpeptide/peptoid-type combinatorial products wherein techniques, such as Edman degradation, would be possible. Different MPS/CPS methods might be envisaged to advance peptide exploratory lead finding vs analog structure–activity development, and some examples of recent research in this area are described below.

3.1. Peptide Lead Discovery Using MPS/CPS Chemistry

3.1.1. Antibody-Targeted Peptide (Epitope or Mimotope) Lead Discovery

The Geysen MPS/CPS "multipin" method *(6)* was the first major approach used to advance peptide epitope mapping *(51)* and mimotope discovery *(52)* in conjunction with antibody screening. With respect to nomenclature, the term epitope may be defined as a continuous sequence or discontinuous array of amino acids of a target ligate (e.g., antigen). Discontinuous epitopes are defined as those consisting of nonadjacent amino acids that are spatially proximate as the result of the secondary/tertiary structural properties of the target ligate. The discovery of synthetic peptides that mimic a discontinuous epitope, termed "mimotopes," is therefore a more randomized strategy as compared to epitope mapping of a ligand of known primary structure. In the case of epitope mapping, the multipin approach or automated MPS instrumentation may be efficiently used to create a complete set of overlapping peptides of desired length related directly to the amino acid sequence of the target ligand. In the case of mimotope identification, a library of peptides can be prepared by randomized *(17)* or partially randomized *(12,14)* MPS/CPS methods (e.g., conserving a dipeptide substructure of known amino acid sequence) to expedite peptide lead(s) optimization.

3.1.2. Receptor-Targeted Peptide (Agonist or Antagonist) Lead Discovery

A recent discovery *(12d)* of a novel series of *N*-acetylated enkephalin antagonists illustrates the application of the split-synthesis and iterative process of MPS/CPS chemistry and biological screening in the discovery of structurally unique receptor ligands. Beyond peptide hormones and neurotransmitters, it is expected that such approaches may be of particular merit to the discovery of peptide ligands that interact with growth factor or cytokine receptors. Several studies using MPS/CPS chemistry to advance peptide analog structure–activity development have been documented *(vide infra)*.

3.1.3. Enzyme-Targeted Peptide (Substrate or Inhibitor) Lead Discovery

A novel hexapeptidyl inhibitor of thrombin has been recently reported *(12c)*, and a split-synthesis MPS/CPS method was used along with iterative enhancement to optimize the primary structure of prototypic peptide leads from an original 400-mixture set of hexapeptides defined at the first two positions (by each of the 20 naturally occurring amino acids) and randomized at the remaining four residues. The discovery of potent inhibitors of HIV protease has been reported *(18)* using a split-synthesis MPS/CPS method and iterative enhancement to optimize the primary structure of prototypic leads from an original 22-mixture set of tetrapeptides defined at the first position and randomized at the remaining three residues. The use of statine as a known P_1-P_1' peptidomimetic (as found in the naturally occurring aspartic proteinase inhibitor, pepstatin) was a key aspect of this study. With respect to optimization of substrates (i.e., substrate mapping), there has been a recent study *(53)* on collagenase in which the P_1' and P_2' sites of a heptapeptide substrate were systematically varied to yield 128 different peptides for determination of their k_{cat}/K_m properties.

3.1.4. Other Examples of Macromolecule-Targeted Peptide Lead Discovery

The discovery of new antimicrobial peptides has been recently described *(12d)* using a split-synthesis MPS/CPS method and iterative enhancement process to optimize the primary structure of prototypic peptide leads from an original 56-mixture set of tetrapeptides defined at the first position (by 19 L-amino acids, 18 D-amino acids, and 19 unnatural amino acids) and randomized at the remaining three residues (by 56 amino acids at each site).

3.2. Peptide Analog Structure–Activity Studies Using MPS/CPS Chemistry

The application of MPS/CPS chemistry methods to advance peptide analog structure-activity studies has been reported for fibrinogen *(54)*, magainin-2 *(55)*, neuropeptide-Y *(56)*, endothelin *(57)*, substance-P *(58)*, tumor necrosis factor *(59)*, and T-cell determinant *(60)*. A recent investigation *(16)* illustrates the preparation of cyclic peptide libraries related to the cyclo(Leu-D-Trp-D-Asp-Pro-D-Val), an endothelin receptor antagonist, and a large-sized (1296 component peptides) cyclic pentapeptide library. No biological data have been reported, but the chemistry strategy provided keen insight into the use of FAB-MS and AAA to determine the possibilities of dimeric products, as well as the synthetic aspects related to orthoganol-protecting groups and sequence-dependent cyclization tendencies, which are of particular significance in the synthesis of cyclic peptide libraries. Finally, it has been recently reported *(61)* that a phosphopeptide library has provided a powerful tool to study the sequence specificity of binding to SH2 domains and development of peptide inhibitors of SH2 protein interactions.

References

1a. Pavia, M. R., Sawyer, T. K., and Moos, W. H. (1993) The generation of molecular diversity. *Bioorg. Med. Chem. Lett.* **3,** 387–396.

1b. Pavia, M. R., Sawyer, T. K., and Moos, W. H. (eds.) (1993) *Bioorg. Med. Chem. Lett.* **3,** Symposium-in-Print on "The Generation of Molecular Diversity."

1c. Moos, W. H., Green, G. D., and Pavia, M. R. (1993) Recent advances in the generation of molecular diversity. *Ann. Rep. Med. Chem.* **28,** 315–324.

2. Dower, W. J. and Fodor, S. P. A. (1991) The search for molecular diversity (II): recombinant and synthetic randomized peptide libraries. *Ann. Rep. Med. Chem.* **26,** 271–280.

3. Jung, G. and Beck-Sickinger, A. G. (1992) Multiple peptide synthesis methods and their applications. *Angew. Chem. Int. Ed. Engl.* **31,** 367–386.

4. Houghten, R. A. (1993) Peptide libraries: criteria and trends. *Trends Genetics* **9,** 235–239.

5. Zuckermann, R. N. (1993) The chemical synthesis of peptidomimetic libraries. *Curr. Opin. Struct. Biol.* **3,** 580–584.

6a. Geysen, H. M., Meloen, R. H., and Barteling, S. J. (1984) Use of a peptide synthesis to probe viral antigens for epitopes to a resolution of a single amino acid. *Proc. Natl. Acad. Sci. USA* **81,** 3998–4002.

6b. Maeji, N. J., Bray, A. M., Valerio, R. M., Seldon, M. A., Wang, J.-X., and Geysen, H. M. (1991) Systematic screening for bioactive peptides. *Pept. Res.* **4,** 142–146.

7a. Bray, A. M., Maeji, N. J., and Geysen, H. M. (1990) The simultaneous multiple production of solution phase peptides: assessment of the Geysen method of simultaneous peptide synthesis. *Tet. Lett.* **31,** 5811–5814.

7b. Bray, A. M., Maeji, N. J., Valerio, R. M., Campbell, R. A., and Geysen, H. M. (1991) Direct cleavage of peptides from a solid support into aqueous buffer. Application in simultaneous multiple peptide synthesis. *J. Org. Chem.* **56,** 6659–6666.

7c. Valerio, R. M., Benstead, M., Bray, A. M., Campbell, R. A., and Maeji, N. J. (1991) Synthesis of peptide analogues using the multipin peptide synthesis method. *Analyt. Biochem.* **197,** 168–177.

7d. Bray, A. M., Maeji, N. J., Jhingran, A. G., and Valerio, R. M. (1991) Gas phase cleavage of peptides from a solid support with ammonia vapor. Application in simultaneous multiple peptide synthesis. *Tet. Lett.* **32,** 6163–6166.

8a. Geysen, H. M. (1987) US Patent 4,708,871.

8b. Geysen, H. M. (1991) Eur. Patent 138,855.

8c. Geysen, H. M. (1990) Int. Patent WO90/9395.

9a. Houghten, R. A. (1985) General method for the rapid solid-phase synthesis of large numbers of peptides: specificity of antigen-antibody interaction at the level of individual amino acids. *Proc. Natl. Acad. Sci. USA* **82,** 5131–5135.

9b. Houghten, R. A., DeGraw, S. T., Bray, M. K., Hoffman, S. R., and Frizzell, N. D. (1986) Simultaneous multiple peptide synthesis: the rapid preparation of large numbers of discrete peptides for biological, immunological and methodological studies. *BioTechniques* **4,** 522–528.

10. Houghten, R. A., Bray, M. K., Degraw, S. T., and Kirby, C. J. (1986) Simplified procedure for carrying out simultaneous multiple hydrogen fluoride cleavages of protected peptide resins. *Int. J. Pept. Prot. Res.* **27,** 675–680.

11. Beck-Sickinger, A. G., Dürr, H., and Jung, G. (1991) Semi-automated T-bag peptide synthesis using 9-fluorenylmethoxycarbonyl strategy and benzotriazo-1-*yl*-tetramethyluronium tetrafluoroborate activation. *Pept. Res.* **4,** 88–94.

12a. Houghten, R. A., Pinalla, C., Blondelle, S. E., Appel, J. R., Dooley, C. T., and Cuervo, J. H. (1991) Generation and use of synthetic peptide combinatorial libraries for basic research and drug discovery. *Nature* **354,** 84–86.

12b. Blondelle, S. E., Takahashi, E., and Houghten, R. A. (1993) Development of new antimicrobial peptides using synthetic peptide combinatorial libraries containing unnatural amino acids. *13th Am. Peptide Symp.,* Edmonton, Canada, Abst. P900.

12c. Cuervo, J. H., Nguyen, D. H., and Houghten, R. A. (1993) Novel thrombin inhibitors determined through the use of synthetic peptide combinatorial libraries. *13th Am. Peptide Symp.,* Edmonton, Canada, Abst. P901.

12d. Dooley, C. T. and Houghten, R. A. (1993) New, potent *N*-acetylated L- and D-amino acid opioid peptides. *13th Am. Peptide Symp.,* Edmonton, Canada, Abst. P903.

13a. Houghten, R. (1986) U.S. Patent 4,631,211.

13b. Houghten, R. A., Cuervo, J. H., Pinilla, C., Appel, J. R., Jr., and Blondelle, S. (1992) Int. Patent WO92/9300.

14a. Furka, Á., Sebestyén, F., Asgedom, M., and Dibó, G. (1988) Cornucopia of peptides by synthesis. *Fourteenth Int. Cong. Biochem.*, vol. 5, Prague, Czechoslovakia, Abst. FR-013, p. 47.
14b. Furka, Á., Sebestyén, F., Asgedom, M., and Dibó, G. (1988) More peptides by less labour. *Tenth Int. Symp. Med. Chem.*, Budapest, Hungary, Abst P-168, p. 288.
14c. Furka, Á., Sebestyén, F., Asgedom, M., and Dibó, G. (1991) General method for rapid synthesis of multi-component peptide mixtures. *Int. J. Pept. Prot. Res.* **37,** 487–493.
14d. Sebestyén, F., Dibó, G., Kovács, A., and Furka, Á. (1993) Chemical synthesis of peptide libraries. *Bioorg. Med. Chem. Lett.* **3,** 413–418.
15. Tatemoto, K., Mann, M. J., and Shimizu, M. (1992) Synthesis of receptor antagonists of neuropeptide-Y. *Proc. Natl. Acad. Sci. USA* **89,** 1174–1178.
16. Darlak, K., Romanovskis, P., and Spatola, A. F. (1993) Cyclic peptide libraries. *Proceedings of the 13th Am. Peptide Symp.* (Hodges, R. and Smith, J., eds.), Escom, B. V., Leiden, Netherlands, in press.
17a. Lam, K. S., Salmon, S. E., Hersh, E. M., Hruby, V. J., Kazmierski, W. M., and Knapp, R. J. (1991) A new type of synthetic peptide library for identifying ligand-binding activity. *Nature* **354,** 82–84.
17b. Lam, K. S., Hruby, V. J., Lebl, M., Knapp, R. J., Kazmierski, W. M., Hersh, E. M., and Salmon, S. E. (1993) The chemical synthesis of large random peptide libraries and their use for the discovery of ligands for macromolecular acceptors. *Bioorg. Med. Chem. Lett.* **3,** 419–424.
18. Owens, R. A., Gesellchen, P. D., Houchins, B. J., and DiMarchi, R. D. (1991) The rapid identification of HIV protease inhibitors through the synthesis and screening of defined peptide mixtures. *Biochem. Biophys. Res. Commun.* **181,** 402–408.
19. Kerr, J. M., Banville, S. C., and Zuckermann, R. N. (1993) Identification of antibody mimotopes containing non-natural amino acids by recombinant and synthetic peptide library affinity selection methods. *Bioorg. Med. Chem. Lett.* **3,** 463–468.
20a. Zuckermann, R. N., Huebner, V., Santi, D. V., and Siani, M. A. (1991) Int. Patent WO91/17823.
20b. Lam, K. S., Salmon, S. E., Hruby, V. J., Hersh, E. M., and Al-Obeidi, F. (1992) Int. Patent WO92/991.
21a. Gutte, B. and Merrifield, R. B. (1971) The synthesis of ribonuclease A. *J. Biol. Chem.* **246,** 1922–1941.
21b. Sarantakis, D., Teichman, J., Lien, E. L., and Fenichel, R. L. (1976) A novel cyclic undecapeptide, WY-40, 770, with prolonged growth hormone release inhibiting activity. *Biochem. Biophys. Res. Commun.* **73,** 336–342.
21c. Konig, W. and Geiger, R. (1970) Eine methode zur synthese von peptiden: Activierung der carbozygruppe mit dicyclohexylcarbodümed unter zusatz von 1-hydroxy-benzotriazolen. *Chem. Ber.* **103,** 788–798.
22. Kaiser, E., Colescott, R. L., Bossinger, C. D., and Cook, P. I. (1970) Color test for detection of free terminal amino groups in the solid-phase synthesis of peptides. *Analyt. Biochem.* **34,** 595–598.

23. Volkmer-Engert, R., Höhne, W., Stigler, R., and Schneider-Mergener, J. (1993) Synthesis of homologous peptide-epitope mixtures on a single resin support and characterization of antibody binding by CE, HPPC, HPLC, and LD-TOF mass spectrometry. *13th Am. Peptide Symp.*, Edmonton, Canada, Abs. P42.
24a. Andrews, P. C., Boyd, J., Loo, R. O., Zhao, R., Zhu, C. Q., Grant, K., and Williams, S. (1993) Synthesis of uniform peptide libraries and methods for physico-chemical analysis. *7th Symp. Protein Soc.*, San Diego, CA, Abst. 363M.
24b. Andrews, P. C., Boyd, J., Loo, R. O., Zhao, R., Zhu, C. Q., Grant, K., and Williams, S. (1993) Synthesis of uniform peptide libraries and methods for physico-chemical analysis, in *Techniques in Protein Chemistry, V* (Crabbe, J., ed.), pp. 485–492.
25. Tjoeng, F. S., Towery, D. S., Bulock, J. W., Whipple, D. E., Fok, K. F., Williams, M. H., Zupec, M. E., and Adams, S. P. (1990) Multiple peptide synthesis using a single support. *Int. J. Pept. Prot. Res.* **35,** 141–146.
26. Fodor, S. P. A., Read, J. L., Pirrung, M. C., Stryer, L., Lu, A. T., and Solas, D. (1991) Light-directed, spatially addressable parallel chemical synthesis. *Science* **251,** 767–773.
27. Pirrung, M. C., Read, J. L., Fodor, S. P. A., and Stryer L. (1992) U.S. Patent 5,143,854.
28. Frank, R. and Döring, R. (1988) Simultaneous multiple peptide synthesis under continuous flow conditions on cellulose paper discs as a segmental solid support. *Tetrahedron* **44,** 6031–6040.
29a. Frank R. (1992) Int. Patent WO92/4366.
29b. Eichler, J., Beyermann, M., Bienert, M., and Hunger, H. D. (1989) German Patent Appl. DD272,856.
30a. Eichler, J., Beyermann, M., and Bienert, M. (1989) Application of cellulose paper as support material in simultaneous peptide synthesis. *Collect. Czech. Chem. Commun.* **54,** 1746–1752.
30b. Frank, R. (1993) Strategies and techniques in simultaneous solid phase synthesis based on the segmentation of membrane type supports. *Bioorg. Med. Chem. Lett.* **3,** 425–430.
30c. Frank, R., Güler, S., Krause, S., and Lindenmaier, W. (1991) Facile and rapid "spot-synthesis" of large numbers of peptides on membrane sheets, in *Peptides 1990* (Giralt, E. and Andreu, D., eds.), Escom, B. V., Leiden, Netherlands, pp. 151,152.
31a. Eichler. J., Furkert, J., Bienert, M., Rohde, W., and Lebl, M. (1991) Multiple peptide synthesis on cotton carriers: elucidation and characterization of an antibody binding site of CRF, in *Peptides 1990* (Giralt, E. and Andreu, D., eds.), Escom, B. V., Leiden, Netherlands, pp. 156,157.
31b. Rinnová, M., Jezek, J., Malon, P., and Lebl, M. (1993) Comparative multiple synthesis of fifty linear peptides: evaluation of cotton carrier vs. T bag-benzhydrylamine resin. *Pept. Res.* **6,** 88–94.
32a. Berg, R. H., Almdal, K., Pedersen, W. B., Holm, A., Tam, J. P., and Merrifield, R. B. (1990) A simple approach to rapid parallel synthesis of multiple peptide analogs, in *Peptides: Chemistry, Structure and Biology* (Rivier, J. E. and Marshall, G. R., eds.), Escom, B. V., Leiden, Netherlands, pp. 1036,1037.

32b. Berg, R. H., Almdal, K., Pedersen, W. B., Holm, A., Tam, J. P., and Merrifield, R. B. (1989) Long-chain polystyrene-grafted polyethylene film matrix: a new support for solid-phase peptide synthesis. *J. Am. Chem. Soc.* **111,** 8024–8026.
33. Berg, R. H., Almdal, K., Pedersen, W. B., Holm, A., and Merrifield, R. B. (1990) Int. Patent WO90/2749.
34. Wolfe, H. R. and Wilk, R. R. (1989) The RaMPS® system: simplified peptide synthesis for life science researchers. *Pept. Res.* **2,** 352–356.
35a. Schnorrenberg, G. and Gerhart, H. (1989) Fully automatic simultaneous multiple peptide synthesis in micromolar scale—rapid synthesis of a series of peptides for screening in biological assays. *Tetrahedron* **45,** 7759–7764.
35b. Schnorrenberg, G., Wiesmüller, K. H., Beck-Sickinger, A. G., Drechsel, H., and Jung, G. (1991) Rapid fully automatic SMPS for epitope mapping of influenza nucleoprotein, in *Peptides 1990* (Giralt, E. and Andreu, D., eds.), Escom, B. V., Leiden, Netherlands, pp. 202,203.
36. Gausepohl, H., Boulin, C., Kraft, M., and Frank, R. W. (1992) Automated multiple peptide synthesis. *Peptide Res.* **5,** 315–320.
37. Nokihara, K., Yamamoto, R., Hazama, M., Wakizawa, O., and Nakamura, S. (1992) Design and applications of a novel simultaneous multiple peptide synthesizer, in *Innovation and Perspectives in Solid Phase Synthesis. Peptides, Polypeptides and Oligonucleotides 1992* (Epton, R., ed.), Intercept, Andover, UK, pp. 445–448.
38a. Groginsky, C. (1990) Independent simultaneous multiple peptide synthesis. *Am. Biotech. Lab.* **8,** 40–43.
38b. Saneii, H. H., Shannon, J. D., Miceli, R. M., Fischer, H. D., and Smith, C. W. (1993) The peptide librarian: fully automated selection and synthesis of peptide libraries. *13th Am. Peptide Symp.,* Edmonton, Canada, Abs. P926.
39. The Symphony Multiplex™ Peptide Synthesizer Owner's Manual, Protein Technologies, Inc., Rainin Instrument Co., Inc. (1993).
40. Hyde, C., Johnson, T., and Sheppard, R. C. (1993) A simple "no compromise" method for multiple peptide synthesis, in *Peptides 1992* (Schneider, C. H. and Eberle, A. N., eds.), Escom, B. V., Leiden, Netherlands, pp. 314,315.
41. Lebl, M., Stierandová, A., Eichler, J., Pátek, M., Pokorny, V., Jehnicka, J., Mudra, P., Zeneísek, K., and Kalousek, J. (1992) An automated multiple solid phase peptide synthesizer utilizing cotton as a carrier, in *Innovation and Perspectives in Solid Phase Synthesis. Peptides, Polypeptides and Oligonucleotides 1992* (Epton, R., ed.), Intercept Ltd., Andover, UK, pp. 251–257.
42. Krchnák, V. and Vágner, J. (1990) Color-monitored solid-phase multiple peptide synthesis under low-pressure continuous flow conditions. *Pept. Res.* **3,** 182–193.
43a. Zuckermann, R. N., Kerr, J. M., Siani, M. A., and Banville, S. C. (1992) Design, construction and application of a fully automated equimolar peptide mixture synthesizer. *Int. J. Pept. Prot. Res.* **40,** 497–506.
43b. Zuckermann, R. N., Kerr, J. M., Siani, M. A., Banville, S. C., and Santi, D. V. (1992) Identification of highest-affinity ligands by affinity selection from equimolar peptide mixtures generated by robotic synthesis. *Proc. Natl. Acad. Sci. USA* **89,** 4505–4509.

44. Neimark, J. and Briand, J.-P. (1993) Development of a fully automated multi-channel peptide synthesizer with integrated TFA cleavage capability. *Pept. Res.* **6,** 219–228.
45. Pinilla, C., Appel, J. R., Blanc, P., and Houghten, R. A. (1992) Rapid identification of ligands using positional scanning synthetic peptide combinatorial libraries. *BioTechniques* **13,** 901–905.
46. Parmley, S. F. and Smith, G. P. (1988) Antibody-selectable filamentous fd phage vectors: affinity purification of target genes. *Gene* **73,** 305–318.
47. Simon, R. J., Kania, R. S., Zuckermann, R. N., Huebner, V. D., Jewell, D. A., Banville, S., Ng, S., Wang, L., Rosenberg, S., Marlowe, C. K., Spellmeyer, D. C., Tan, R., Frankel, A. D., Santi, D. V., Cohen, F. E., and Bartlett, P. A. (1992) Peptoids: a modular approach to drug discovery. *Proc. Natl Acad. Sci. USA* **89,** 9367–9371.
48. Brenner, S. and Lerner, R. A. (1992) Encoded combinatorial chemistry. *Proc. Natl. Acad. Sci. USA* **89,** 5381–5383.
49. Kerr, J. M., Banville, S. C., and Zuckermann, R. N. (1993) Encoded combinatorial peptide libraries containing non-natural amino acids. *J. Am. Chem. Soc.* **115,** 2529–2531.
50. Nikolaiev, V., Stierandová, A., Krchnák, V., Seligmann, B., Lam, K. S., Salmon, S. E., and Lebl, M. (1993) Peptide-encoding for structure determination of non-sequenceable polymers within libraries synthesized and tested on solid-phase supports. *Pept. Res.* **6,** 161–170.
51. Geysen, H. M., Barteling, S. J., and Meloen, R. H. (1985) Small peptides induce antibodies with a sequence and structural requirement for binding antigen comparable to antibodies raised against the native protein. *Proc. Natl. Acad. Sci. USA* **82,** 178–182.
52. Geysen, M. H., Rodda, S. M., and Mason, T. J. (1984) A priori delineation of a peptide which mimics a discontinuous antigenic determinant. *Mol. Immunol.* **23,** 709–715.
53. Berman, J., Green, M., Sugg, E., Anderegg, R., Millington, D. S., Norwood, D. L., McGeehan, J., and Wiseman, J. (1992) Rapid optimization of enzyme substrates using defined substrate mixtures. *J. Biol. Chem.* **267,** 1434–1437.
54. Ruggeri, Z. M., Houghten, R. A., Russell, S. R., and Zimmerman, T. S. (1986) Inhibition of platelet function with synthetic peptides designed to be high affinity antagonists of fibrinogen binding to platelets. *Proc. Natl. Acad. Sci. USA* **83,** 5708–5712.
55. Cuervo, J. H., Rodriguez, B., and Houghten, R. A. (1990) Synthesis and antimicrobial activity of magainin alanine substitution analogs, in *Peptides: Chemistry, Structure and Biology* (Rivier, J. E. and Marshall, G. R., eds.), Escom, B. V., Leiden, Netherlands, pp. 124–126.
56. Beck-Sickinger, A. G., Gaida, W., Schnorrenberg, G., Lang, R., and Jung, G. (1990) Neuropeptide Y: identification of the binding site. *Int. J. Peptide Protein Res.* **36,** 522–530.
57. Spellmeyer, D. C., Brown, S., Stauber, G. B., Geysen, H. M., and Valerio, R. (1993) Endothelin receptor ligands. Replacement net approach to SAR determination of potent hexapeptides. *Bioorg. Med. Chem. Lett.* **3,** 519–524.

58. Wang, J., DiPasquale, A. J., Bray, A. M., Maiji, N. J., and Geysen, H. M. (1993) Study of stereo-requirements of substance P binding to NK1 receptors using analogues with systematic D-amino acid replacements. *Bioorg. Med. Chem. Lett.* **3,** 451–456.
59. Rathjen, D. A. and Aston, R. (1993) Selective enhancement of tumour necrosis factor activity: mapping regions with monoclonal antibodies. *Bioorg. Med. Chem. Lett.* **3,** 457–462.
60. Mutch, D. A., Rodda, S. J., Benstead, M., Valerio, R. M., and Geysen, H. M. (1991) Effects of end groups on the stimulatory capacity of minimal length T cell determinant peptides, *Pept. Res.* **4,** 132–137.
61. Songyang, Z., Shoelson, S. E., Chaudhuri, M., Gish, G., Pawson, T., Haser, W. G., King, F., Roberts, T., Ratnofsky, S., Lechleider, R. J., Neel, B. G., Birge, R. B., Fajardo, J. E., Chou, M. M., Hanafusa, H., Schaffhausen, B., and Cantley, L. C. (1993) SH2 domains recognize specific phosphopeptide sequences. *Cell* **72,** 767–778.

Index